**Advances in
Electrochemical Science
and Engineering**

Volume 14
Electrocatalysis

Advances in Electrochemical Science and Engineering

Advisory Board

Philippe Allongue, Ecole Polytechnique, Palaiseau, France
A. Robert Hillman, University of Leicester, Leicester, UK
Tetsuya Osaka, Waseda University, Tokyo, Japan
Laurence Peter, University of Bath, Bath, UK
Lubomyr T. Romankiw, IBM Watson Research Center, Yorktown Heights, USA
Shi-Gang Sun, Xiamen University, Xiamen, China
Esther Takeuchi, SUNY Stony Brook, Stony Brook; and Brookhaven National Laboratory, Brookhaven, USA
Mark W. Verbrugge, General Motors Research and Development, Warren, MI, USA

In collaboration with the International Society of Electrochemistry

Advances in Electrochemical Science and Engineering

Volume 14
Electrocatalysis

Edited by
*Richard C. Alkire, Dieter M. Kolb[†], Ludwig A. Kibler,
and Jacek Lipkowski*

Verlag GmbH & Co. KGaA

The Editors

Richard C. Alkire
Department of Chemical and
Biomolecular Engineering
University of Illinois
Urbana, IL 61801
United States

Ludwig A. Kibler
Institute of Electrochemistry
University of Ulm
Albert-Einstein-Allee 47
89081 Ulm
Germany

Dieter M. Kolb[†]
Institute of Electrochemistry
University of Ulm
Albert-Einstein-Allee 47
89081 Ulm
Germany

Jacek Lipkowski
Department of Chemistry
University of Guelph
N1G 2W1 Guelph, Ontario
Canada

■ All books published by **Wiley-VCH** are carefully produced. Nevertheless, authors, editors, and publisher do not warrant the information contained in these books, including this book, to be free of errors. Readers are advised to keep in mind that statements, data, illustrations, procedural details or other items may inadvertently be inaccurate.

Library of Congress Card No.: applied for

British Library Cataloguing-in-Publication Data
A catalogue record for this book is available from the British Library.

Bibliographic information published by the Deutsche Nationalbibliothek
The Deutsche Nationalbibliothek lists this publication in the Deutsche Nationalbibliografie; detailed bibliographic data are available on the Internet at http://dnb.d-nb.de.

© 2013 WILEY-VCH Verlag GmbH & Co. KGaA, Weinheim Germany

All rights reserved (including those of translation into other languages). No part of this book may be reproduced in any form – by photoprinting, microfilm, or any other means – nor transmitted or translated into a machine language without written permission from the publishers. Registered names, trademarks, etc. used in this book, even when not specifically marked as such, are not to be considered unprotected by law.

Print ISBN: 978-3-527-33227-4

ePDF ISBN: 978-3-527-68046-7

ePub ISBN: 978-3-527-68045-0

mobi ISBN: 978-3-527-68044-3

oBook ISBN: 978-3-527-68043-6

Cover Design Schulz Grafik-Design, Fubgonheim

Typesetting Thomson Digital, Noida, India

Printing and Binding Markono Print Media Pte Ltd, Singapore

Printed in Singapore
Printed on acid-free paper

In Memoriam

Prof. Dr. Dieter M. Kolb (1942–2011)

Dieter M. Kolb has been the Head of the Institute of Electrochemistry at the University of Ulm, Germany, from 1990 to 2010, and has served as Co-Editor of this series since 1997. It is with deep sadness to report that Professor Kolb passed away on October 4, 2011. He is well known for his contributions to the fundamental understanding of electrochemical phenomena. In particular, he was among the founders of electrochemical surface science by combining classical electrochemical understanding with atomic-level information obtained by modern surface science techniques like the *in-situ* scanning tunneling microscopy (STM). These scientific activities have stimulated not only related research in other electrochemical laboratories but also studies of non-electrochemists in the field of interfacial electrochemistry.

His scientific accomplishments include initial stages of metal deposition together with underpotential deposition of foreign metals on single crystal substrates, nanostructuring of electrode surfaces by metal clusters generated at the tip of an STM, surface reconstruction of gold electrodes, metallization of organic layers, and electrocatalysis. By using well-defined electrode surfaces, he obtained unprecedented insights into the nature of elementary processes, especially the influence of surface structure on electrochemical reactions.

Contents

Preface *XIII*
List of Contributors *XV*

1   **Multiscale Modeling of Electrochemical Systems** *1*
Jonathan E. Mueller, Donato Fantauzzi, and Timo Jacob
1.1   Introduction *1*
1.2   Introduction to Multiscale Modeling *3*
1.3   Electronic Structure Modeling *6*
1.3.1   Modern Electronic Structure Theory *6*
1.3.1.1   Quantum Mechanical Foundations *6*
1.3.1.2   Born–Oppenheimer Approximation *8*
1.3.1.3   Single-Electron Hamiltonians *9*
1.3.1.4   Basis Sets *10*
1.3.1.5   Enforcing the Pauli Principle *11*
1.3.1.6   Electron Correlation Methods *12*
1.3.1.7   Density Functional Theory *14*
1.3.2   Applications of Electronic Structure to Geometric Properties *16*
1.3.2.1   Geometry Optimization *16*
1.3.2.2   Transition State Searches *17*
1.3.3   Corrections to Potential Energy Surfaces and Reaction Pathways *18*
1.3.3.1   Energy and Entropy Corrections *18*
1.3.3.2   Thermodynamic State Functions *20*
1.3.3.3   Reaction Energies and Rates *21*
1.3.4   Electronic Structure Models in Electrochemistry *22*
1.3.4.1   Modeling the Electrode Surface: Cluster versus Slab *23*
1.3.4.2   Modeling the Solvent: Explicit versus Implicit *24*
1.3.4.3   Modeling the Electrode Potential *25*
1.3.5   Summary *26*
1.4   Molecular Simulations *27*
1.4.1   Energy Terms and Force Field Parameters *27*
1.4.1.1   Covalent Bond Interactions *27*
1.4.1.2   Non-Covalent Interactions *32*
1.4.2   Parametrization and Validation *34*
1.4.3   Atomistic Simulations *35*

1.4.3.1	Monte Carlo Methods 35
1.4.3.2	Molecular Dynamics 36
1.4.3.3	QM/MM 37
1.4.4	Sampling and Analysis 39
1.4.5	Applications of Molecular Modeling in Electrochemistry 39
1.4.6	Summary 40
1.5	Reaction Modeling 40
1.5.1	Introduction 40
1.5.2	Chemical Kinetics 41
1.5.3	Kinetic Monte Carlo 41
1.5.3.1	System States and the Lattice Approximation 41
1.5.3.2	Reaction Rates 42
1.5.3.3	Reaction Dynamics 43
1.5.3.4	Applications of kMC in Electrochemistry 45
1.5.4	Summary 45
1.6	The Oxygen Reduction Reaction on Pt(111) 46
1.6.1	Introduction to the Oxygen Reduction Reaction 46
1.6.2	Preliminary Considerations 46
1.6.3	DFT Calculations 48
1.6.4	Method Validation 49
1.6.5	Reaction Energies 49
1.6.6	Solvation Effects 51
1.6.7	Free Energy Contributions 52
1.6.8	Influence of an Electrode Potential 53
1.6.9	Modeling the Kinetic Rates 55
1.6.10	Summary 58
1.7	Formic Acid Oxidation on Pt(111) 59
1.7.1	Introduction to Formic Acid Oxidation 59
1.7.2	Density Functional Theory Calculations 60
1.7.3	Gas Phase Reactions 60
1.7.4	Explicit Solvation Model 61
1.7.5	Eley–Rideal Mechanisms and the Electrode Potential 63
1.7.6	Kinetic Rate Model of Formic Acid Oxidation 65
1.7.7	Summary 66
1.8	Concluding Remarks 66
	References 67
2	**Statistical Mechanics and Kinetic Modeling of Electrochemical Reactions on Single-Crystal Electrodes Using the Lattice-Gas Approximation** 75
	Marc T.M. Koper
2.1	Introduction 75
2.2	Lattice-Gas Modeling of Electrochemical Surface Reactions 76
2.3	Statistical Mechanics and Approximations 79
2.3.1	Static System 79
2.3.2	Dynamical System 83

2.4	Monte Carlo Simulations *84*	
2.5	Applications to Electrosorption, Electrodeposition and Electrocatalysis *85*	
2.5.1	Electrosorption and Electrodeposition *85*	
2.5.2	Electrocatalysis *92*	
2.6	Conclusions *96*	
	References *96*	

3	**Single Molecular Electrochemistry within an STM** *99*	
	Richard J. Nichols and Simon J. Higgins	
3.1	Introduction *99*	
3.2	Experimental Methods for Single Molecule Electrical Measurements in Electrochemical Environments *101*	
3.3	Electron Transfer Mechanisms *103*	
3.3.1	Tunneling *106*	
3.3.2	Resonant Tunneling *109*	
3.3.3	Hopping Models *111*	
3.4	Single Molecule Electrochemical Studies with an STM *115*	
3.4.1	Adsorbed Iron Complexes *115*	
3.4.2	Viologens *118*	
3.4.3	Osmium and Cobalt Metal Complexes *122*	
3.4.4	PyrroloTTF (pTTF) *125*	
3.4.5	Perylene Tetracarboxylic Diimides *128*	
3.4.6	Oligo(phenylene ethynylene) Derivates *130*	
3.5	Conclusions and Outlook *131*	
	References *134*	

4	**From Microbial Bioelectrocatalysis to Microbial Bioelectrochemical Systems** *137*	
	Uwe Schröder and Falk Harnisch	
4.1	Prelude: From Fundamentals to Biotechnology *137*	
4.2	Microbial Bioelectrochemical Systems (BESs) *137*	
4.2.1	The Archetype: Microbial Fuel Cells (MFCs) *137*	
4.2.2	Strength Through Diversity: Microbial Bioelectrochemical Systems *139*	
4.3	Bioelectrocatalysis: Microorganisms Catalyze Electrochemical Reactions *140*	
4.3.1	Energetic Considerations of Microbial Bioelectrocatalysis *141*	
4.3.1.1	Case Study: The Anodic Acetate Oxidation by *Geobacteraceae* *142*	
4.3.1.2	Case Study: The Cathodic Hydrogen Evolution Reaction (HER) *143*	
4.3.2	Microbial Electron Transfer Mechanisms *144*	
4.3.2.1	Direct Electron Transfer (DET) *144*	
4.3.2.2	Mediated Electron Transfer (MET) *146*	
4.3.2.3	Cathodic Electron Transfer Mechanisms *148*	
4.3.3	Microbial Interactions: Ecological Networks *148*	

4.3.3.1	Interspecies Electron Transfer and "Scavenging" of Redox-Shuttles *148*
4.4	Characterizing Anodic Biofilms by Electrochemical and Biological Means *149*
4.4.1.1	Case Study: On the use of Cyclic Voltammetry *154*
4.4.1.2	Case Study: Raman Microscopy *155*
	References *156*
5	**Electrocapillarity of Solids and its Impact on Heterogeneous Catalysis** *163*
	Jörg Weissmüller
5.1	Introduction *163*
5.2	Mechanics of Solid Electrodes *164*
5.2.1	Outline – Surface Stress and Surface Tension *164*
5.2.2	Solid Versus Fluid *167*
5.2.3	Free Energy of Elastic Solid Surfaces *167*
5.2.4	Deforming a Solid Surface *170*
5.2.5	Case Study: Thought Experiment in Electrowetting *172*
5.2.6	Capillary Equations for Fluids and Solids *175*
5.2.7	Case Study: Molecular Dynamics Study of Surface-Induced Pressure *177*
5.3	Electrocapillary Coupling at Equilibrium *177*
5.3.1	Outline – Polarizable and Nonpolarizable Electrodes *177*
5.3.2	Lippmann Equation and Electrocapillary Coupling Coefficient *179*
5.3.3	Case Study: Cantilever-Bending Experiment in Electrolyte *181*
5.3.4	Important Maxwell Relations for Electrocapillarity *183*
5.3.5	Electrocapillary Coupling During Electrosorption *184*
5.3.6	Coupling Coefficient for Adsorption from Gas *185*
5.3.7	Coupling Coefficient for the Langmuir Isotherm *186*
5.3.8	Case Study: Strain-Dependent Hydrogen Underpotential Deposition *187*
5.3.9	Coupling Coefficient for Potential of Zero Charge and Work Function *190*
5.3.10	Empirical Data for the Electrocapillary Coupling Coefficient *193*
5.4	Exploring the Dynamics *198*
5.4.1	Outline *198*
5.4.2	Cyclic Cantilever-Bending Experiments *199*
5.4.3	Dynamic Electro-Chemo-Mechanical Analysis *200*
5.5	Mechanically Modulated Catalysis *203*
5.5.1	Outline *203*
5.5.2	Phenomenology; Distinguishing Capacitive from Faraday Current *204*
5.5.3	Rate equations: Butler–Volmer kinetics *206*
5.5.4	Rate Equations: Heyrowsky Reaction *207*
5.6	Summary and Outlook *212*
	References *215*

6	**Synthesis of Precious Metal Nanoparticles with High Surface Energy and High Electrocatalytic Activity** *221*
	Long Huang, Zhi-You Zhou, Na Tian, and Shi-Gang Sun
6.1	Introduction *221*
6.2	Shape-Controlled Synthesis of Monometallic Nanocrystals with High Surface Energy *224*
6.2.1	Electrochemical Route *224*
6.2.1.1	Platinum *224*
6.2.1.2	Palladium *228*
6.2.2	Wet-Chemical Route *230*
6.2.2.1	Platinum *230*
6.2.2.2	Palladium *232*
6.2.2.3	Gold *234*
6.3	Shape-Controlled Synthesis of Bimetallic NCs with High Surface Energy *235*
6.3.1	Surface Modification *236*
6.3.1.1	Bi-Modified THH Pt NCs *237*
6.3.1.2	Ru-Modified THH Pt NCs *239*
6.3.1.3	Au-Modified Pt THH NCs *241*
6.3.1.4	Pt-Modified Au Prisms with High-Index Facets *243*
6.3.2	Alloy NCs *245*
6.3.2.1	THH PdPt Alloy *245*
6.3.2.2	HOH PdAu Alloy *247*
6.3.3	Core–Shell Structured NCs *248*
6.4	Concluding Remarks and Perspective *254*
	References *256*
7	**X-Ray Studies of Strained Catalytic Dealloyed Pt Surfaces** *259*
	Peter Strasser
7.1	Introduction *259*
7.2	Dealloyed Bimetallic Surfaces *262*
7.3	Dealloyed Strained Pt Core–Shell Model Surfaces *264*
7.4	X-Ray Studies of Dealloyed Strained $PtCu_3(111)$ Single Crystal Surfaces *266*
7.5	X-Ray Studies of Dealloyed Strained Pt-Cu Polycrystalline Thin Film Surfaces *269*
7.6	X-Ray Studies of Dealloyed Strained Alloy Nanoparticles *273*
7.6.1	Bragg Brentano Powder X-Ray Diffraction (XRD) *273*
7.6.2	*In Situ* High Temperature Powder X-Ray Diffraction (XRD) *274*
7.6.3	Synchrotron X-Ray Photoemission Spectroscopy (XPS) *276*
7.6.4	Anomalous Small Angle X-Ray Scattering (ASAXS) *277*
7.6.5	Anomalous Powder X-Ray Diffraction (AXRD) *277*

7.6.6	High Energy X-Ray Diffraction (HE-XRD) and Atomic Pair Distribution Function (PDF) Analysis *281*
7.7	Conclusions *284*
	References *285*

Index *293*

Preface

This book is devoted to the rapidly growing field of electrocatalysis and aspects of modern electrochemical surface science. Recent progress is reviewed with a particular emphasis on methodological developments that are driving electrochemical surface science research and its relation to kinetics of electrode reactions across a broad range of fundamental topics and applications. These key fundamental research developments include modern experimental methods with well-defined model electrodes, as well as achievements in theoretical electrochemistry.

Mueller, Fantauzzi, and Jacob provide a comprehensive introduction to multiscale modeling in Chapter 1, which is not only designed for theorists but also for experimentalists with limited background in modern computational chemistry. The use of density functional theory for a basic understanding of elementary steps in oxygen reduction and formic acid oxidation on Pt(111) is highlighted. Experimental results for well-defined single-crystal electrodes are compared with kinetic modeling based on a statistical-mechanical description of electrochemical surface reactions in Chapter 2 by Koper. Nichols and Higgins present in Chapter 3 new insights into charge transfer across electrical junctions by single-molecule electrochemical studies with the tip of a scanning tunneling microscope. The field of bioelectrocatalysis is addressed in Chapter 4 by Schröder and Harnisch, who describe microbial electron transfer mechanisms and the catalysis of electrochemical reactions by microorganisms. In Chapter 5, Weissmüller presents the fundamentals of the theory of electrocapillarity of solids and their application in the field of strain-dependent catalysis. In Chapter 6, Huang, Zhou, Tian, and Sun review recent progress in shape-controlled synthesis of metal nanoparticles with high-energy facets and their applications in electrocatalysis. The geometric and electronic structure of strained dealloyed bimetallic Pt catalysts as studied by X-ray measurements is shown by Strasser in Chapter 7.

Several chapters address applications for which electrocatalysts efficiently convert electrical energy into chemical bonds, or the reverse, converting chemical energy from fuels into electrical energy. Prominent among these applications is the task of catalyzing electrode reactions in fuel cells.

This book was initiated under the editorial leadership of Prof. Dr. D.M. Kolb. Upon his untimely passing, the lead editorial tasks were taken on by Guest Editor Dr. L.A. Kibler, with the assistance of Prof. R.C. Alkire.

The combined experimental and theoretical approach is of interest to chemists, physicists, biochemists, surface and materials scientists, and engineers. The opportunities for impact in this field are far greater than the current researchers trained in electrochemistry can accomplish. By providing up-to-date reviews with deep level of coverage of key background topics, this book is adapted to students and professionals entering the field, as well as experienced researchers seeking to expand their scope and mastery.

Ulm, Germany   *L.A. Kibler*
Urbana, IL, USA   *R.C. Alkire*

List of Contributors

Donato Fantauzzi
Universität Ulm
Institut für Elektrochemie
Albert-Einstein-Allee 47
89081 Ulm
Germany

Falk Harnisch
TU Braunschweig
Institute of Environmental and
Sustainable Chemistry
Hagenring 30
38106 Braunschweig
Germany

and

Helmholtz Centre for Environmental
Research - UFZ
Department of Environmental
Microbiology
Permoserstraße 15
04318 Leipzig
Germany

Simon J. Higgins
University of Liverpool
Department of Chemistry
Donnan and Robert Robinson
Laboratories
Crown Street
Liverpool L69 7ZD
UK

Long Huang
Xiamen University
State Key Laboratory of Physical
Chemistry of Solid Surfaces
College of Chemistry and Chemical
Engineering
Department of Chemistry
South Siming Road 422
Xiamen 361005
China

Timo Jacob
Universität Ulm
Institut für Elektrochemie
Albert-Einstein-Allee 47
89081 Ulm
Germany

Marc T.M. Koper
Leiden University
Leiden Institute of Chemistry
Einsteinweg 55
2333 CC Leiden
The Netherlands

and

Hokkaido University
Catalysis Research Center
Kita21, Nishi10, Kita-ku
Sapporo 001-0021
Japan

Jonathan E. Mueller
Universität Ulm
Institut für Elektrochemie
Albert-Einstein-Allee 47
89081 Ulm
Germany

Richard J. Nichols
University of Liverpool
Department of Chemistry
Donnan and Robert Robinson
Laboratories
Crown Street
Liverpool L69 7ZD
UK

Uwe Schröder
TU Braunschweig
Institute of Environmental and
Sustainable Chemistry
Hagenring 30
38106 Braunschweig
Germany

Peter Strasser
Technical University Berlin
Department of Chemistry
Chemical Engineering Division
Strasse des 17. juni 124
10623 Berlin
Germany

Shi-Gang Sun
Xiamen University
State Key Laboratory of Physical
Chemistry of Solid Surfaces
College of Chemistry and Chemical
Engineering
Department of Chemistry
South Siming Road 422
Xiamen 361005
China

Na Tian
Xiamen University
State Key Laboratory of Physical
Chemistry of Solid Surfaces
College of Chemistry and Chemical
Engineering
Department of Chemistry
South Siming Road 422
Xiamen 361005
China

Jörg Weissmüller
Technische Universität Hamburg-
Harburg
Institut für Werkstoffphysik und
Werkstofftechnologie
Eißendorfer Straße 42
21073 Hamburg
Germany

and

Helmholtz-Zentrum Geesthacht
Institut für Werkstoffforschung
Werkstoffmechanik
Max-Planck-Straße 1
21502 Geesthacht
Germany

Zhi-You Zhou
Xiamen University
State Key Laboratory of Physical
Chemistry of Solid Surfaces
College of Chemistry and Chemical
Engineering
Department of Chemistry
South Siming Road 422
Xiamen 361005
China

1
Multiscale Modeling of Electrochemical Systems
Jonathan E. Mueller, Donato Fantauzzi, and Timo Jacob

1.1
Introduction

As one of the classic branches of physical chemistry, electrochemistry enjoys a long history. Its relevance and vitality remain unabated as it not only finds numerous applications in traditional industries, but also provides the scientific impetus for a plethora of emerging technologies. Nevertheless, in spite of its venerability and the ubiquity of its applications, many of the fundamental processes, underlying some of the most basic electrochemical phenomena, are only now being brought to light.

Electrochemistry is concerned with the interconversion of electrical and chemical energy. This interconversion is facilitated by transferring an electron between two species involved in a chemical reaction, such that the chemical energy associated with the chemical reaction is converted into the electrical energy associated with transferring the electron from one species to the other. Taking advantage of the electrical energy associated with this electron transfer for experimental or technological purposes requires separating the complementary oxidation and reduction reactions of which every electron transfer is composed. Thus, an electrochemical system includes an electron conducting phase (a metal or semiconductor), an ion conducting phase (typically an electrolyte, with a selectively permeable barrier to provide the requisite chemical separation), and the interfaces between these phases at which the oxidation and reduction reactions take place.

Thus, the fundamental properties of an electrochemical system are the electric potentials across each phase and interface, the charge transport rates across the conducting phases, and the chemical concentrations and reaction rates at the oxidation and reduction interfaces. Traditional experimental techniques (e.g., cyclic voltammetry) measure one or more of these continuous observables in an effort to understand the interrelationships between purely electrochemical phenomena (e.g., electrode potential, current density). While these techniques often shed light on both fundamental (e.g., ionic charge) and statistical (e.g., diffusion rates) properties of the atoms and ions that make up an electrochemical system, they provide little insight into the detailed atomic structure of the system.

Electrocatalysis: Theoretical Foundations and Model Experiments, First Edition.
Edited by Richard C. Alkire, Ludwig A. Kibler, Dieter M. Kolb, and Jacek Lipkowski.
© 2013 Wiley-VCH Verlag GmbH & Co. KGaA. Published 2013 by Wiley-VCH Verlag GmbH & Co. KGaA.

In contrast, modern surface science techniques (e.g., STM, XPS, SIMS, LEISS) typically probe the atomistic details of the interface regions, and support efforts to gain insight into the atomistic processes underlying electrochemical phenomena. Indeed, these methods have been applied to gas–solid interfaces with resounding success, elucidating the atomistic structures underlying macroscopic phenomena [1]. Unfortunately, the presence of the electrolyte at the electrode surface hampers the application of many of these surface science techniques. Because the resulting solid–electrolyte interface is an essential component of any electrochemical system, electrochemistry has not yet fully experienced the atomistic revolution enjoyed by other departments of surface science, although these techniques are increasingly making their way into electrochemistry [2].

The dramatic increases in computing power realized over the past decades coupled with improved algorithms and methodologies have enabled theorists to develop reliable, atomistic-level descriptions of surface structures and processes [3]. In particular, periodic density functional theory (DFT) now exhibits a degree of efficiency and accuracy which allows it not only to be used to explain, but also to predict experimental results, allowing theory to take a proactive, or even leading, role in surface science investigations. A prime example of this is the design of a new steam reforming catalyst based on a combination of theoretical and experimental fundamental research [4].

The application of DFT to electrochemical systems is not as straightforward as it is to the surface–vacuum interfaces of surface sciences. There have indeed been promising efforts in this direction [5–7], and there is a growing interest in theoretical electrochemistry [8–10]; however, proper treatments of the electrolyte and electrode potential provide novel challenges for which there are not yet universally agreed upon solutions. Nevertheless, there are already success stories, such as the theoretical prediction [11,12] and experimental confirmation [13] of the nonmonotonic dependence of the electrocatalytic activity of the hydrogen evolution reaction (HER) on the thickness of Pd overlayers on Au(111).

Common to both the experimental and theoretical approaches mentioned above is the existence of two regimes – the macroscopic and the atomistic – and the importance of relating these in order to obtain a comprehensive picture of an electrochemical system. Statistical mechanics provides the necessary framework for relating the discrete properties and atomistic structures of the atomistic regime to the continuous variable controlled or observed in the macroscopic regime. The fundamental assumption underlying this relationship is what Richard Feynman called the "atomic hypothesis", which we rephrase in terms of electrochemistry as follows: "there is nothing that *electrochemical* systems do that cannot be understood from the point of view that they are made up of atoms acting according to the laws of physics" [14].

Modern computational methods, based on the principles of quantum mechanics, provide a means of probing the atomistic details of electrochemical systems, as do the techniques of modern surface science techniques. The concepts of statistical mechanics are critical for extending the results of these molecular scale models to macro-scale descriptions of electrochemical systems. Such a procedure creates a multiscale model of an electrochemical system, built up from the atomistic details of the quantum regime to a description of the electrochemical phenomena observed in macroscopic systems.

This chapter is intended to serve as an introduction to multiscale modeling for electrochemists with minimal background in the methods of modern computational chemistry. Thus, the fundamentals of some of the most important methods are presented within the framework of multiscale modeling, which integrates diverse methods into a single multiscale model, which then spans a wider range of time and length scales than is otherwise possible. The physical ideas underlying the methods and the conceptual framework used to weave them together are emphasized over the specific how-to details of running simulations. Thus Section 1.2 gives an overview of the multiscale modeling and Sections 1.3–1.5 present three different levels of theory used as components in many multiscale models: electronic structure modeling methods, molecular modeling methods, and chemical reaction modeling methods. The development of appropriate models for simulating electrochemical systems at each level of theory is the key outcome of each of these sections. To illustrate the application of some of the methods of multiscale modeling to electrochemistry, two concrete examples are presented in detail. In Section 1.6 a detailed mechanistic study of the oxygen reduction reaction on Pt(111) is presented. In Section 1.7 a similar study of formic acid oxidation illustrates additional approaches and modeling techniques. In both cases the focus is on the methods and modeling techniques used rather than the particular conclusions reached in each study.

1.2
Introduction to Multiscale Modeling

Electrochemical phenomena can be viewed over a wide range of time and length scales, ranging from electronic transfer processes which take place over distances on the scale of nanometers in times of the order of femtoseconds, to large-scale industrial processes involving moles of atoms occupying spaces best measured in meters and lasting hours, days or even years. Bridging these time and length scales is one of the central tasks of modern theoretical electrochemistry. This is the case for both scientists seeking to further our fundamental understanding of electrochemistry, and engineers developing applications of electrochemical processes and systems. Thus, the former continue the hard work of uncovering the atomistic processes underlying macroscopic electrochemical processes [2], while the latter seek to bring together interconnected phenomena spanning many time and length scales to design a product with the desired functionality [15]. In both cases a multiscale framework is needed to interrelate phenomena at the relevant time and length scales.

Traditionally, computational chemistry uses a single computational tool to model a given system at a particular time and length scale. Several of the major categories that these computational tools fit into, along with the approximate time and length scales, to which they can be applied, are shown in Figure 1.1. The physical laws appropriate for the system components or building blocks at each time and length scale govern the models developed at that level and determine which system

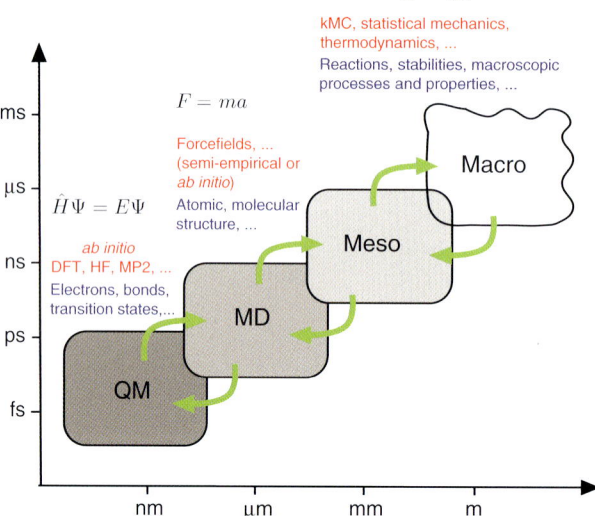

Figure 1.1 Schematic representation of various categories of simulation methods used in multiscale modeling according to the time and length scales of the models to which they are most applicable.

properties we can obtain directly. Thus each level of theory focuses on the system under a single aspect. Multiscale modeling aims at stitching these various aspects together into a unified whole, such that macroscopic properties emerge from underlying microscopic phenomena.

Two strategies are available for stitching methods of differing scales together into a single, coherent multiscale model. In the first, known as concurrent coupling, the various levels of simulation are incorporated into a single multiscale model. Thus, as illustrated in Figure 1.2, a single simulation makes direct use of various levels of theory and explicitly describes phenomena taking place at a range of time and length scales. Concurrent coupling is typically realized by dividing the system into various regions, each of which is treated using a different level of theory. Defining the boundaries between these regions and then determining how the regions interact with each other is the primary challenge in concurrent coupling. The key disadvantage, is that, because the time propagation of the system dynamics is limited by the process with the smallest time step, there are often only limited gains in the time scales which can be achieved using concurrent coupling schemes. Nevertheless, significant gains in the length scales treated are very realizable. A common example of concurrent coupling is QM/MM modeling in which an electronic structure method is used to describe a small reactive portion of a system, which is otherwise described using a molecular force field. We discuss this approach in greater detail in Section 1.4.3.3.

The second strategy, known as sequential coupling, uses results from modeling at one level as the inputs for a model at another level. This often entails fitting the parameters that define a model at one level to results derived from another level, and

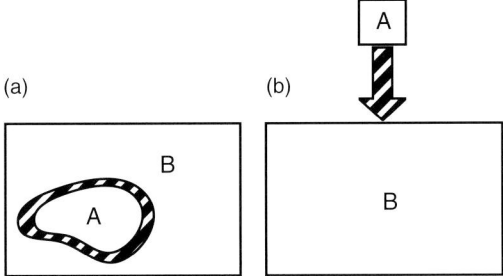

Figure 1.2 Schematic representation of (a) concurrent and (b) serial coupling approaches to integrating two diverse levels of theory into a single, multiscale modeling framework. In concurrent coupling different parts of a single model system are treated using methods describing the system at two different levels of detail. Here a small region treated with a more detailed method (A) is typically embedded within a larger region described using a more course-grained method (B). The shaded overlap region where the fibers from these distinct computational methods are woven together into a single, continuous model is critical to the successful implementation of a concurrent coupling approach. In serial coupling results obtained using a more fundamental method (A) are used as the basis for optimizing the parameters needed to define a course-grained method (B), which can then be applied to larger model systems for larger time and length scale simulations.

is often referred to as parameter passing. Thus, as shown in Figure 1.2, one method is derived from another, such that subsequent simulations carried out using the derived method are not constrained by the time and length scale limitations of the parent model(s). The aim in this procedure is often to extend the atomistic details, and thus presumably accuracy, of smaller scale methods to systems which would normally be too large for them to treat. Of course, there is a price to be paid for these gains in computational efficiency. Larger scale methods can only be derived from smaller (typically more exact) methods by making approximations and simplifying assumptions. To verify the validity of these approximations, it is important to make a direct comparison between results from the derived method and results from its parent method(s) for cases which were not included in the derivation. It is also possible that errors are propagated from the parent method(s) in the process of derivation. Comparison with experiment, where possible, is an important means of locating such errors. A common example of sequential modeling is force field parameter development and parameter optimization using results obtained from electronic structure calculations. This strategy is described in greater detail in Section 1.4.2.

Thus, communication between various levels of theory is at the heart of multiscale modeling. In the case of concurrent coupling this involves the continuous translation and shuttling of information across the boundaries between regions modeled using different methods in order to maintain the unity of the overall model. In the case of sequential coupling, information is first transferred from a parent to a daughter method, as the former method is derived or parametrized from the latter. However, subsequent validation requires the reverse flow of information as results from the derived method are returned to the parent method for comparison.

1.3
Electronic Structure Modeling

Multiscale modeling in electrochemistry typically begins at the atomic level, where the interactions of the electrons and nuclei which make up the electrochemical system are described in the language of quantum mechanics. Indeed, modern physics claims that the quantum mechanical description of such a system, in the form of a space- and time-dependent wave function, $\Psi(\mathbf{r}, t)$, is exact, and contains all that there is to know about the system. However, we are unable to obtain exact wavefunctions for all but the most trivial systems. Thus, Paul Dirac once famously noted: "The underlying physical laws necessary for the mathematical theory of a large part of physics and the whole of chemistry are thus completely known, and the difficulty is only that the exact application of these laws leads to equations much too complicated to be soluble. It therefore becomes desirable that approximate practical methods of applying quantum mechanics should be developed, which can lead to an explanation of the main features of complex atomic systems without too much computation" [16].

For several decades Dirac's prescription to avoid excessive computation was necessary for making headway in the effort to understand chemistry from quantum mechanical, first principles, and primarily conceptual rather than strictly numerical connections between quantum mechanics and chemical phenomena were developed. However, since the advent of modern computers and the subsequent exponential increase in their computing power, a new avenue for applications of quantum mechanics to chemistry has opened up which takes advantage of the enormous computing power available today.

1.3.1
Modern Electronic Structure Theory

Even with all of the computational resources available today, exact solutions to the Schrödinger equation are, in all but the simplest cases, unattainable. Thus, much effort has been dedicated to developing numerical methods for obtaining approximate wavefunctions that provide reliable descriptions of systems of interest. The aim of this section is to briefly present the conceptual underpinnings of modern electronic structure theory and methods and the conceptual bridges that can be used to link their results to macroscopic electrochemical systems. More detailed and complete treatments of quantum mechanics [14–18], quantum chemistry [19–21], statistical mechanics [22,23], chemical kinetics [24] and electronic structure methods [25–27] are readily available in many of the works referenced in this chapter, as well as in many more we have not had occasion to cite.

1.3.1.1 Quantum Mechanical Foundations
The central premise of quantum mechanics is that every physical system is completely described by a wavefunction, $\Psi(\mathbf{r}, t)$, such that all system properties can be obtained by consulting the wavefunction. [One must bear in mind that the wavefunction, $\Psi(\mathbf{r}, t)$, is defined over the configuration space of the system, rather than over three-dimensional space. In the following, we shall take $\mathbf{r}$ to denote a point (or state of the system) in this configuration space.] Because the absolute square of the wavefunction, $\Psi(\mathbf{r}, t) \cdot \Psi^*(\mathbf{r}, t)$, in a region of configuration space is proportional to the probability

density of finding the system in the same region of space (or equivalently, in the state corresponding to that region of configuration space), the wavefunction of a system must be normalizable. Furthermore, physically meaningful wavefunctions are finite, single-valued, continuous and continuously differentiable over all space.

Operators (e.g., $\hat{A}$) and their corresponding eigenvalues (a_i) take the place of the dynamical variables used in classical mechanics. The eigenfunction solutions, ϕ_is, of the corresponding eigenvalue equation:

$$\hat{A}\phi_i = a_i\phi_i \tag{1.1}$$

form a possible basis set for expressing the system states, which are represented as wavefunctions, $\Psi = \sum_i c_i\phi_i$. A single measurement on any state of the system always yields a single eigenvalue (a_i). The probability that a particular eigenstate is observed is proportional to the absolute square of its coefficient, c_i^2, in the linear combination of eigenfunctions making up the initial state of the system, $\Psi(\mathbf{r}, t)$. However, once a particular eigenvalue, a_i, is observed the system remains in the corresponding eigenstate, ϕ_i, until it is further perturbed.

A collection of measurements on an ensemble of systems or particles with identical initial wavefunctions, results in an average value, that is, the expectation value ($\langle \hat{A} \rangle$) of the system.

$$\langle \hat{A} \rangle = \int \Psi^*(\mathbf{r}, t)\hat{A}\Psi(\mathbf{r}, t)d\mathbf{r} = \int \Psi^*(\mathbf{r}, t)\hat{A}\Psi(\mathbf{r}, t)d\mathbf{r}$$
$$= \sum_i \sum_j a_i c_i c_j \int \phi_j^* \phi_i \, d\tau = \sum_i a_i c_i^2 \tag{1.2}$$

The dynamics of a non-relativistic, quantum mechanical system are governed by the time-dependent Schrödinger equation:

$$i\hbar \left(\frac{\partial \Psi(\mathbf{r}, t)}{\partial t} \right) = \hat{H}\Psi(\mathbf{r}, t) \tag{1.3}$$

When the energy of the system (i.e., $\hat{H}$) has no explicit time dependence, we can derive and make use of the time-independent Schrödinger equation:

$$\hat{H}\Psi(\mathbf{r}) = E\Psi(\mathbf{r}) \tag{1.4}$$

The Hamiltonian operator, $\hat{H}$, which operates on the wavefunction to extract the system energy, E, contains both potential energy terms for the interactions of particles in the system with each other or any external fields, $\hat{V}$, and a kinetic energy term, $\hat{T}$, for each particle, which for a particle with mass m is written in atomic units[1] as:

$$\hat{T} = \frac{-\nabla^2}{2m} \tag{1.5}$$

1) Atomic units simplify the appearance of many of the most important equations in quantum mechanics by setting the following constants as fundamental atomic units equal to 1: the unit of mass is the rest mass of an electron ($m_e = 1$ atomic unit of mass), the unit of charge is the elementary charge ($e = 1$ atomic unit of charge), the unit of angular momentum is the reduced Planck's constant ($\hbar = 1$ atomic unit of angular momentum), the atomic unit for the electric constant is the Coulomb force constant ($\frac{1}{4\pi\varepsilon} = 1$ atomic unit). Atomic units for other dimensions are readily derived from these.

For a molecule composed of N nuclei with nuclear masses and charges m_ν and Z_ν, respectively, and M electrons, $\hat{H}$ can be written as:

$$\hat{H} = \sum_{\nu=1}^{N} \frac{-\nabla_\nu^2}{2m_\nu} + \sum_{i=1}^{M} \frac{-\nabla_i^2}{2} - \sum_{i=1}^{M}\sum_{\nu=1}^{N} \frac{Z_\nu}{|\mathbf{r}_i - \mathbf{R}_\nu|} + \frac{1}{2}\sum_{\substack{i,j=1 \\ i \neq j}}^{M} \frac{1}{|\mathbf{r}_i - \mathbf{r}_j|}$$

$$+ \frac{1}{2}\sum_{\substack{\nu,\mu=1 \\ \nu \neq \mu}}^{N} \frac{Z_\nu Z_\mu}{|\mathbf{R}_\nu - \mathbf{R}_\mu|} = \sum_{\nu=1}^{N} \hat{T}_\nu + \sum_{i=1}^{M} \hat{T}_i + \hat{V}$$

(1.6)

The form of the potential energy terms describing the electromagnetic interactions between each pair of particles couples the motions of the particles, barring analytical solutions for all but the most trivial systems. Nevertheless, useful approximate solutions are within reach for many systems of interest, in part due to the variational principle, which states that because all solutions of the Schrödinger equation are linear combinations of the eigenfunction solutions, that the ground state (i.e., lowest energy) solution forms the lower limit of the system energy. Thus, the energy of any solution we generate will be greater than or equal to the ground state energy, and we can optimize any approximation of the ground state by minimizing its energy.

1.3.1.2 Born–Oppenheimer Approximation

The Hamiltonian for a molecule includes kinetic and potential energy terms for both electrons and nuclei, and operates on a wavefunction describing both electrons and nuclei. However, because the kinetic energy of the nuclei is small relative to the electronic kinetic energy, it can be ignored. This allows the electronic wavefunction to be calculated based on localized nuclear coordinates, rather than a delocalized nuclear wavefunction. Thus, the following electronic Schrödinger equation can be separated out of the time-independent Schrödinger equation:

$$\hat{H}_e \Psi_e(\mathbf{r}_i, \{\mathbf{R}_\nu\})$$

$$= \left(-\sum_{i=1}^{M} \frac{1}{2}\nabla_i^2 + \frac{1}{2}\sum_{\substack{i,j=1 \\ i \neq j}}^{M} \frac{1}{|\mathbf{r}_i - \mathbf{r}_j|} - \sum_{i=1}^{M}\sum_{\nu=1}^{N} \frac{Z_\nu}{|\mathbf{r}_i - \mathbf{R}_\nu|} \right) \Psi_e(\mathbf{r}_i, \{\mathbf{R}_\nu\})$$

(1.7)

$$= E_e(\mathbf{R}_\nu) \Psi_e(\mathbf{r}_i, \{\mathbf{R}_\nu\})$$

where $\Psi_e(\mathbf{r}_i \sigma_i, \{\mathbf{R}_\nu\})$ are the electronic wave functions, and $\{\mathbf{R}_\nu\}$ denotes that nuclear spatial coordinates are parameters and not variables.

In this approximation, first introduced by Max Born and J. Robert Oppenheimer in 1927 [28], the heavy nuclei are thought of as fixed relative to the rapid motion of the quickly moving electrons, allowing the electrons to fully equilibrate to fixed nuclear positions. The equilibrated electronic energy as a function of the nuclear coordinates then forms a potential energy surface, with local minima (stable structures) and saddle points (transition states). Thus, obtaining the electronic structure and the corresponding system energy as a function of the nuclear positions is important in computational chemistry. The Born–Oppenheimer approximation is appropriate for the vast majority of cases; however, it breaks down for nuclear configurations where there are solutions to the electronic Schrödinger equation with similar energies.

1.3.1.3 Single-Electron Hamiltonians

While the Born–Oppenheimer approximation simplifies the Schrödinger equation by separating out the motion of the nuclei, the wavefunctions for the electrons are still coupled through their electrostatic interactions, making the resulting equations very difficult to solve. Of course there is no such difficulty for a single-electron system.

$$\left[\frac{-\nabla^2}{2} + \sum_{\nu=1}^{N} \frac{-Z_\nu}{|\mathbf{r} - \mathbf{R}_\nu|}\right]\psi_e = E_e \psi_e \quad (1.8)$$

This suggests that a crude, but readily soluble, approximation for a multi-electron system might be made by neglecting the electron–electron interactions altogether, so that the wavefunction for each electron ψ_i is solved for independently.[2] In this case, the total electronic wavefunction, Ψ_e, would be written as the product of single electron wavefunctions ψ_a. The total wavefunction is known as the Hartree product:

$$\Psi_e = \prod_{i,j} \psi_i(\mathbf{r}_j) \quad (1.9)$$

A more reasonable approximation than non-interacting electrons is to decouple the motions of the electrons, so that each electron interacts with the field associated with the average charge density of the other electrons rather than directly with the wavefunctions describing the other electrons. The resulting single-electron hamiltonian is known as the Hartree hamiltonian [29–31] (here for the ith-electron):

$$h_i = -\frac{\nabla_i^2}{2} + \sum_{\nu=1}^{N} \frac{-Z_\nu}{|\mathbf{r}_i - \mathbf{R}_\nu|} + \sum_{j=1}^{M} \int \psi_j^*(\mathbf{r}') \frac{1}{|\mathbf{r}_i - \mathbf{r}'|} \psi_j(\mathbf{r}') d\mathbf{r}' \quad (1.10)$$

Because the final potential energy term contains the charge density of the other electrons, the electronic wavefunctions are now dependent on each other. Thus, to evaluate the Hartree hamiltonian for one electron, the wavefunctions for all of the other electrons (or at least their net charge density distribution) must first be known. The way out of this chicken and egg problem, is to start with a trial wavefunction. This initial guess provides the background charge density for solving for each single-electron Hartree hamiltonian individually. These new solutions then provide the initial guess for a new set of solutions. The process can be repeated iteratively until the solutions converge (i.e., the wavefunctions remain relatively unchanged over the

---

2) Here we use ψ instead of Ψ to indicate single-particle wavefunctions of a multi-particle system.

course of a single iteration in which each wavefunction is optimized against the other current wavefunctions in turn) in what is known as a self-consistent field (SCF) method. The energy for such a system can be written as the sum of the energies of the single-electron hamiltonians; however, because each electron–electron interaction is fully accounted for twice (i.e., once in the single electron hamiltonian of each electron involved) we must compensate for the double counting by subtracting the total electron repulsion energy from the sum of individual electron energies. Thus, the total energy of the system can be written as:

$$E = \sum_{i=1}^{M} \varepsilon_i - \frac{1}{2} \sum_{j \neq i} \iint \frac{\rho_i \rho_j}{r_{ij}} d^3r_i d^3r_j \tag{1.11}$$

where ρ_i is the self-consistent charge density distribution corresponding to the single-electron wavefunctions, $\psi_i(r_i)$.

1.3.1.4 Basis Sets

The single-electron wavefunctions (ψ_i) that form the Hartree product total electronic wavefunctions are not known analytically, but rather must be formed from some other basis set (ϕ_β).

$$\psi_i = \sum_\beta c_\beta \phi_\beta \tag{1.12}$$

While, from a purely mathematical point of view, any complete, and thus infinite, basis set will do, in practice we are limited to finite basis sets. Furthermore, computational considerations favor small basis sets, whose functions, when substituted into the various HF equations, result in integrals which can be efficiently computed. At the same time, the resulting single-electron wavefunctions should be as accurate as possible. This is best accomplished by choosing basis functions that resemble the well-known solutions to atomic hydrogen.

1.3.1.4.1 Slater and Gaussian Type Orbitals

An obvious choice along these lines is Slater type orbitals [32] (STOs), centered on the atomic nuclei. This basis set mirrors the exact orbitals for the hydrogen atom, and naturally forms a minimum basis set, which means that each electron is described by only one basis function.

$$\Phi_{z,n,l,m}(r, \Theta, \phi) = N Y_{l,m}(\Theta, \phi) r^{n-1} \exp(-Zr) \tag{1.13}$$

Unfortunately, all of the requisite integrals cannot be evaluated analytically, thus Gaussian type orbitals [33,34] (GTOs) provide an alternative basis, with easy to evaluate integrals.

$$\Phi_{z,n,l,m}(r, \Theta, \phi) = N Y_{l,m}(\Theta, \phi) r^{2n-2-l} \exp(-Zr^2) \tag{1.14}$$

However, GTOs lack a cusp at the origin and decay too rapidly as they move away from the origin. This second deficiency can be remedied by using a linear combination of GTOs in place of each STO, which has been fit to reproduce the STO it replaces. Linear combinations of three GTOs have been found to optimize the relationship between accuracy and computational expense. This is known as the STO-3G basis set. Results can be further improved by moving beyond a minimum basis set, so that each electron orbital is described by not just one basis set orbital, but two, three or more. Such decompression results in so called double-ζ, triple-ζ, *and so on* basis sets. Additionally, polarization functions, consisting of basis functions with one quantum number higher angular momentum (i.e., p for s and d for p orbitals), can be included to give more flexibility to valence electrons, and diffuse functions included to better describe weakly bound electrons.

1.3.1.4.2 Plane Waves
For periodic systems, particularly those with delocalized electrons, such as metals, plane waves provide a natural choice for the basis set:

$$\phi_k(\mathbf{r}) = e^{-i\mathbf{k}\mathbf{r}} \tag{1.15}$$

where for a unit cell vector $\mathbf{r}$, $\mathbf{k}$ is restricted to the first Brillouin zone:

$$\mathbf{r} \cdot \mathbf{k} = 2\pi n \tag{1.16}$$

for n is any integer.

The size of the basis set required for the calculations to converge (i.e., the highest value of n) must be tested for each system.

1.3.1.4.3 Effective Core Potentials
A common strategy for reducing the expense of modeling heavy elements, which contain many electrons, is to replace each nucleus and its core electrons with an effective core potential (ECP). The valence electrons, which are primarily responsible for chemical interactions, are then explicitly modeled in the presence of the ECP (or pseudopotential as it is called in the physics community). Besides reducing computational expenses, ECPs also provide a means of implicitly incorporating the relativistic effects which can be important for accurately describing the core electrons of heavy elements into a simulation without further complicating the rest of the treatment of valence electrons with unimportant relativistic contributions. While it is possible to include all core electrons in the ECP, better results are often obtained by modeling some of the highest energy core electrons explicitly alongside the valence electrons.

1.3.1.5 Enforcing the Pauli Principle
As fermions, electrons are not allowed to share identical states; rather, they are indistinguishable particles which must occupy distinct states in such a way that the sign of their combined wavefunction is reversed when they are exchanged. The Pauli principle requires that the total electronic wavefunction is antisymmetric with respect to the interchange of any two electrons. This implies two conditions. The first is that all single-electron spatial orbitals must be orthonormal:

$$\int \phi_i^* \phi_j dr = \delta_{ij} \tag{1.17}$$

The second is that the overall wavefunction, $\Psi(\ldots, i, \ldots, j, \ldots)$, is antisymmetric with respect to the exchange of any two electrons i and j:

$$\Psi(\ldots, j, \ldots, i, \ldots) = -\Psi(\ldots, i, \ldots, j, \ldots) \tag{1.18}$$

These rules are typically enforced by writing the total wavefunction in the form of a Slater determinant, whose terms are products of single-electron spin orbitals, $\psi_i = \phi_i \sigma_i$, which are each composed of a spatial orbital, ϕ_i, and a spin component, σ_i. For an M-electron system the Slater determinant has the following form:

$$\Psi_e = \frac{1}{\sqrt{M!}} \begin{vmatrix} \psi'_1(r_1\sigma_1) & \psi'_1(r_2\sigma_2) & \cdots & \psi'_1(r_M\sigma_M) \\ \psi'_2(r_1\sigma_1) & \psi'_2(r_2\sigma_2) & \cdots & \psi'_2(r_M\sigma_M) \\ \vdots & \vdots & \ddots & \vdots \\ \psi'_M(r_1\sigma_1) & \psi'_M(r_2\sigma_2) & \cdots & \psi'_M(r_M\sigma_M) \end{vmatrix}. \quad (1.19)$$

Because electrons are indistinguishable, every electron appears in every orbital. Furthermore, electrons have two spin possibilities. Thus for every spatially distinct single-electron wavefunction there are actually a pair of orbitals, one with up (α) and one with down (β) spin. An appropriately anti-symmetrized pair of spin orbitals (α and β) have the form:

$$\frac{\alpha(1)\beta(2) - \alpha(2)\beta(1)}{\sqrt{2}} \quad (1.20)$$

1.3.1.6 Electron Correlation Methods

An important property of spin is that when the interaction of two same-spin orbitals is evaluated, an extra electron correlation, known as exchange, appears, due to the ability of same-spin electrons to avoid each other. Thus the expression for the potential energy of interaction between two same-spin electrons occupying orbitals ψ_i and ψ_j is:

$$V_{ee} = \iint \left(\frac{\psi_i(r_1)\psi_j(r_2)\psi_i^*(r_1)\psi_j^*(r_2)}{|r_1 - r_2|} - \frac{\psi_i(r_1)\psi_j(r_2)\psi_i^*(r_2)\psi_j^*(r_1)}{|r_1 - r_2|} \right) d^3r_1 d^3r_2 \quad (1.21)$$

$$= J_{ij} - K_{ij}$$

While for opposite-spin electrons, it is simply:

$$V_{ee} = \iint \frac{\psi_i(r_1)\psi_j(r_2)\psi_i^*(r_1)\psi_j^*(r_2)}{|r_1 - r_2|} d^3r_1 d^3r_2 = J_{ij} \quad (1.22)$$

Nevertheless, this electron exchange is the only electron correlation accounted for in the Hartree–Fock method. Accounting for dynamical correlation due to explicit electron–electron interaction and non-dynamical correlation due to wavefunction contributions from higher energy electron configurations requires more advanced methods.

The Hartree–Fock equations approximate the Schrödinger equation by ignoring all correlation, with the exception of exchange. Nevertheless, by reintroducing this ignored electron correlation, they can be used as a stepping stone for returning to the full Schrödinger equation. How the Hartree–Fock method can be corrected to recapture the electron correlation that has been lost, is the subject of this section.

1.3.1.6.1 Configuration Interaction

By solving the Hartree–Fock equations we arrive at an optimal HF wavefunction, Ψ_{HF}. Furthermore, we are also able to solve for excited states of the system, within the Hartree–Fock approximation, Ψ_{HF}^n. Any one of these solutions can be substituted into the exact Schrödinger equation (i.e., operated on using the exact many-electron Hamiltonian operator to yield the system energy), such that the variational principle still applies. However, further improving any one of these solutions is hampered by the coupling of the electronic motions, which is the original problem the HF equations were designed to avoid. Nevertheless, if a set of Hartree–Fock configurations $\{\Psi_{HF}^n\}$ exists we can write the wavefunction as a linear combination of Hartree–Fock configurations and optimize this linear combination according to the variational principle. This procedure is known as the configuration interaction (CI) method. Thus, finding approximate solutions to the exact Hamiltonian is made tractable by writing the wavefunction as a linear combination of Hartree–Fock configurations. The exact solution is approached in the limit of an infinite basis set (i.e., all possible Hartree–Fock configurations); however, in practice, computational considerations limit us to finite basis sets.

1.3.1.6.2 Perturbation Theory

The strategy in CI methods is to make use of solutions to the Hartree–Fock equations as the basis set for a linear combination of configurations that can be optimized for the exact, multi-electron hamiltonian within the context of the variational principle. By contrast, perturbation theory rewrites the multi-electron hamiltonian in terms of an analytically solvable component (H_0) and a correction factor (H_1). Møller–Plesset (MP) perturbation theory, takes H_0 to be the sum of the Hartree–Fock one-electron operators, excluding the electron–electron repulsion terms. The perturbation term (H_1) is then taken to be the electron–electron repulsion terms (taking into account the double counting of these interactions in the Hartree–Fock equations). This choice is appropriate insofar as the Schrödinger equation with H_0 is solvable; however, the electron–electron repulsion term is not always the relatively small perturbation that the theory assumes it to be.

The uncorrected energy from MP includes no electron–electron interactions, and the first-order correction recovers the Hartree–Fock result. Additional correction terms may offer further improvements to the Hartree–Fock energy; however, these terms become increasingly complex and computationally expensive to calculate. Furthermore, the nature of these correction factors means that there is no guarantee that the approximation at any level (beyond the first) provides either an upper or a lower bound to the real energy.

1.3.1.6.3 Coupled Cluster

Coupled cluster methods provide a way to systematically add various types of excitations to the wavefunction. Thus, the first-order terms include all single-electron excitations, the second-order terms all double excitations, and so on. Coupled cluster methodology resembles the CI approach insofar as it writes the wavefunction as a linear combination of configurations. It resembles perturbation theory in systematically ordering corrections factors from least to most significant.

1.3.1.6.4 Electron Correlation in Electrochemistry
At present, methods that explicitly include electron correlation are too expensive for use on most systems of interest in electrochemistry. Nevertheless, they occasionally find application calculating dispersion interactions or van der Waals forces, which depend heavily on electron correlations

1.3.1.7 Density Functional Theory
An alternative approach to wavefunction methods, is to work instead with the electron density. Not only is electron density more intuitive, but it also provides computational advantages, which have enabled density functional theory (DFT) to develop into a powerful and widely used methodology.

1.3.1.7.1 The Hohenberg–Kohn Theorems
Two theorems, first proved by Hohenberg and Kohn [35,36], provide the foundation for DFT. The first of these, the Hohenberg–Kohn existence theorem, states that the non-degenerate, ground-state electron density of a system uniquely determines the external potential of the system (and thus the hamiltonian) within a constant. This means that the system energy can be expressed as a functional of the electron density:

$$E_0 = f[\varrho(\mathbf{r})] \tag{1.23}$$

The second theorem, the Hohenberg–Kohn variational theorem, justifies the application of variational methods to the density. Thus, the correct ground state electron density distribution is the density distribution corresponding to the energy minimum. Thus

$$\frac{\partial f[\varrho(\mathbf{r}, \lambda)]}{\partial \lambda} = 0 \tag{1.24}$$

leads to the optimal ground state electron density $\varrho_0(\mathbf{r}, \lambda)$.

1.3.1.7.2 The Energy Functional
Unfortunately $f[\varrho_0(\mathbf{r})]$ is unknown; however, if we consider the electron density to be made up of non-interacting electrons (this is allowed, provided we really have the same density), we can write the energy as a sum of single energy operators:

$$E[\rho(r)] = \sum_{i}^{M} \left(\left\langle \chi_i \left| \frac{-\nabla_i^2}{2} \right| \chi_i \right\rangle - \left\langle \chi_i \left| \sum_{v}^{N} \frac{Z_v}{|\mathbf{r}_i - \mathbf{R}_v|} \right| \chi_i \right\rangle \right) \\ + \sum_{i}^{M} \left\langle \chi_i \left| \frac{1}{2} \int \frac{\rho(\mathbf{r'})}{|\mathbf{r}_i - \mathbf{r'}|} d^3 r' \right| \chi_i \right\rangle + E_{\text{XC}}[\rho(r)] \tag{1.25}$$

The first three terms correspond to the electronic kinetic energy, the electron–nucleus attractions, and the electron–electron repulsions, respectively. The form of these terms is well known; however, the form of the final term, E_{XC} which includes the electron exchange and correlation energies, along with a correction to the kinetic energy term, is unknown. From here, a single-electron hamiltonian is easily extracted:

$$h_i^{\text{KS}} = -\frac{\nabla_i^2}{2} - \sum_{v}^{N} \frac{Z_v}{|\mathbf{r}_i - \mathbf{R}_v|} + \int \frac{\rho(\mathbf{r'})}{|\mathbf{r}_i - \mathbf{r'}|} d^3 r' + V_{\text{XC}} \tag{1.26}$$

and

$$V_{XC} = \frac{\partial E_{XC}}{\partial \rho} \tag{1.27}$$

As in the case of the HF method, the energies obtained with this single-electron hamiltonian for each electron in the system can be summed to arrive at the total system energy. However, in contrast to HF, the result is the exact system energy, rather than an approximate energy. The difficulty is that, in contrast to HF, the exact form of V_{XC} is unknown. Thus, utilizing DFT requires approximating this term.

1.3.1.7.3 Exchange–Correlation Functionals A vast number of approximations for E_{XC} have been developed. These (and the DFT methods they are utilized within) can be conveniently classed in rungs on "Jacob's ladder" (Figure 1.3), which span the gap between Hartree results and chemical accuracy [37,38]. On the bottom rung of the ladder are methods which employ the localized density approximation (LDA). Here, the value of E_{XC} at each point is typically borrowed from the value of E_{XC} for a uniform electron gas with the same density as the local density of the point of interest. In any case, the value of E_{XC} depends only on $\rho(r)$.

The next level up is known as the generalized gradient approximation (GGA). Here, E_{XC} depends not only on $\rho(r)$, but also on its gradient, $\nabla \rho(r)$. Including either the Laplacian of the density ($\nabla^2 \rho(r)$), or the local electronic kinetic energies is known as the meta-generalized gradient approximation (MGGA).

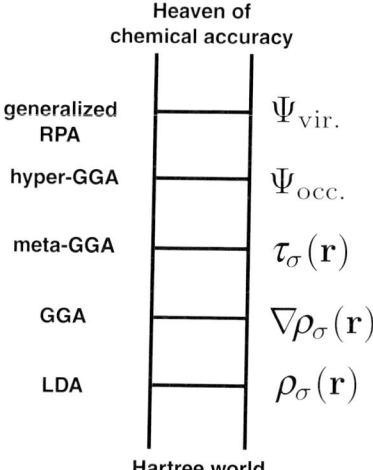

Figure 1.3 Jacob's ladder illustrating the hierarchy of approximations used to construct exchange-correlation functionals. Abreviated categorizations of methods are shown to the left of the ladder and the new level of dependence added at each level is shown on the right. Thus, the exchange-correlation functional at any given rung of the ladder will be a function of not only the quantity or construct directly to its right, but also of all quantities and constructs below it.

The hyper-generalized gradient approximation (hyper-GGA), uses the Kohn–Sham (KS) orbitals to calculate the exact HF exchange. The total E_{XC} is then written as a linear combination of the HF exchange and the E_{XC} from a LDA and/or GGA method.

Finally, at the highest level of "Jacob's ladder" sit generalized random phase methods (RPM). Not only are occupied KS orbitals included, but also virtual (i.e., unoccupied) KS orbitals.

1.3.2
Applications of Electronic Structure to Geometric Properties

The system energy is the most basic output of electronic structure calculations. Due to the Born–Oppenheimer approximation, this calculated system energy corresponds to the energy associated with a particular set of nuclear coordinates (i.e., a particular system geometry). Although the multi-dimensional configuration space containing all possible geometries is often quite large, only a few select geometries, namely local minima and the saddlepoints along the minimum energy pathways (MEP) connecting the local minima, are typically required for an accurate description of the chemical characteristics and behavior of the system. Methods for finding these local minima and saddlepoints, that is, those corresponding to the resting states (RS) and transition states (TS) of a chemical system, are indispensable for successfully applying electronic structure methods to chemical systems.

1.3.2.1 Geometry Optimization

Geometry optimization procedures begin with an initial geometry guess, typically provided by the user based on previous calculations, experimental results or chemical intuition. In addition to the energy of this initial structure, $U(\mathbf{r})$, additional quantities may be calculated in order to aid the following optimization procedure. The first of these is the energy gradient, $g = -\nabla U(\mathbf{r}) = \sum_i -\frac{\partial U(\mathbf{r})}{\partial r_i}$, which corresponds to the net forces exerted on each atom and is expressed as a vector in configuration space, $g = -\sum_i F_i$. In some cases, the components of the gradient are best evaluated analytically from the function used to compute the system energy, $U(r)$. However, in other cases it is either more efficient or necessary to evaluate them numerically, by displacing the geometry along the various directions in configuration space and computing the energy change that results from each displacement. By definition the gradient at a RS point (or any stationary point including a TS) is zero, thus, the gradient can also be used to help identify the end point of the calculation.

Another useful metric is the second derivative matrix, also known as the Hessian. The matrix components correspond to all possible pairs of coordinates i and j (including $i=j$) in configuration space, and have the value: $\left(\frac{\partial^2 U(\mathbf{r})}{\partial r_i \partial r_j}\right)$. At a stationary point, all of the off-diagonal terms of the Hessian are zero. In addition, the diagonal terms are all positive at a RS, and all are positive except for one, which is negative, at a TS. Thus the Hessian can not only be used to verify that a RS has been reached, but can also be used to distinguish between various types of RSs.

The most basic optimization strategy is to optimize the variables (i.e., coordinates in configuration space) one at a time. Because the variables are not independent there is no guarantee that the local minimum will be approached with this procedure, and even should it be approached this is likely to require many optimization cycles. Thus, for systems with more than a handful of variables (i.e., the vast majority of molecular geometries) this method is neither practical nor reliable. Fortunately other strategies are available.

Steepest descent methods optimize the structure at each optimization step along the direction of the gradient of the starting geometry for that step. Convergence is guaranteed using steepest descent methods; however, their efficiency is hampered by their intrinsic requirement that adjacent optimization steps be in perpendicular directions. Conjugate gradient methods attempt to remedy this and improve their efficiency by forming the new search direction from a linear combination of the previous search direction and the gradient of the current geometry.

Newton–Raphson methods take advantage of not only the gradient but also the Hessian to advance the optimization yet more quickly toward the nearest stationary point. Because this stationary point could be a minimum, maximum or saddlepoint, it is important that the initial guess is made within the desired region of configuration space.

1.3.2.2 Transition State Searches

Methods for locating saddle points can conveniently be categorized by the type of input or initial guess that they require. Local methods require nothing more than an initial geometry guess near the transition state [39]. As we have already seen Newton–Raphson methods are available for optimizing an initial input structure to the nearest stationary point (in this case a saddle point). An analysis of which internal coordinates (bond distances, angles, etc.) are significantly different in the products compared with the reactants sometimes provides insight into the region of configuration space in which the saddle point is most likely to be found, suggesting an initial guess. However, this is often not the case, making an acceptable initial guess hard to come by. This difficulty, combined with the high computation cost of calculating the Hessian at each step, makes local methods an impractical sole means for finding most saddle points. At the very least a different preliminary method is needed to arrive at a reasonable initial guess, before turning to local optimization methods. Interpolation methods are able to fill this role. Unlike local optimization methods, they require the minima on either side of the saddle point being sought as inputs [40]. Various schemes can then be used to trace out and optimize a MEP between the given minima. The crest of this MEP then corresponds to the saddle point being sought. In some cases system coordinates can be chosen such that the reaction coordinates are seen to clearly correspond to just one or two of these system coordinates. Constrained minimization, in which these reactive coordinates are stepped from their reactant to their product values, provides an intuitive means of mapping out a reaction pathway, in what is known as coordinate driving. Linearly

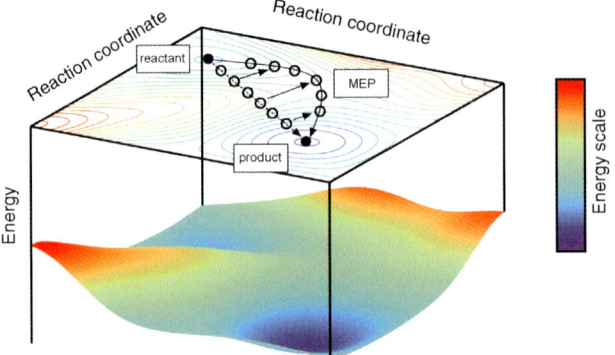

Figure 1.4 Illustration of how the NEB method would be applied to a two-dimensional potential energy surface to find a transition state. A typical initial guess used to initiate a NEB procedure, and the final MEP, at which the application of the NEB method arrives, are shown on the potential energy surface.

interpolating the reactant and product cartesian coordinates is another interpolation strategy, with a variety of strategies for optimizing the interpolated image(s) and driving one of them to the saddle point being available [41]. Particularly popular are variations of the nudged elastic band (NEB) method (Figure 1.4), which uses a spring constant to connect and evenly distribute images over the MEP [42].

1.3.3
Corrections to Potential Energy Surfaces and Reaction Pathways

While local minima and saddle points are the critical locations on a potential energy surface (PES) for describing chemical reactions, there is a collection of system states associated with each of these stationary points on the PES. These associated states include the vibrational, rotational, translational and electronic states of the molecules within the system, as well as configurational states of the system components. Taking into consideration the energetic and entropic contributions of these states is often vital for accurately modeling chemical reactions that might occur in a system.

1.3.3.1 Energy and Entropy Corrections
Vibrational states of molecules are typically treated within a harmonic approximation. In this case, the normal modes, $\mathbf{w}_\nu$, and corresponding frequencies, ω_ν, are found by solving the eigenvalue equation given by the Hessian:

$$\mathbf{H}\mathbf{w}_\nu = \omega_\nu^2 \mathbf{w}_\nu \tag{1.28}$$

Once the normal modes are known, the allowed energies and states for each mode are determined by the quantum mechanical treatment of the individual harmonic oscillators. Thus the energy states for a vibrational mode with frequency ω are given by:

$$\varepsilon_n = \hbar\omega\left(n + \frac{1}{2}\right), n = 0, 1, 2, 3, \ldots \tag{1.29}$$

The total partition function for all vibrational states ($n = 0, 1, 2,...$), from which their entropic contributions at temperature T can be extracted, is given by:

$$q_{vib} = \frac{\exp\left[-\frac{\hbar\omega}{2k_B T}\right]}{1 - \exp\left[-\frac{\hbar\omega}{k_B T}\right]} \tag{1.30}$$

In cases where gas phase molecules consisting of more than a single atom are present (i.e., where molecules are free to rotate), rotational degrees of freedom should be considered. These are typically accounted for within a rigid rotor approximation, which assumes that the internal geometry of the molecule is fixed. For a diatomic molecule the energy levels obtained from the quantum mechanical solution of a rigid rotor for two masses, m_1 and m_2, lying distances r_1 and r_2, respectively, from their combined center of mass, leading to a moment of inertia $I = m_1 r_1^2 + m_2 r_2^2$, are:

$$\varepsilon_j = \frac{\hbar^2}{2I}j(j+1), j = 0, 1, 2, 3, \ldots \tag{1.31}$$

When we take into account the symmetry index, σ (which is 1 for a heterogeneous and 2 for a homogeneous diatomic molecule), we can write the partition function encompassing all rotational states as:

$$q_{rot} = \frac{2 I k_B T}{\sigma \hbar^2} \tag{1.32}$$

For more complex molecules with three moments of inertia, I_1, I_2 and I_3, the partition function is:

$$q_{rot} = \frac{\sqrt{\pi I_1 I_2 I_3}}{\sigma} \left(\frac{2 k_B T}{\hbar^2}\right)^{\frac{3}{2}} \tag{1.33}$$

In the case of gas phase molecules, not only rotational but also translational motion should be considered. The expected energy contribution from translational degrees of freedom for an ideal gas particle with mass m at temperature T is given by:

$$\langle \varepsilon_{trans} \rangle = \frac{3}{2} k_B T \tag{1.34}$$

and the partition function for such a particle constrained in a volume V is:

$$q_{trans} = \left(\frac{m k_B T}{2\pi \hbar^2}\right)^{\frac{3}{2}} V \tag{1.35}$$

While there are cases where excited electronic states are low enough in energy to play a significant role, in the majority of cases only the ground electronic state is accessible and thus the ground state energy computed using an electronic structure method provides a good approximation for the electronic energy of the system, E_{elec}. The partition function for this electronic state, q_{elec} is given by its degeneracy, g_0, that is, its spin multiplicity.

The total system energy can now be computed by summing these various energy contributions:

$$E_{tot} = E_{elec} + E_{vib} + E_{rot} + E_{trans} \tag{1.36}$$

Similarly the overall partition function is formed by taking the product of the various contributing partition functions:

$$Q = q_{elec} \cdot q_{vib} \cdot q_{rot} \cdot q_{trans} \tag{1.37}$$

Armed with these energy correction terms and their corresponding partition functions, we are able to calculate a variety of useful thermodynamic quantities, such as the zero-point energy, entropy, enthalpy and Gibbs free energy.

1.3.3.2 Thermodynamic State Functions

A stationary state, and indeed any point on a PES, corresponds to a set of fixed nuclear coordinates. However, the Heisenberg uncertainty principle does not allow for fixed nuclear coordinates, because the positions and momenta of the nuclei would then be simultaneously, exactly determined. Similarly, the absence of a zero-energy state among the vibrational states means that regardless of the temperature (i.e., even as absolute zero is approached), molecules continue to vibrate, albeit only in their lowest vibrational states. Thus meaningful comparison with experiment requires correcting system energies to include these zero-point corrections. Adding these zero-point corrections to the "bottom of the well" energy, E_{elec}, yields the zero-point-corrected energy of the system:

$$E_0 = E_e + \frac{1}{2} \sum_{\omega} \hbar \omega \tag{1.38}$$

This zero-point energy then corresponds to the effective energy in experiments conducted at temperatures approaching 0 K.

Making additional comparisons with macroscopic quantities such as pressure p, temperature T, and volume V requires us to use the tools of statistical mechanics to derive appropriate thermodynamic quantities from the overall partition function, we found above. Thus, the temperature-dependent internal energy of a system at a particular temperature T, $U(S, V)$, can be calculated from its partition function, Q:

$$U(S, V) = k_B T^2 \left(\frac{\partial \ln Q}{\partial T} \right)_{N,V} \tag{1.39}$$

An often more relevant quantity is the enthalpy, $H(S, p)$, since it corresponds to experimentally determined heats of formation:

$$H(S, p) = U(S, V) + pV \tag{1.40}$$

The entropy, $S(U, T)$, of a system can also be calculated directly from the partition function:

$$S(U, V) = k_B \ln Q + k_B T \left(\frac{\partial \ln Q}{\partial T} \right)_{N,V} \tag{1.41}$$

Using this entropy both the Helmholtz, $A(T, V)$, and the Gibbs, $G(T, p)$, free energies can also be computed:

$$A(T, V) = U(S, V) - TS \qquad G(T, p) = H(S, p) - TS \qquad (1.42)$$

The Gibbs free energy is particularly useful as it corresponds to the most common experimental conditions (fixed temperature and pressure). Finally, it should be noted that these thermodynamic quantities can not only be computed for individual points on the PES, but that these points can then be used to construct new energy surfaces, corresponding to the thermodynamic quantity of interest.

1.3.3.3 Reaction Energies and Rates

The raw energy from a single electronic structure calculation typically has little meaning on its own for several reasons. First, the use of pseudo-potentials masks the energetic contributions of core electrons. Second, the energy to form the system from infinitely separated electrons and nuclei (or pseudo-potentials when they are used) is almost never directly relevant to experimental measurables. Finally, systematic errors associated with the use of any particular method and basis set tend to result in relatively large errors for a given individual calculation.

If, instead, relative energies are calculated, the situation changes considerably. Approximations and systematic errors often cancel out, yielding much improved accuracy. Furthermore, reaction energies and heats of formation are standard experimentally measured quantities, which are readily available for many reactions and substances.

The heat of formation, $\Delta_f H°_{298}$, of a system composed of i-indexed atoms with stoichiometric factors, ν_i, and standard state enthalpies, $H°_{298}(i)$, computed from electronic structure calculations as shown above, and an electronic-structure-calculation-derived system enthalpy of $H°_{298\text{sys}}$ is:

$$\Delta_f H°_{298} = H°_{298\text{sys}} - \sum_i \nu_i H°_{298,i} \qquad (1.43)$$

Reaction energies, enthalpies and free energies can also be calculated analogously. Thus, the Gibbs free energy for a reaction is given by:

$$\Delta G° = G°_{\text{products}} - G°_{\text{reactants}} \qquad (1.44)$$

The Gibbs free energy (or thermodynamic quantity for another relevant ensemble) can then be used to calculate the equilibrium constant for the reaction of interest:

$$K_{eq} = \exp\left[\frac{-\Delta G°}{RT}\right] \qquad (1.45)$$

This is all the information needed to characterize a system at equilibrium; however, non-equilibrium systems, and in particular reaction rates, are often of interest in electrochemistry. The rate of an elementary reaction step can be written as the product of the concentrations of the reactants, each raised to the power of its stoichiometric factor, times a rate constant:

$$\text{Rate} = k_{\text{rate}} \prod_i [i]^{\nu_i} \tag{1.46}$$

Transition state theory provides an approximate means for calculating the rate constant from the initial state (i.e., reactants) and transition state separating the reactants and products on the PES:

$$k_{\text{rate}} = \frac{k_B T}{h} \exp\left[\frac{-\Delta G^{\ddagger}}{RT}\right] \tag{1.47}$$

where $\Delta G^{\ddagger}$ is the free energy difference between the reactants and the transition state. The critical assumption here is that the system begins in the reactant state and then samples nearby states until at last it samples the transition state, at which point it proceeds directly to the product state. Thus, the exponential term can be interpreted as the Boltzmann factor for finding the system at the transition state, and the forefactor gives the rate at which new states are sampled.

1.3.4
Electronic Structure Models in Electrochemistry

Traditional electrochemical experiments involve systems composed of around 10^{26} atoms. However, with current computational technology, routine electronic structure calculations are only feasible for around 10^2 atoms. Thus, modeling macroscopic-scale experimental systems with nano-scale electronic structure calculations requires the three following assumptions or approximations. First, the assumption that the experimental system can be broken up into various spatially localized processes (e.g., electrochemical reactions at the interface or ion diffusion in the electrolyte). Secondly, that the macroscopic properties of these processes can not merely be derived from the sum of all underlying atomistic processes, but rather, that they can be derived from the properties of a limited number of representative exemplars of the atomistic processes (rather than summing them exhaustively). Finally, that there are models involving no more than 10^2 atoms which reliably represent the critical atomistic processes in the macroscopic system and can thus be used as the exemplars that form the starting point for derivations of macroscopic properties.

The most commonly modeled processes and regions using electronic structure calculations are the chemical reactions at the interface between the ion- and electron-conductors. Typically, this is a solid–liquid interface. The number of atoms explicitly involved in an individual electrochemical reaction at such an interface is typically within the limits of the number of atoms that can be explicitly treated with electronic structure methods. The electrode surface, as well as the surrounding electrolyte, often plays a crucial role in surface reactions; however, both extend well beyond the nano-scale limitations of electronic structure calculations into the macroscopic scale. Thus, the application of electronic structure methods to electrochemical surface reactions, requires restricting the extent to which the solvent and surface are treated explicitly (i.e., modeled as individual atoms).

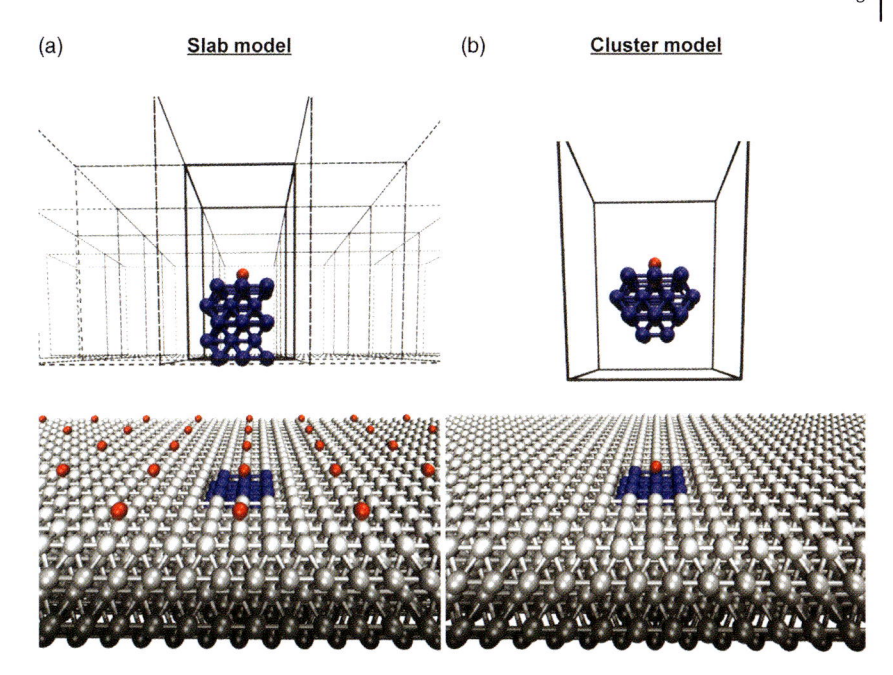

Figure 1.5 Illustrations of slab (a) and cluster (b) methods for modeling an electrode surface. In the slab approximation, the electrode surface is modeled using a periodically infinite slab. This infinite slab (with a periodically recurring adsorbate) is shown below, while the contents of a single simulation supercell are shown above among the outlines of neighboring supercells. In the cluster approximation a cluster is conceptually lifted out of the electrode surface, and taken to be representative of the whole on the basis of convergence tests, where various cluster sizes and geometries are tested.

1.3.4.1 Modeling the Electrode Surface: Cluster versus Slab

There are two basic approaches for modeling extended electrode surfaces: the cluster approach and the slab/supercell approach (Figure 1.5) [27,43–45]. In the cluster approach a cluster of atoms surrounding the active surface site, where the reaction is to take place, is cut out and used to model the electrode surface. In determining which particular cluster model to use, tests should be performed to verify that the cluster chosen well approximates the chemical activity of the active site for the reaction of interest. Because computational capabilities were once much more limited than they are today, clusters were a popular model choice, because a cluster can be formed with as few as a single atom. However, many of these early, small cluster models have been found to result in inadequate, if not altogether false, descriptions of the surface chemistry. Indeed, it has been shown that clusters consisting of 20 to more than 50 atoms are required to reliably model the simplest metal surfaces [46]. Thus it is important to verify that one's results are converged with respect to cluster size before adopting a cluster model. Cluster models are particularly appropriate for studying isolated reactions, in which no neighboring adsorbates are involved. In contrast to cluster models, which limit the extent of the electrode in all directions, slab/supercell methods limit only the depth of the electrode. Infinite extent in the directions perpendicular to the surface is procured by introducing periodicity. Convergence should be tested to verify that the slab is sufficiently thick to model the surface of a bulk system. The convergence for unit cell lengths perpendicular to the surface is also important to verify when low surface coverage situations are being modeled. Otherwise these unit cell lengths can be modified to correspond to the surface coverage being modeled.

1.3.4.2 Modeling the Solvent: Explicit versus Implicit

The simplest way to deal with the electrolyte or surrounding solvent molecules in electronic structure calculations is to assume that they do not exist. Such gas phase models are appropriate for modeling many surface science experiments, especially under ultra high vacuum (UHV) conditions, however; solvent effects often exert significant influence on surface chemistry in electrochemical environments. Two strategies are used to include these important solvent effects in electronic structure calculations. They are known as explicit and implicit solvation methods (Figure 1.6). In explicit solvation methods a limited number of explicit solvent molecules directly surrounding the solvated molecule model solvation. This procedure is particularly important when one of these solvent molecules plays a chemical role in the reaction, such as temporarily donating a proton. Nevertheless, explicit solvation is not only computationally expensive, but cumbersome. Each additional explicit solvent molecule not only increases the number of electrons in the system, but increases the number of possible solvent configurations that must be tested in order to find the optimal solvent configuration. Furthermore, because DFT does not well describe the van der Waals forces, which are often important in solvation interactions, there is no guarantee that even a large number of optimally placed solvent molecules in a DFT calculation will adequately describe solvation effects.

(a) **Explicit solvation model**   (b) **Implicit solvation model**

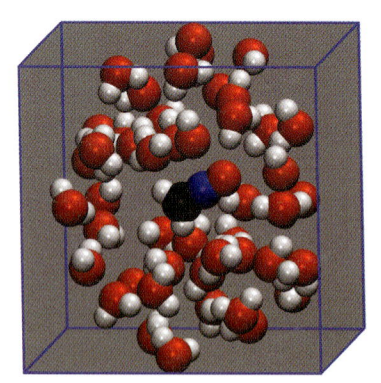

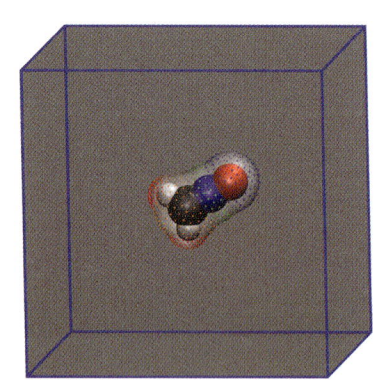

Figure 1.6 Illustrations of explicit (a) and implicit (b) treatments of solvation. In an explicit solvent model, the solvent is modeled using a limited number of explicit solvent molecules. In an implicit solvent model, the effects of the solvent are typically included in the form of an electric field surrounding the solute molecule(s).

To avoid the computational expense of explicit solvation methods, implicit methods have been developed. These methods assume that solvation effects primarily result from electrostatic interactions, and that the effect of the solvent on the solvated molecule can thus be reasonably represented as an electric field, known as the reaction field. To do this the solvent is modeled as a continuous dielectric with a cavity in the region of the solvated molecule. The polarization of the solvent can then be solved for based on the charge distribution of the solute in the cavity, yielding an electric field in the cavity which acts on the solvated molecule. A self-consistent solution can be approached by solving for the solvent polarization and then its subsequent effect on the solvated molecule as a part of each SCF iteration.

1.3.4.3 Modeling the Electrode Potential

The electrode potential is a critical element in an electrochemical system and distinguishes electrochemistry from simple catalysis. The electrode potential affects electrochemical reactions in two ways. First, because it relates to the chemical potential difference of electrons at the anode and cathode, it determines the energy for transferring an electron between electrodes following each oxidative process at the anode and each reductive process at the cathode. Thus, the electrode potential determines the chemical potential of electrons involved in electrochemical half-cell reactions. Electronic structure calculations typically consider neutral systems and thus do not include the electron transfer between electrodes. However, this can be included in the free energy of the half-cell reaction by including an additional potential-dependent term. Thus the free energy of a reaction that produces z electrons at an electrode at potential U is:

$$\Delta G = G_{\text{products}} - G_{\text{reactants}} - zeU \tag{1.48}$$

where e is the charge of an electron.

Secondly, the electrode potential attracts oppositely charged counter ions, which form a double layer region above the surface, across which an electric field is induced. This electric field influences the structures and energies of surface adsorbates and reactions.

There are three general approaches for handling this induced electric field. The first is to ignore it altogether and treat the electrode potential as solely affecting the chemical potential of electrons (i.e., the fermi level of the system). The second approach is to include an applied external field in the calculations. This is typically carried out within the supercell approach by introducing a dipole layer into the vacuum region between slabs. The primary difficulty here is knowing how strong the electric field, $\mathbf{E}$, induced by the electrode potential, U, is [47]. If the double layer thickness, d, can be estimated [48], then this can be approximated (assuming a plate-capacitor) as:

$$|\mathbf{E}| = \frac{U}{d} \tag{1.49}$$

The third approach is to place a net charge on the electrode. In the case of a slab model (i.e., periodic boundary conditions) counter charges are required to keep the cell neutral. They can be introduced in the form of a uniform background charge, a localized counter electrode or explicit counter ions. Some care should be taken in doing this, as artifacts are known to arise, particularly in the case of a uniform background charge.

Once the slab is allowed to take on a net charge, it is not only possible to set this charge explicitly by fixing the number of electrons, N_e, but also implicitly by fixing the chemical potential of the electrons associated with the slab, μ_e. While the latter approach more directly models a fixed electrode potential, calculations with a fixed number of electrons are typically sufficiently faster that it is more efficient to vary N_e over several calculations until the value corresponding to the desired μ_e is found, than to restrain μ_e directly within the electronic structure calculations.

1.3.5
Summary

This section began by reviewing some of the fundamental principles of quantum mechanics. Central to the application of these principles is the Schrödinger equation, which is typically too complicated to allow for exact solution. A series of approximations was then outlined, which led to a range of methods for calculating approximate electronic wavefunctions (or electron densities in the case of DFT) and electronic energies. The first of these approximations was the Born–Oppenheimer approximation which, by separating the nuclear and electronic components of the wavefunction, leads to a potential energy surface defined by the electronic energy corresponding to each set of possible nuclear coordinate values. Optimization procedures can then be used to locate local minima and saddle points on this potential energy surface, which correspond to stable molecular geometries and transition states, respectively. Vibrational modes can be derived from further analysis of these stationary points, which can then be used to derive a variety of thermodynamic quantities for the chemical species represented by stationary points on the potential energy surface. Finally, the application of transition state theory allows derivation of kinetic reaction rates from these thermodynamic quantities. Thus, electronic structure methods can be used to provide both thermodynamic and kinetic properties of chemical species and reactions to which they are applied.

Electronic structure methods are most commonly used in electrochemistry to model chemical reactions on electrode surfaces. In addition to the atoms directly involved in the reaction of interest there are three system components which need to be considered in developing a reliable model: the electrode surface, the solvating electrolyte, and the electrode potential. The electrode surface is modeled using either a finite cluster or an infinite slab. In either case it is crucial to verify that the cluster or slab used is sufficiently large and well-formed to reliably model the catalytic properties of a macroscopic electrode. Depending on its anticipated role in the reaction, the electrolyte solvent can be modeled either implicitly, using an applied electric field to simulate its electrostatic influence on the system, or explicitly by including a small number of solvent molecules in the calculation. Finally, the influence of the electrode potential can be accounted for by adjusting the fermi level of calculations after their completion. Additionally, it is possible to include the effects of the electrode potential in forming a double layer, either by applying an appropriate external electric field to the calculation directly, or by placing a net charge on the electrode, which induces a similar effect.

1.4
Molecular Simulations

The equations of quantum mechanics are complex and computationally expensive to solve, even in approximate forms. This drastically limits both the size of the systems which it is feasible to treat quantum mechanically, as well as the number of energy calculations that can be performed on a given system. Force field methods use classical-like forces to approximate the quantum mechanical interactions between atoms and molecules in order to decrease the computational expense of calculating these forces by many orders of magnitude. Thus, at the molecular level, intuitive concepts of chemical bonding are used to embed quantum mechanical effects and phenomena into a classical framework.

The aim of this section is twofold: first, to briefly describe the conceptual foundations and practical development of force fields typically used in molecular simulations; second, to consider various ways in which force fields can be applied to electrochemical systems in terms of both simulation techniques and model systems. Strategies for realizing both sequential and concurrent coupling with the electronic structure methods presented in the previous section are also discussed. More complete treatments of molecular simulation methods [25,26,49,50] and additional details relating to particular methods or applications are available in the references cited in the chapter, as well as in many others.

1.4.1
Energy Terms and Force Field Parameters

A force field is a set of equations for calculating the potential energy of a system of atoms. The system energy is computed by summing energy contributions derived from various types of interatomic interactions, as shown in Eq. (1.50).

$$V_{sys} = V_{bond} + V_{angle} + V_{torsion} + V_{coulomb} + V_{vdW} + \ldots \tag{1.50}$$

These potential energy contributions are typically calculated as functions of the system geometry. Parameters in the energy expressions are typically fit to experimental or computational data. In spite of the wide variety of functional forms used by various force fields to compute the individual energy terms, the widespread agreement among them regarding the types of energy contributions that need to be included reveals an underlying conceptual unity.

1.4.1.1 Covalent Bond Interactions
Energy terms in most force fields can be divided into two classes: those associated with covalently bonded atoms (i.e., with covalent bond dependence) (Figure 1.7) and those associated with all pairs of atoms (i.e., no covalent bond dependence).

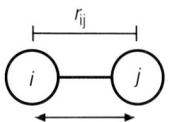

Figure 1.7 Schematic representation of covalent bond interactions. The bond energy is primarily a function of the internuclear bond distance, r_{ij}, between atoms i and j.

1.4.1.1.1 Reactive versus Non-Reactive

Whether a force field is "reactive" or not (i.e., describes changes in chemical bonding) is the most fundamental distinction between force field types. Reactive force fields compute a covalent bond order as a function of interatomic distances, and allow it to continuously vary between zero and full bond order (1, 2 or 3 depending on the atoms involved), such that a physically reasonable description of bond formation and cleavage is given [51–53]. Furthermore, other bond-dependent quantities scale with this continuous bond order as it varies between zero and full bond order. Because chemical reactions may result in an atom converting from one type or configuration to another (e.g., one orbital hybridization or oxidation state to another), a universal description of each atom within the force field must adequately model all of these possible forms.

In the case of non-reactive force fields whole number covalent bond orders are predetermined in the input or computed from initial bond distances. These covalent bond assignments are then maintained for the duration of the simulation. Thus, different atom types (i.e., force field descriptions) can be used to describe the same element in different chemical bonding configurations.

1.4.1.1.2 Covalent Bond Strength

The most basic function of covalent bond energy terms is to describe the energy associated with stretching or compressing the bond. For the small changes associated with molecular vibrations, a harmonic oscillator provides a simple model of the energy associated with stretching or compressing the bond (Figure 1.8). Thus, the simplest non-reactive force fields calculate the energy associated with stretching or compressing covalent bonds using a harmonic oscillator, in which case the bonding can be described by Eq. (1.51):

$$V_{\text{bond-harmonic}} = \sum_{i,j \neq i} \frac{1}{2} k_b \left(r_{ij} - r_0 \right)^2. \tag{1.51}$$

This equation is known as Hooke's law, where r_0 denotes the equilibrium distance and k_b the force constant of the bond. The total bond potential is the sum over all bonds.

Describing covalent bond energies using harmonic oscillators forbids dissociation, because the energy increases quadratically as the bond is stretched, whereas dissociation requires the covalent energy function to level off at the dissociation energy within a couple of angstroms of the equilibrium bond distance. Nevertheless, further excessive bond compression should result in high energies due to nuclear–nuclear repulsions. Thus, a function describing the covalent bond energy must be asymmetric, with one side of a harmonic-like well climbing steeply to represent

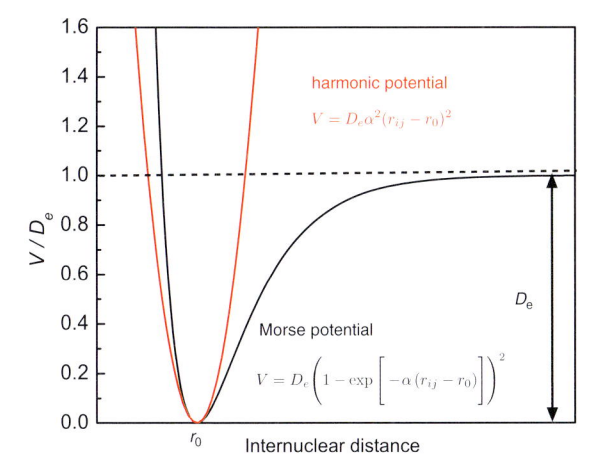

Figure 1.8 Graphic representation of common expressions used to compute covalent bond energies or energy deviations. A harmonic potential can be used to compute the energy for displacing covalently bonded atoms from their equilibrium distance within a non-reactive context. A Morse potential can be used to describe covalent bond formation/cleavage in a reactive force field. $\vec{r}_{ij}$ between two atoms i and j, and the force constant k_b.

internuclear repulsions, while the other side of the well gradually flattens out as the atoms dissociate. The Morse potential is a simple function meeting these criteria, and is one of the simplest choices for a reactive internuclear covalent energy expression (Figure 1.8). It can be written in terms of a minimum interatomic potential energy, D_e, a corresponding equilibrium distance, r_0, and a factor, α, related to the well stiffness and hence the allowed vibrational modes.

$$V_{\text{bond-Morse}} = \sum_{i,j \neq i} D_e (1 - \exp[-\alpha(\vec{r}_{ij} - r_0)])^2 \tag{1.52}$$

In both reactive and non-reactive force fields it is convenient to reference the energy at the bottom of the covalent bond well to the bond energy relative to some standard state of the atoms. This means that the computed energies have some intrinsic, independent meaning, and not simply relative values for comparison with other structures one explicitly calculates. A further requirement for reactive force fields is that the depth of the covalent bond energy function (i.e., the energy at the bottom of the well compared with the energy at the dissociation limit) must correspond to the bond energy relative to the dissociated atoms.

1.4.1.1.3 Angles

Describing geometries of molecules involving more than two atoms involves not just bond distances, but also bond angles involving three bonded atoms (two atoms each bonded to a third, central atom). Angle energy terms are typically computed as penalties based on the difference between the actual value of the angle and its optimal equilibrium value. Thus they correct the energy for deviations from the optimum geometry. Equation (1.53) shows how energy contributions related to angles might be computed within a harmonic

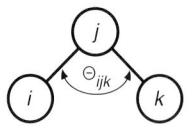

Figure 1.9 Schematic representation of the angle contribution. This contribution is described by the force constant k_Θ and the angle Θ_{ijk}.

approximation for displacements from an equilibrium angle.

$$V_{\text{angle}} = \sum_{i,j,k} \frac{1}{2} k_\Theta (\Theta_{ijk} - \Theta_0)^2 \tag{1.53}$$

k_Θ denotes the force constant, Θ_{ijk} the angle between the bonds as shown in Figure 1.9 and Θ_0 the equilibrium angle.

1.4.1.1.4 Torsion

The energy associated with rotation about a chemical bond is given by a torsional energy term which, like angle terms, is computed as an energy penalty for deviations from an optimal, equilibrium dihedral angle θ_0 (Figure 1.10). Unlike angle terms, where only small deviations from equilibrium are expected, torsional angle terms need to be able to account for all possible dihedral angles, since full rotation is common. Thus a periodic function is needed. The following abbreviated Fourier series provides a simple expression for the torsional energy:

$$V_{\text{torsion}} = \sum_{\text{Quadrupletts}} \sum_{n=0}^{3} \frac{1}{2} V_n [1 + \cos(n\phi - \delta)] \tag{1.54}$$

Which terms in the n summation are non-zero will depend on the geometry of the rotation. For example, rotation about the double bond in ethene has a periodicity of 180° and so only the $n = 2$ term is non-zero. Analogously, rotation about the single bond of ethane has a periodicity of 120°, so that only the $n = 3$ term is needed. If one of the hydrogen atoms on each carbon is replaced with another substituent, then the $n = 1$ term is needed to differentiate between the various minima. Additional

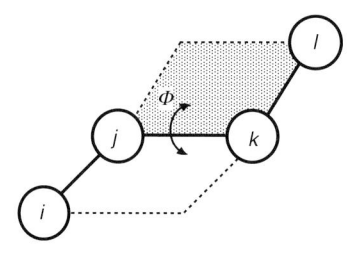

Figure 1.10 Schematic representation of the torsional deformation. This deformation is described by the angle of intersection between the two planes, which are defined by the atoms i, j, k and j, k, l, and the force constant V_n.

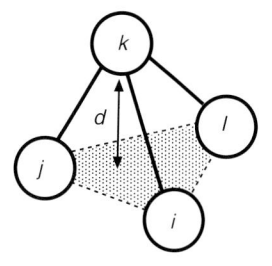

Figure 1.11 Schematic representation of out-of-plane energy correction for central atom k with respect to the plane defined by atoms i, j and l.

combinations and/or higher order terms may be used to model more complex situations.

In the case of reactive force fields it is important for these correction terms to vanish as any one of the covalent bonds involved in the torsion or angle dissociates. Otherwise the penalty will be falsely imposed on groups of non-bonded atoms. Polynomial expressions are typically used to model angle displacements, and trigonometric functions are most often used for torsions.

1.4.1.1.5 Inversion An sp^2 atom with three planar substituents should suffer relatively large energetic losses for displacement from the plane its substituents occupy. Because the angle correction terms are not strong enough to account for these forces, an additional inversion energy term is often included for such atoms. The simplest way to express the inversion energy is as a harmonic function for displacements measured in terms of the distance, d of the central atom from the plane defined by its substituents and an empirically chosen spring constant, α. See Figure 1.11.

$$V_{inversion} = \sum_k \frac{1}{2}\alpha d^2 \qquad (1.55)$$

1.4.1.1.6 Under/Over Coordination In reactive force fields penalties for under- and over-coordination (i.e., an atom forming too many or too few covalent bonds) are sometimes used to enforce proper covalency. To do this the total bond order associated with an atom is computed by summing the bond orders of the covalent bonds the atom is involved in. Then an energy penalty is assigned depending on how much this total bond order falls below or surpasses the expected, optimal total bond order. It is also possible to partially enforce the expected atom bond orders by correcting the bond orders so that bond orders involving atoms whose valencies are not fully filled (based on preliminary bond orders) are enhanced, while bond orders involving over-coordinated atoms are diminished.

1.4.1.1.7 Cross Terms Additional terms can also be introduced to correct for the interactions between the various terms. For example, the equilibrium C—H bond distance in methane, should be dependent on the H—C—H angles. The introduction of such cross terms can be useful in reducing the number of different atom types required to adequately represent a given element.

1.4.1.1.8 Resonance An extra energy term is required to stabilize resonant structures in conjugated systems. This energy can be added into non-reactive force field descriptions, either by developing an additional atom type to correspond to each bonding combination that results in resonance or by adding correction values from crude electronic structure calculations. For reactive force fields this terms comes into play when successive bonds have bond orders near 1.5.

1.4.1.2 Non-Covalent Interactions

Beside interactions associated with covalent bonds, several interactions between atoms which are not associated with covalent bonds are important. These interactions can be defined as occurring only between non-bonded pairs of atoms. Alternatively, when the strengths of the covalent bond interactions are adjusted to compensate for their presence, they can be evaluated for every pair of atoms.

1.4.1.2.1 Electrostatics Electrostatic interactions are typically modeled by approximating individual atoms with point charges q, whose energy of interaction is given by:

$$V_{coulomb} = \sum_i \sum_{j<i} \frac{1}{4\pi\varepsilon_{ij}} \frac{q_i q_j}{r_{ij}} \tag{1.56}$$

The partial charges associated with the atoms are either assigned based on the results of experiments or theoretical measurements, or are computed within the force field based on the system geometry and relative electronegativities of the atoms. The permittivity (ε_{ij}) is often assigned empirically to match experimental data or higher level calculations rather than being derived from first principles. See Figure 1.12.

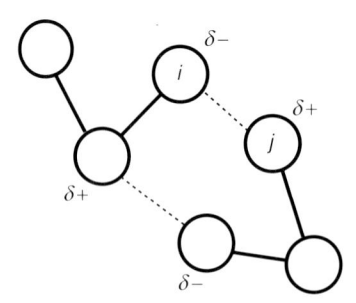

Figure 1.12 Schematic representation of electrostatic interactions between atoms i and j, carrying partial charges q_i and q_j, respectively. The interaction energy is then calculated based on these partial charges, the distance between i and j, r_{ij}, and the permittivity of the medium, ε_{ij}.

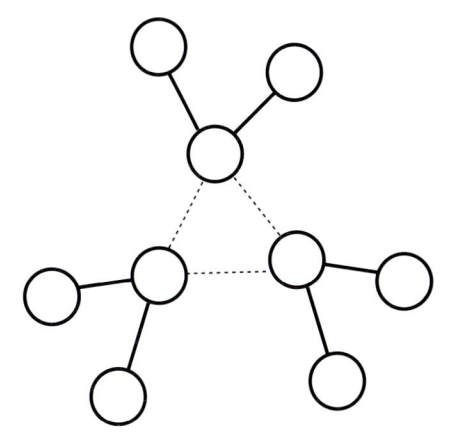

Figure 1.13 Schematic representation of the Van der Waals contribution. This contribution is described by a Lennard-Jones potential.

1.4.1.2.2 Van der Waals A second important class of non-bond interactions are van der Waals interactions. These forces result from temporarily induced dipoles, as two atoms approach another (Figure 1.13). A variety of potential energy expressions can be used to describe van der Waals interactions. The Lennard-Jones potential (also known as a 12-6 potential, where the numbers refer to the repulsive and attractive exponents which make up the potential) is the one most frequently used (Figure 1.14). It is given in Eq. (1.57):

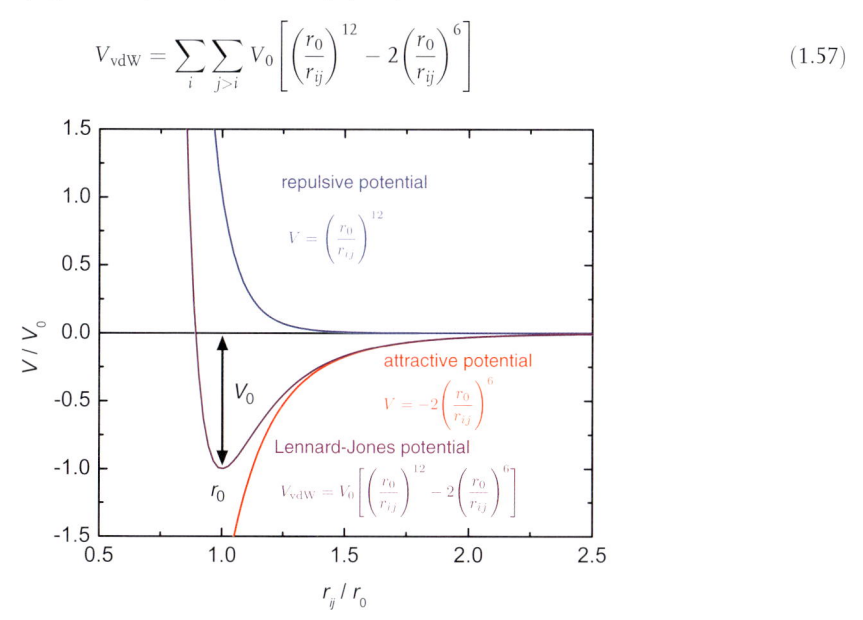

$$V_{\text{vdW}} = \sum_i \sum_{j>i} V_0 \left[\left(\frac{r_0}{r_{ij}}\right)^{12} - 2\left(\frac{r_0}{r_{ij}}\right)^6 \right] \tag{1.57}$$

Figure 1.14 Graphical representation of the Lennard-Jones potential, showing its attractive and repulsive components.

Thus, van der Waals forces are attractive at large values, but repulsive for small values of the internuclear separation r_{ij}.

1.4.1.2.3 Hydrogen Bonds

Hydrogen bonding is often partially accounted for in the van der Waals, and electrostatic terms; however, an additional correction factor is useful for improving the description of hydrogen bonds. This attractive interaction between hydrogen H_i and an atom X_j to which it is hydrogen bonded, is often modeled as:

$$V_{Hbond} = \sum_i \sum_j V_0 \left[5 \left(\frac{r_0}{r_{ij}} \right)^{12} - 6 \left(\frac{r_0}{r_{ij}} \right)^{10} \right] \tag{1.58}$$

This closely resembles the Lennard–Jones potential used to model van der Waals forces, except that the attractive region disappears more quickly as the interatomic distance increases.

1.4.2 Parametrization and Validation

The many equations that make up a force field involve a number of parameters. These parameters must be obtained somehow. One possibility is fitting them to experimental data; however, the amount of relevant experimental data available for fitting force field parameters is often miniscule compared with the number of parameters that need to be established. Thus, a second alternative is to use data from electronic structure calculations to find suitable force field parameters. This second approach is a classic example of serial multiscale modeling. Results obtained from electronic structure methods provide the input for a molecular model.

Parameters are typically fit by minimizing some measure of the error they result in when results they produce are compared with the training set data. This can be accomplished by either fitting individual parameters to particularly relevant parts of the training set, or by fitting groups of parameters to larger subsets of the training set. A variety of mathematical tools and computational algorithms is available for optimizing force field parameters, however, it is important to keep the physical nature of the parameters in mind (i.e., magnitude and qualitative behavior of the energies they produce), so that the various energy terms continue to correspond to the chemical concepts they were designed to represent. The physical reasonableness of a force field formulation thus provides an important justification for applying the force field to systems which are not explicitly included in the training set. Nevertheless, it is important to know the limitations of an individual force field and to use caution when applying it to a system in which the atomic interactions deviate significantly from those contained in the training set.

A given parametrization can be validated using one of three different types of tests. The first type of test uses the parameters to reproduce data that were not included in the training set, but which are of the same type (e.g., energies of formation from DFT) as that contained in the training set. If the parameters are able to reproduce this data, it suggests that their validity extends beyond the immediate confines of the data they were trained against. A second way of using QM results to validate a parametrization is to extract novel structures or reaction pathways, which are observed during the course of simulations making use of the parameters, and verify that QM calculations indeed predict similar structures and energies for these snapshots. Returning to a previous level of modeling in order to verify the reliability of the newly established level is an important, although often overlooked, step in building a robust, multiscale framework.

The third type of test involves reproducing larger scale macroscopic properties, which are known experimentally, but cannot be directly reproduced using the electronic structure methods used to produce all or most of the training set (e. g., the melting point of a solid). This final validation procedure is useful in verifying that the parameters describe not only the atomistic phenomena predicted by electronic structure methods, but also higher order experimentally observable phenomena they are designed to model on the basis of the atomistic interactions.

1.4.3
Atomistic Simulations

While the behavior of some systems can be accurately characterized by exploring only the few points in phase space which correspond to selected local minima and the saddle-points connecting them, broader regions of phase space must be explored to reliably model other systems. Small molecules under UHV provide a good example of the former type because the vast majority of intermolecular interactions involve only a pair of molecules. Liquids provide a typical example of the latter type, because the relatively weak, but nevertheless significant interactions between molecules support a vast number of molecular configurations with similar energies.

The relevant local energy minima and transition states between these are too numerous to make an explicit, systematic search a practical computational tool. Furthermore, average thermodynamic properties of such systems, rather than detailed snapshots, are typically what we want to extract from the simulations. A representative sample of configuration space is sufficient for this task. Atomistic simulation techniques provide us with strategies for obtaining such representative samples.

Because a large number of structures must be sampled to provide a representative sampling of a region of phase space, less computationally intensive methods are generally preferred for computing the necessary system properties for each sampled configuration. Thus atomistic force fields provide the most common PES description used in atomistic simulations. Nevertheless, methods, such as Car–Parrinello (CPMD) [54], Born–Oppenheimer (BOMD) [55] and Ehrenfest molecular dynamics [56] have been developed to utilize electronic structure methods in molecular simulations.

1.4.3.1 Monte Carlo Methods
Monte Carlo (MC) methods [49,57] provide a means of randomly exploring phase space starting in the region of an initial structure. For each simulation

step a random number determines how to perturb the system. The energy of the perturbed system is then calculated and compared with the energy of the system prior to the perturbation. If the energy of the system has decreased, then the new structure is automatically accepted. If the energy of the system increases by ΔE then the Boltzmann factor $e^{\frac{-\Delta E}{k_B T}}$ is compared with a random number between 0 and 1. If the Boltzmann factor is larger than this random number, then the new structure is accepted. This enables a thermodynamically weighted sampling of configuration space at the chosen temperature, T, while still allowing for the possibility of overcoming energy barriers to sample configuration space beyond the valley in which the structure of the initial guess lies.

One of the advantages of MC methods is that nearly any conceivable internal or external system coordinate can be used as the variable to be perturbed in generating new structures. Cartesian coordinates, bond distances and angles can be perturbed, atom positions can be exchanged, or atoms can even be added to or removed from the system.

1.4.3.2 Molecular Dynamics

The ergodic hypothesis implies that, regardless of the initial state we choose, if a thermodynamic variable is monitored as we follow the dynamical trajectory of the system, its value will converge to the appropriate statistical, expectation value as the time approaches infinity. In practice this means that, if followed over a sufficiently long time period, the system dynamics provide a thermodynamically representative sampling of a system. Typically this is accomplished by applying Newton's second law to determine the motion of the atoms.

$$m\frac{\partial^2 \mathbf{r}}{\partial t^2} = -\nabla V \tag{1.59}$$

Here, V represents the potential energy surface, which can be mapped out piece-wise using a force field, electronic structure or other method for computing system energies. For all but the most trivial systems, analytical solutions to these equations are not readily available. Thus the dynamics are propagated by using a variety of numerical algorithms (e.g., Verlet, leap-frog or velocity Verlet [58]) to estimate the particle positions and velocities across finite time intervals. The smaller these time intervals or time steps, the better simulation approximates the true trajectory; however, smaller time steps mean that more simulation time is required to explore phase space. Meaningful trajectories typically require time steps which are at least an order of magnitude shorter than the fastest process in the system. Typically these are vibrational frequencies with an order of magnitude of $10^{14} s^{-1}$, so that time steps of the order of 10^{-15} are necessary.

Because the application of Newton's second law of motion (Eq. (1.59)) conserves energy, the natural ensemble for molecular dynamics (MD) simulations is the NVE ensemble, in which the number of each type of particle (N), the system volume (V) and the energy of system (E) are held constant. However, algorithms for simulating other ensembles, in which the chemical potential (μ) might be held constant instead of N, or the external pressure (P) instead of the system volume, or the system temperature (T) instead of E, have also been developed (e.g., NVT, NPT, μVT) [59].

1.4.3.3 QM/MM

While force field parameters are often arrived at by means of a serial approach to multiscale modeling, it is also possible to combine molecular and electronic structure approaches to coherently model a single system using a concurrent approach to multiscale modeling, known as QM/MM modeling [60]. This approach is suited for systems in which properties better described by electronic structure methods (e.g., reactivity, electron density) are important in only a small portion of a large system. This smaller subsystem is modeled using an electronic structure method (i.e., using the principles of quantum mechanics, QM), while the rest of the system is described by an atomistic force field (i.e., using molecular modeling, MM). The system energy is then the sum of the energies of the isolated QM and MM regions in addition to their interaction energy (see Figure 1.15).

The way in which a system is divided into QM and MM regions and the manner in which these two regions are subsequently coupled are the defining characteristics for any QM/MM method. Location (as determined either by cartesian coordinates or chemical connectivity) often plays a decisive role in assigning atoms and molecules to either the QM or MM region, such that only atoms located in the vicinity of the reaction (or other phenomenon of interest) are treated with QM. This region can be either fixed or allowed to vary throughout the simulation as the reaction spreads or migrates. The assignment of atoms to either the QM or MM region can either take place once for all at the beginning of the simulation, or the atoms can be reassigned at regular intervals based on their up-to-date locations.

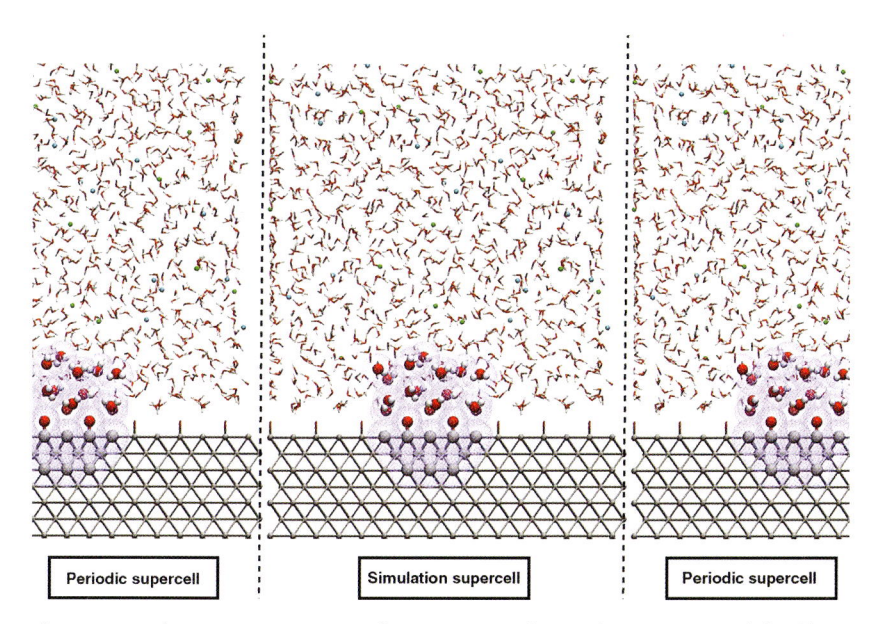

Figure 1.15 Schematic representation of a QM/MM simulation. The QM-region is defined by balls and sticks, the MM-region in licorice. The QM/MM boundary region is represented by a blue net surrounding the QM region.

Methods for coupling the QM and MM regions can be divided into two basic categories: mechanical embedding (ME) schemes and electrostatic embedding (EE) schemes. These two general approaches differ in the degree and manner of electrostatic coupling between the two regions. In ME schemes, the QM region is first treated without reference to the surrounding MM region. The electronic structure results for the QM system are then reinterpreted into MM terms and, as such, embedded into the molecular description of the MM region. Thus, perturbations in the electronic structure of the QM region from electrostatic interactions with the MM region are ignored. A second potential problem with ME is that it is often difficult to find an appropriate MM description of the QM region, as the lack of such a description was likely one of the motivations for turning to a QM/MM approach. The primary advantage of the ME approach is its relative simplicity.

Instead of being ignored (as is the case in ME approaches) EE schemes include the electrostatic influence of the MM region on the QM region by introducing point charges, and possibly higher order multipoles (indeed, finding a way to do this efficiently and accurately is a key challenge in the implementations of EE schemes), into the electronic structure calculations, representing the electrostatic influence of the MM region. Thus, polarization of the electronic structure of the QM region by the electrostatic influence of the MM region is enabled. In turn the QM region should be allowed to polarize the MM – and, indeed, there are self-consistent methods available for reaching electrostatic equilibrium – however, this effect is typically ignored. Nevertheless, EE schemes are able to avoid the key disadvantages of ME schemes; however, the simplicity and computational efficiency of the ME approach is lost.

Depending on the system geometry, it may be necessary to draw the QM–MM boundary such that it separates covalently bonded atoms. This is problematic for the atom left in the QM region, because it is left with dangling bonds, which need to be incorporated into covalent bonds for a reliable description of its electronic structure. This is typically accomplished by adding either link atoms or so-called "localized orbitals" to rebuild covalent bonds severed by the QM–MM boundary. The link atom approach typically replaces the missing atom in a severed covalent bond with a hydrogen atom, which is positioned to recreate, as best as possible, the original covalent bond. This approach is popular because it is fairly straightforward and easy to generalize. When an EE scheme is used, care must be taken to ensure that nearby point charges (representing the missing atoms from the severed covalent bonds) do not induce artificial polarization of the covalent bond.

Localized orbital approaches, on the other hand, are more fundamental, but also more complicated to implement. In one version, strictly localized bond orbitals (SLBOs) are determined from calculations of analogous bonds in small molecules. The SLBOs are then included in the electronic structure calculations, but not allowed to mix with any other orbitals during SCF optimization. Yet another option is to replace the missing atoms with complete sets of $s + 3p$ orbitals. Their hybridization (sp^3, sp^2+p_z, etc.) is determined by the local geometry/coordination and only active orbitals (i.e., those involved in the severed covalent bond) are included in SCF optimization. The key disadvantage with these localized orbital approaches is that a new localized orbital must be developed to accurately describe each unique covalent bond.

1.4.4
Sampling and Analysis

Abstracting reliable information from molecular simulations depends on having a representative sample with respect to the properties or behaviors of interest. This typically requires equilibrating the system by performing dynamics until any biases in the initial configuration of the system have been randomized away. Following the equilibration period, measurements can be made on the further dynamics of the system. This further portion of the simulation must be extensive enough for the relevant regions of phase space to be adequately explored for satisfactory representation in the property extracted. Performing multiple simulations with independent initial configurations provides an alternative strategy for obtaining a statistically meaningful sampling of configuration space.

A vast assortment of information, ranging from macroscopic thermodynamic properties to average structural properties to individual atomistic steps in reaction mechanisms, can be extracted from appropriate molecular simulations. Thermodynamic properties, such as temperature and pressure, are easily calculated from the atomic positions and momenta using their statistical mechanical definitions. Structural motifs can be extracted by calculating radial distribution functions for various atom types in the system, or assigning molecular bonding based on interatomic distances. Entropy and related properties can be extracted from the auto-correlation function [61–64].

1.4.5
Applications of Molecular Modeling in Electrochemistry

Molecular methods have not enjoyed the same level of popularity for applications in electrochemistry enjoyed by electronic structure methods. A large part of this is probably due to the inability of the vast majority of force fields to describe the reactive events which are at the heart of electrochemistry. Nevertheless, several non-reactive aspects of electrochemical systems have been modeled extensively using molecular force fields. The structure of metal–electrolyte interfaces has been one popular subject for such studies [50]. Another has been the solvation of ions in the electrolyte solution and their adsorption on the electrode surface [50]. The structure and stability of the electrochemical double layer that forms above the electrode surface has been a third [65]. In all of these cases statistical averaging over molecular dynamic trajectories is typically used to arrive at a description of the interfacial structure.

The recent advent of force fields capable of describing chemical reactions opens the door to studying catalytic events, similar to those currently investigated using electronic structure methods, only over a significantly larger time and length scale. This has indeed already been realized to a limited extent [66]; however, there are many more systems to which this approach might yet be fruitfully applied.

1.4.6
Summary

This section began with an overview of various energy terms which can contribute to force field descriptions of molecular systems. The use of results from electronic structure calculations to optimize force field parameters in the process of force field development is a quintessential example of serial coupling. Monte Carlo and molecular dynamic simulations are two strategies for sampling the potential energy surface defined by a force field in order to extract information about the chemical system it describes. It is also possible to use molecular force field and electronic structure methods to simultaneously describe different regions of the same system in what are known as QM/MM methods. These are a classical example of a concurrent coupling approach to multiscale modeling. Molecular modeling techniques have been applied to aspects of electrochemistry, especially the electrochemical double layer and other electrode–electrolyte interfaces. The ability to employ reactive force fields to study electrochemical reactions traditionally confined to electronic structure methods has recently opened up exciting opportunities in this arena.

1.5
Reaction Modeling

1.5.1
Introduction

The system descriptions used for the molecular level simulations described in the previous section were primarily physical, rather than chemical. Chemical properties, such as covalency, are indeed typically included in force fields; however, they are included under a physical, as opposed to a chemical, guise. In other words, force fields are written in the language of Newtonian forces, rather than chemical reactivities, and thus it is these physical forces (rather than their chemical basis) that are explicitly used to determine system dynamics. In contrast to these physical force field descriptions, the methods presented in this section employ primarily chemical descriptions of systems, with chemical reactivity providing the driving force for exploring the chemical dynamics of a system. The primary advantage of turning from a physical to a chemical description of a reactive system is quite simple. The vast majority of physical processes and events (e.g., collisions, rotations and vibrations) are not reactive events (i.e., chemical processes) and thus in and of themselves are of no interest for obtaining a chemical description of a system. Yet because of their enormous frequency (often many orders of magnitude greater) compared with chemical events, the vast majority of the computational resources required for a physical simulation of a system are used to simulate uninteresting physical processes. Only a very small fraction of the employed resources are then actually used to directly simulate the reactive events of interest. A chemical description of a system, on the other hand, only considers chemical processes explicitly and thus avoids the tremendous expense of explicitly simulating uninteresting physical processes.

1.5.2
Chemical Kinetics

A chemical description of a system is typically given in terms of discrete chemical states, which are connected by chemical reactions. Mathematically, this means that a chemical system can be modeled by a so-called Markov process. The dynamics of such a system (i.e., a system limited to discrete states with probabilistic rules governing the transitions between these various states) is governed by the so-called Master equation:

$$\frac{dP(\sigma)}{dt} = \sum_{\sigma' \neq \sigma} P(\sigma')W(\sigma' \to \sigma) - \sum_{\sigma' \neq \sigma} P(\sigma)W(\sigma \to \sigma') \qquad (1.60)$$

System states are denoted as σ, and the probability of finding the system in a particular state, σ, is $P(\sigma)$. The probability per unit time of a transition from state σ to state σ' (assuming the system is initially found in state σ) is represented by $W(\sigma \to \sigma')$. If the system can be broken down into a collection of lattice sites with index i, then the overall state of the system can be written as a conjunction of the states of the individual sites and the evolution of the state of a particular site i can be written as:

$$dP(\sigma_i) = \sum_k w_{ik}^+(\sigma)dt - w_{ik}^-(\sigma)dt \qquad (1.61)$$

where k is an index over all processes leading to or from the σ-state and where the transition probabilities, w_{ij}, for the processes are functions of the overall state of the system. The Master equation, in either form, is deterministic; however, its high dimensionality makes an analytical solution impossible to find for all but the most trivial systems, and the vast number of states and processes to consider makes a deterministic, numerical solution equally impractical in most cases. Thus stochastic methods are typically employed to simulate the chemical dynamics described by the Master equation. The remainder of this section is devoted to describing how a family of approaches, known as kinetic Monte Carlo (kMC) methods, can be employed to provide stochastic solutions of the Master equation.

1.5.3
Kinetic Monte Carlo

Kinetic Monte Carlo methods extend the time scale of reactive simulations by using statistical mechanics to fast forward the dynamics to the next reactive event, rather than explicitly simulating long periods of quasi-equilibrium dynamics between reactions [67–70].

1.5.3.1 System States and the Lattice Approximation
In order to apply the Master equation to a chemical system we must first characterize it in terms of discrete states. For crystal structures or surfaces the lattice

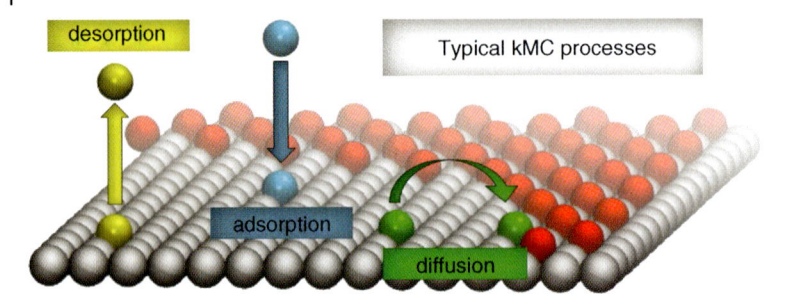

Figure 1.16 Schematic representation of typical processes included in kMC models: adsorption, desorption and diffusion.

approximation provides an easy and intuitive way of doing this. In the lattice approximation the system is spatially divided into discrete sites, with each site being allowed to be either wholly unoccupied or occupied by one of the possible chemical species present in the system. Typically, three types of processes, the first two of which are illustrated in Figure 1.16, are then considered. First, adsorption/desorption of a species at a site. Second, diffusion of an adsorbate from one site to an adjacent site. Third, reactions between species on adjacent sites. The reactive events involve two neighboring sites at most, or perhaps just one site and an implicitly included gas or liquid phase molecule. As a result only the local configuration (i.e., the state of nearby sites) is necessary for determining which reactive processes are possible at a given lattice site. The main disadvantages of the lattice approximation are that it is only applicable to strictly crystalline or crystal-like systems, and that it typically only allows for simple species and processes involving at most one or two lattice sites. To avoid these disadvantages, off-lattice kMC approaches have been developed [71–74]. Nevertheless, the vast majority of kMC calculations still make use of the lattice approximation.

1.5.3.2 Reaction Rates

Once the discrete states of the system have been defined the possible transitions between these states and their corresponding transition rates need to be determined. Within the on-lattice approximation, reactions typically only involve one or two lattice sites directly and are typically only influenced by the states of nearby sites (rather than the configuration of the entire system). As such, all possible reactions are easily categorized into a much smaller number of prototypical reactions, each of which has an associated reaction rate which is valid for all reactions of its type. The rates of these prototypical reactions are then extracted from either experimental data or higher level calculations making use of transition state theory, as discussed in Section 1.3.3.3. The use of an electronic structure or reactive force field method to obtain the reaction parameters needed to define a kMC model is a good example of a multiscale model that can be developed using either serial or concurrent coupling [75]. In the serial-coupling approach all reaction parameters need to fill in a preconceived reaction table (whose contents are determined by the developer) and obtained from the higher level method independent of the kMC model they are used to parametrize. The primary benefit of this approach is the simplicity of its implementation. In contrast, concurrent-coupling approaches try to take advantage of automated reaction and transition state search algorithms to locate and characterize reactions on-the-fly, that is, during the course of a kMC simulation. The possibility of discovering novel reactions or reaction pathways has thus far been the primary motivation for the development of on-the-fly methods [71,76]. Another possible benefit for systems with large numbers of possible reactions would be the computational savings resulting from only calculating the barriers for reactions actually needed to run the kMC simulation rather than all conceivable reactions.

1.5.3.3 Reaction Dynamics

Now that the system states and probabilities of transitions between states have been determined, it remains to be discussed how to stochastically step from one state to another in such a way that the reactive dynamics of the system are meaningfully

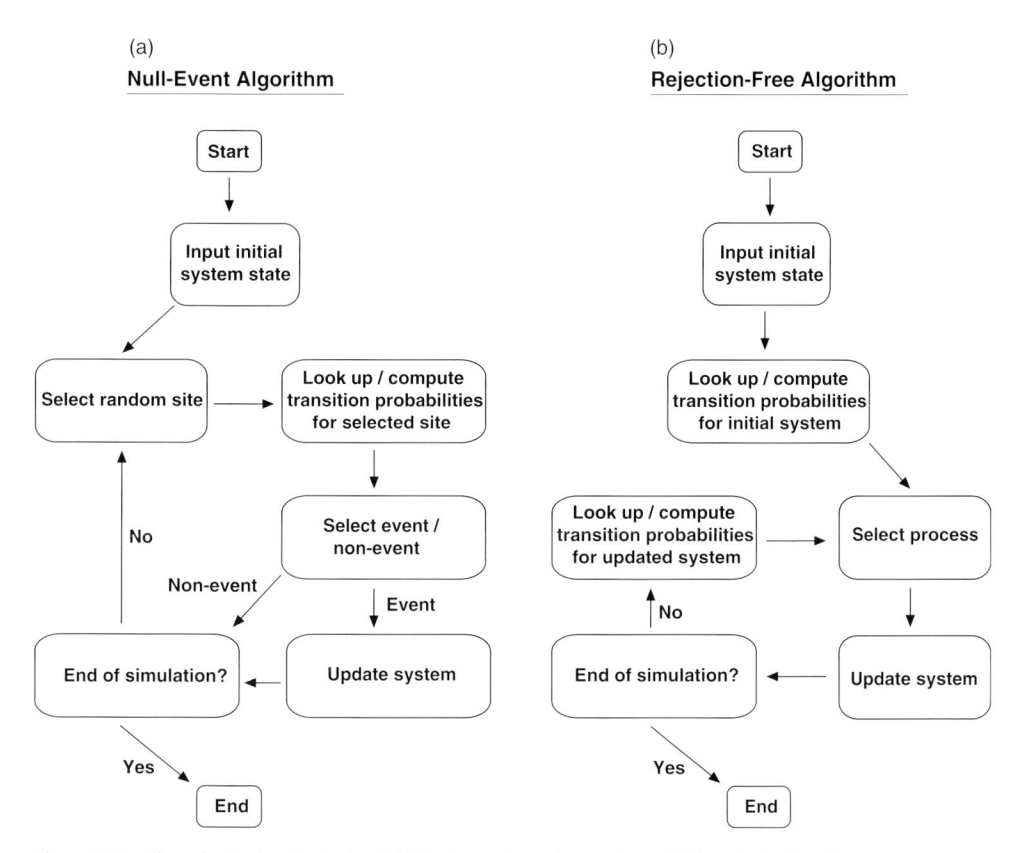

Figure 1.17 Flow charts showing typical kMC schemes based on null-event (a) and rejection-free (b) algorithms.

modeled. Given an initial state of the system, there are two basic strategies for selecting the reactive event that should lead to the next system state (see Figure 1.17).

In null-event algorithms (e.g., the well known Metropolis algorithm [77]) a site is randomly selected (with each site having an equal probability of being selected) and the probability, w_i, for each possible reaction at the site is computed (or looked up in a reaction table). These probabilities are then normalized using the total reaction probability, w_{tot}, for the most reactive site possible in the system (note that w_{tot} is computed once at the beginning for the simulations and then remains constant throughout the simulation, regardless of which site is currently being sampled). Thus, the sum of the normalized probabilities, $k_i = \frac{w_i}{w_{tot}}$, for the reactions associated with the selected site is less than one. A random number, x, is now generated and the jth process is selected for execution when:

$$\sum_{i=1}^{j-1} k_i < x \leq \sum_{i=j}^{n} k_i \qquad (1.62)$$

If $x > \sum_i k_i$ no event takes place. Otherwise, the system is updated in accordance with the event which has been carried out before the simulation proceeds to the next step, where another site is selected and an event associated with this site is carried out if the Monte Carlo procedure described above (i.e., the comparison of the process probabilities with a new random number) is met. Thus, the cycle of steps continues until the desired number of iterations or reactive events has been reached. A major advantage of a null-event algorithm is that the reaction rates for only a very limited number of precesses need to be computed at each iteration. Thus null-event algorithms are well suited for being concurrently coupled with a higher level of theory, which calculates the required reaction barriers and rates as needed on-the-fly. However, a major disadvantage of null-event algorithms can be that not every step results in a successful event. Thus there are wasted Monte Carlo steps, whose computational expense does not contribute directly to advancing the reaction dynamics of the system. In contrast, rejection-free algorithms execute a reactive process at every Monte Carlo step; however, the price they pay for this is that the probability of every process must be taken into account at every step. A typical rejection-free algorithm proceeds as follows. The probability, w_i, of every possible process that the system in its current state might undergo is calculated (or looked up in a table) and these are summed to give a total probability or reaction rate for the entire system:

$$w_{tot} = \sum_{i=1}^{n} w_i \qquad (1.63)$$

This total probability is then used to normalize the individual probabilities and a random number is used to choose one of these, as in the case of the null-event algorithm. However, here, because the sum of the normalized probabilities is exactly one, an event is guaranteed to occur. An additional advantage of rejection-free algorithms is that each kMC iteration can be converted into real time based on the rates for all possible reactions in the system. If the rates, w_i, of all possible reactions (i.e., all reactions that might be chosen) in the system are known, then there is a direct connection between the kMC iteration and real time, such that real time is advanced by Δt at each iteration:

$$\Delta t = \frac{-\ln(x)}{\sum_{i=1}^{n} w_i} \quad (1.64)$$

where x is a random number between 0 and 1 [78]. All that now remains is to update the system based on the executed event. This entails changing the state of affected lattice sites and the reaction processes and rates associated with them along with the overall system rate. Because only a small fraction of the system has typically been affected by the occurrence of a single process it is typically more efficient to only update these parts of the system, rather than the system as a whole. In any case, a variety of data structures, ranging from simple arrays or lists to binary trees, can be chosen to better facilitate the efficient or convenient search (e.g., simple linear, n-level linear or binary) and update (e.g., local vs global) strategies [69].

1.5.3.4 Applications of kMC in Electrochemistry

The lattice approximation makes kMC particularly well suited for simulations of crystal growth [79], making it an obvious choice for simulations of metal deposition under electrochemical conditions [15]. For similar reasons it is sometimes well suited for studies of surface catalysis [80]; however, the large number of possible surface reactions coupled with the possibility of adsorbates occupying more than one surface site can make such applications somewhat cumbersome and involved.

Finally, a common difficulty with kMC should be mentioned. The original goal of the kMC method was to separate fast uninteresting physical processes (such as vibrations) out of the dynamics and leave only interesting chemical processes implicitly in the simulations. Unfortunately, for many surface simulations the vast majority of kMC processes are simple diffusion steps, which in and of themselves are not very interesting. Various approaches to further accelerate kMC simulations have been developed [69]; however, the first passage method, which separates out and effectively fast forwards diffusion by wrapping up collision-free diffusion processes into single Monte Carlo events is particularly worth mentioning [81].

1.5.4
Summary

This section began by recounting the role of the Master equation in chemical kinetics theory. A common approach to stochastically propagating the dynamics prescribed by the Master equation is the use of kinetic Monte Carlo methods. Traditionally kMC simulations make use of pre-defined reaction tables of allowed chemical events and the lattice approximation, which restricts all species to discrete lattice sites. Thus, kMC is particularly well suited for crystal growth and metal deposition simulations. Nevertheless, the development of off-lattice and on-the-fly kMC approaches opens the door to more exotic applications as well.

1.6
The Oxygen Reduction Reaction on Pt(111)

1.6.1
Introduction to the Oxygen Reduction Reaction

The oxygen reduction reaction (ORR) is a fundamental electrochemical reaction which plays a critical role in such varied phenomena as combustion, corrosion and cell respiration, and finds application in many energy conversion and/or storage technologies [82–85]. Of particular interest is its central role in powering polymer-electrolyte (or proton-exchange) membrane fuel cells (PEMFCs), which is driving the search for improved ORR catalysts. One of the most commonly used ORR catalysts is Pt(111). Although the ORR on Pt(111) should ideally generate 1.23 V per electron (see Figure 1.18), an operating potential of at most 0.90 V is experimentally realized. A detailed atomistic-level understanding of the ORR mechanism on Pt(111) could provide important insights toward understanding the cause of this ~ 0.3 V over-potential and developing strategies to reduce or eliminate it.

In this section we highlight an application of some of the techniques discussed in the previous sections to study the mechanism of the ORR on Pt(111). Concurrent coupling between DFT, which is at the heart of the approach presented here, and a Poisson–Boltzmann continuum solvation model are employed to obtain a detailed picture of the relevant energy landscape. Eyring's canonical transition state theory is then used, by way of serial coupling, to develop a kinetic rate model of the system.

1.6.2
Preliminary Considerations

Before applying any of our methods to this system, it is important to think through exactly what we want to model and what appropriate starting points or initial guesses we need to pursue. In this case, the reaction we want to model (shown in Figure 1.18) involves breaking an O—O bond and forming four O—H bonds, two to each O atom. Now we might begin by asking in which order these five individual steps take place. In other words, when does the O—O bond break and when do the O—H bonds each form? The first possibility is that the O—O bond dissociation might take place before any O—H bonds are formed. In this case O_2 is the dissociating species

$$2 H_{2(g)} \longrightarrow 4 H^+_{(aq)} + 4 e^- \qquad E^\circ = 0 \text{ V (anode)}$$

$$O_{2(g)} + 4 H^+_{(aq)} + 4 e^- \longrightarrow 2 H_2O_{(l)} \qquad E^\circ = 1.229 \text{ V (cathode)}$$

$$2 H_{2(g)} + O_{2(g)} \longrightarrow 2 H_2O_{(l)} \qquad E^\circ = 1.229 \text{ V}$$

Figure 1.18 Hydrogen oxidation (anode) and oxygen reduction (cathode) at fuel-cell electrodes.

and so we name this the O_2-dissociation mechanism. The second possibility is that a single H is added to O_2 to form OOH before O—O bond dissociation, so that OOH is the dissociating species. We call this mechanism the OOH-dissociation mechanism. Finally, if a H is added to each O to form HOOH before the O—O bond dissociates, then we call this a H_2O_2-dissociation mechanism. The addition of more than one H to either O prior to O—O bond dissociation is not considered because it would involve either over-filling the valency of oxygen (even if we ignore bonding to the surface) or require simultaneous dissociation of the O—O bond, both of which seem unlikely. Thus, we consider these three general reaction mechanisms, which are illustrated in Figure 1.19.

Because the ORR is catalyzed by the Pt(111) surface, we expect its critical step(s) to involve adsorbates. Thus, we anticipate that the O—O bond cleavage takes place on the Pt(111) surface, along with the H-addition steps. While this means that the oxygen species being protonated is expected to be adsorbed to the catalyst surface throughout the reaction, there are two possible immediate sources for the protons added during the H-addition steps. These might either first adsorb to the Pt(111) surface and then react with the oxygen species, or the protons might come directly from the electrolyte and thus never adsorb to the surface directly. The former type of reaction is known as a Langmuir–Hinshelwood-type (LH-type) reaction, and the latter as an Eley–Rideal-type (ER-type) reaction. A schematic of these is presented in Figure 1.20. Both cases should be considered within each of the three dissociation

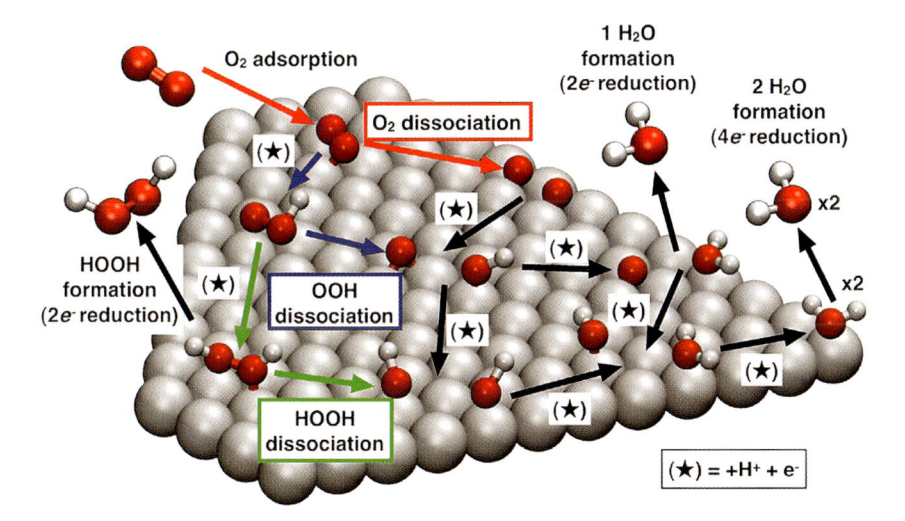

Figure 1.19 Schematic representation of the three investigated ORR mechanisms. Highlighted with a red box and red arrows is the O_2 dissociation mechanism, in blue the OOH dissociation mechanism and in green the HOOH dissociation mechanism. The black arrows denote the subsequent reactions after the first O—O bond dissociation. A star denotes a coupled proton electron transfer (CPET) process, which can take place through either a LH-type or ER-type mechanism.

Langmuir–Hinshelwood (LH) mechanism

Eley–Rideal (ER) mechanism

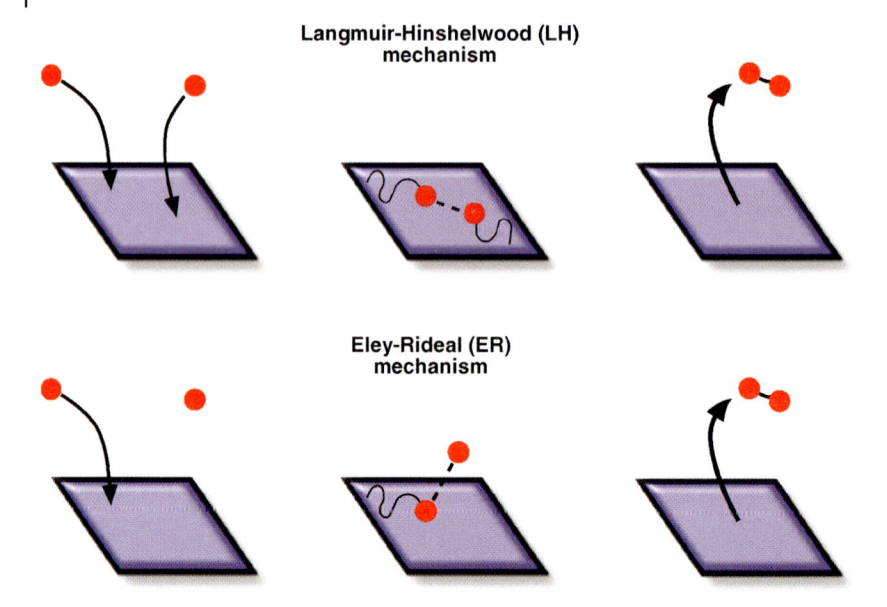

Figure 1.20 Schematic representation of Langmuir–Hinshelwood (LH)- and Eley–Rideal (ER)-type reaction mechanisms.

mechanisms discussed above, and, indeed, both have been considered [86,87], however, we will limit most of our discussion here to LH-type reactions.

1.6.3
DFT Calculations

The geometries for the reactants, products, intermediates and transition states involved in each dissociation pathway were optimized within a DFT description of the system. In particular, the B3LYP flavor of DFT [88–91] in conjunction with an LACVP** basis set [92–96] was used as implemented in the Jaguar program [97] to calculate the energies and forces involved in obtaining the optimal geometry for each system state, and it corresponding "raw" system energy, E^{SCF}. In all cases, not only converged energies and geometries, but also converged spin states are used in arriving at the final computed energies, a detail which is often essential for accurately reproducing experimental results [98]. Using an appropriate model for the electrode surface is even more vital for reliable calculations; however, the development of such a model can be an arduous task [86, 99–101]. The results presented in this section were obtained using a Pt_{35} cluster model. The cluster contains three layers of atoms, most of which are fixed at their experimental bulk positions. Only four atoms in the center of the (111) surface are allowed to relax. Although this cluster is significantly larger than many of the metal clusters commonly used to model catalytic surface reactions, extensive cluster-size convergence studies have shown that a 3-layer Pt_{28} cluster is the smallest, shallowest cluster capable of reliably reproducing adsorption energies [99]. The Pt_{35} cluster has been shown to provide a reliable model for an extended Pt(111) surface, such as is found on a Pt electrode.

1.6.4
Method Validation

While computational convergence demonstrates a level of consistency within our DFT model, it is unable to identify systematic errors which might invalidate converged results. Thus, it is good practice to check a new model/level of theory by comparing results obtained using it to experimental results. The binding energies of O, O_2 and H_2O to Pt(111) under low-coverage, low-pressure and low-temperature conditions have been measured experimentally using thermal desorption spectroscopy (TDS) [102–104] and electron energy-loss spectroscopy (EELS) [105–107], and found to be in the ranges 3.47–3.73 eV, 0.3–0.5 eV, and 0.43–0.65 eV, respectively. The O_2 dissociation barrier has also been measured to be in the range 0.3–0.5 eV and STM imaging shows that while O_2 preferentially binds at a bridge site, dissociation involves a different intermediate binding site. Using the DFT methods and models presented directly above, the binding energies for O, O_2 and H_2O were calculated to be 3.25, 0.49 and 0.60 eV, respectively and O_2 is found to have a dissociation barrier of 0.65 eV, which involved migration to an fcc site. Thus, the experimental results are well reproduced within an acceptable degree of accuracy by the theoretical approach taken here. Unfortunately, a direct comparison under electrochemical conditions is not yet possible because traditional ORR experiments measure overall activity rather than individual binding energies.

1.6.5
Reaction Energies

Based on the raw SFC energies (E^{SCF}) computed for relevant reactants, products, intermediates, transition states, and reference molecules using the methods and model outlined above, reaction energies and barriers can be computed for each step in the three ORR reaction mechanisms considered here. Gas phase H_2, O_2 and the bare Pt_{35} cluster are taken as reference states, so that the reaction energy (ΔE_{gas}) associated with each reactant, product, intermediate or transition state is calculated as follows:

$$\Delta E^{gas} := E^{SCF}_{sys} - E^{SCF}_{Pt_{35}} - \frac{1}{2} N_H E^{SCF}_{H_2} - \frac{1}{2} N_O E^{SCF}_{O_2} \qquad (1.65)$$

These energies are plotted in Figure 1.21, which shows the energetics of three different ORR mechanisms using the LH mechanism for all H-addition steps. Excepting the removal of water at the last step, all individual reaction steps (in all three mechanisms) are downhill. Nevertheless, there are barriers to slow down the kinetic rates of many steps. Thus, the initial dissociation of O—O has a barrier of

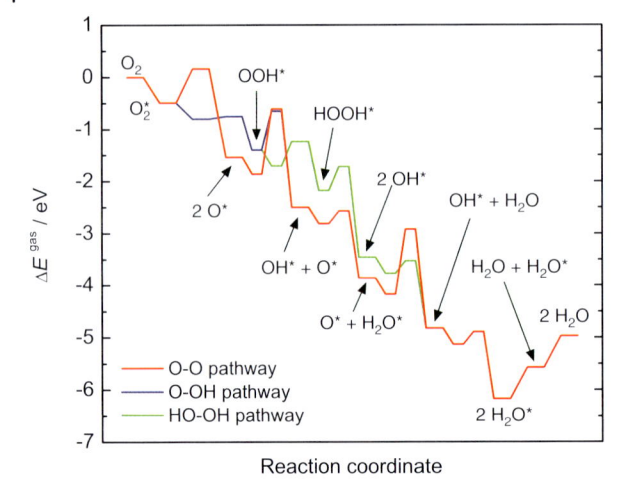

Figure 1.21 Energy profiles of LH-type mechanisms for the surface ORR under gas phase conditions. Only the adsorbates containing O atoms are denoted for clarity. Unlabeled potential energy wells correspond to the preceding intermediate +H*.

0.65 eV, and the addition of a first H to O to form OH via a LH-type mechanism is the largest barrier: 1.25 eV. The barrier for H-addition to form H_2O is small (0.24 eV), and desorption of the final product (H_2O) has a similar barrier (0.60 eV, which is the same as the reaction energy in this case) to the initial O—O bond dissociation step. Thus, the addition of an initial H to a lone O to form OH is the rate-limiting reaction within the O_2-dissociation mechanism. The OOH mechanism introduces a new H-addition step at its commencement ($O_2 + H \rightarrow OOH$), whose barrier is negligible (0.05 eV). The subsequent OOH dissociation step (OOH $\rightarrow$ O + OH) has a barrier (0.74 eV) which is slightly higher than the dissociation barrier in the O_2 dissociation mechanism. Subsequent steps follow the same pathway as is found in the O_2 dissociation mechanism, and so once again the H-addition to form OH is the reaction step with the highest barrier (1.25 eV). The H_2O_2-dissociation mechanism avoids this high-barrier step by adding an H to each O in O_2 before the O—O bond is cleaved. The barrier for adding this second H to form HOOH using a LH mechanism is 0.47 eV, and is almost identical to the subsequent barrier for breaking the O—O bond in the thus formed H_2O_2. Thus, the energy required to desorb the final product (H_2O) from the surface (0.60 eV) is the highest barrier for this mechanism. Because it is less than half the value of the highest barriers in the other two mechanisms, we anticipate that the H_2O_2 dissociation mechanism will dominate under these conditions (i.e., zero pressure and temperature and no zero-point-energy effects!). As should be evident from the conditions mentioned above, there are a number of factors which need to be corrected for before applying our model to electrochemical systems. These include solvation effects, free energy contributions, and the electrode potential. We now explain the inclusion of each of these in turn.

1.6.6
Solvation Effects

Solvation energies were calculated using an implicit solvation model. The solvent molecules are modeled as a continuous dielectric material, surrounding a cavity, specifically fitted to the solvated molecule(s). Assuming a Boltzmann-type distribution of solvent ions, the Poisson–Boltzmann equation gives the electrostatic potential in terms of which the interaction energy between the solvent and solute can be computed using:

$$\nabla[\varepsilon(\mathbf{r}) \cdot \nabla V_{es}(\mathbf{r})] = -4\pi \left[\rho^f(\mathbf{r}) + \sum_i c_i^\infty z_i \lambda(\mathbf{r}) \exp\left(\frac{-z_i V_{es}(\mathbf{r})}{k_B T} \right) \right] \quad (1.66)$$

The terms in the above Poisson–Boltzmann equation are the following: $\varepsilon(\mathbf{r})$ is the dielectric constant of the solvent, V_{es} is the electrostatic potential, $\rho^f(\mathbf{r})$ is the charge density of the solute, c_i^∞ is the bulk concentration of solvent ion-type i, which has a net charge of z_i and $\lambda(\mathbf{r})$ is the accessibility that the solvent has to position $\mathbf{r}$. The free energy of solvation is then the sum of the energy required to form the solute cavity in the dielectric and the energy required for charging the solvent near the cavity surface. The solvation energy was thus calculated using the Jaguar [97] Poisson–Boltzmann solver for each step of the SCF energy procedure and included in the SCF energy at each iteration. Nevertheless, the solvation energy (E^{solv}) is easily extracted again, so that reaction energies, which now include solvation effects can be written as:

$$\Delta E^{solv} = E^{gas} + (E^{solv}_{products} - E^{solv}_{reactants}) \quad (1.67)$$

These solvation energies are plotted in Figure 1.22. There we see that solvation dramatically increases the attraction of O_2 to the surface, so that the adsorption of O_2

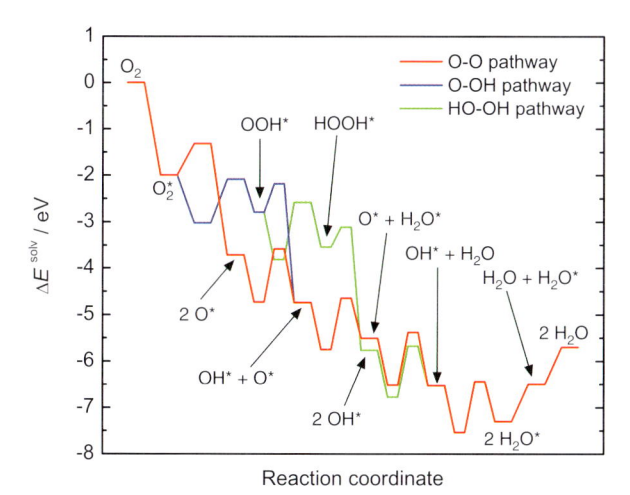

Figure 1.22 Energy profiles of LH-type mechanisms for the surface ORR under the influence of implicit water solvation. Only the adsorbates containing O atoms are denoted for clarity. Unlabeled potential energy wells correspond to the preceding intermediate $+H^*$.

is now −1.71 eV (instead of −1.05 eV) downhill. The barriers for O—O dissociation (0.69 eV instead of 0.65 eV) and the first H-addition steps (1.14 eV instead of 1.25 eV) show only minor changes. On the other hand, the barrier for water formation increases dramatically (1.10 eV instead of 0.24 eV) and is now uphill with a reaction energy of 0.24 eV (was −1.05 eV downhill). Finally, the desorption of water is more uphill than it previously was (0.83 eV instead of 0.60 eV). Although the inclusion of solvation effects dramatically changes some of the barriers, the initial H-addition step still has the highest barrier, which has undergone only a small decrease.

Similar trends are observed when we consider the effects of solvation on the OOH-dissociation mechanism. The first H-addition step (OOH formation) is now uphill (0.22 eV instead of −0.60 eV) and has a significant barrier (0.94 eV instead of 0.05 eV). In contrast, the OOH-dissociation barrier remains relatively unchanged (0.62 eV instead of 0.74 eV). Thus OH formation remains the process with the highest barrier (1.14 eV instead of 1.25 eV) following the inclusion of solvation effects.

In the case of the H_2O_2-dissociation mechanism, we once again see solvation primarily resulting in increased H-addition barriers, while leaving the O—O bond dissociation step relatively untouched. Thus the second H-addition to form H_2O_2 is an uphill reaction when solvated and now has the highest barrier within the mechanism (1.23 eV instead of 0.47 eV), while the H_2O_2 dissociation barrier remains almost constant (0.43 eV instead of 0.46 eV).

Thus, now that solvation effects have been included the H_2O_2-dissociation mechanism has the highest overall barrier (1.23 eV vs. 1.14 eV for each of the other two mechanisms) instead of the lowest, as was the case before including solvation (0.60 eV vs. 1.25 for each of the other two mechanisms).

1.6.7
Free Energy Contributions

Free energies were calculated within an ideal gas approximation. Vibrational states (i.e., normal modes) were extracted from the Hessian of each structure and then used to calculate the zero-point-energy correction (E^{ZPE}) the temperature-dependent enthalpy correction, $H(T)$, and the temperature-dependent entropy, $S(T)$. The Gibbs free energy in solvent, ΔG_T^{solv}, was then calculated as:

$$\Delta G_T^{solv} := \Delta E^{solv} + \Delta E^{ZPE} + \Delta H_T^{vib} - T\Delta S_T^{vib} \tag{1.68}$$

where ΔE^{solv}, ΔE^{ZPE}, ΔH_T^{vib} and ΔS_T^{vib} indicate the difference of these respective quantities between the products and reactants under consideration. The Gibbs free energy for the three LH reaction mechanisms at 298 K are shown in Figure 1.23.

The inclusion of these thermal contributions decreases the favorability of O_2 dissociation substantially from 1.71 to 1.35 eV; however, the barrier is unchanged (0.68 eV). The OH formation reaction energy and barrier both show similar small increases (from 0.0 to 0.20 eV and from 1.14 to 1.19 eV, respectively). A similar trend is also observed for H_2O formation as the overall reaction energy increases from

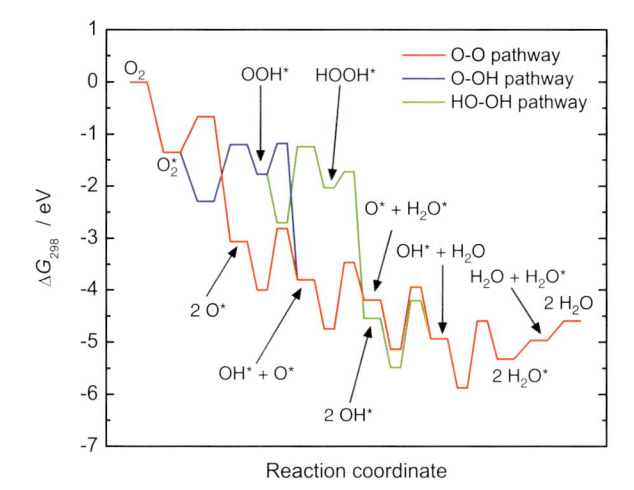

Figure 1.23 Energy profiles of LH-type mechanisms for the surface ORR under gas phase conditions. Only the adsorbates containing O atoms are denoted for clarity. Unlabeled potential energy wells correspond to the preceding intermediate $+H^*$.

0.24 to 0.55 eV and the reaction barrier from 1.10 to 1.27 eV. Thus, with the inclusion of thermal contributions at 298 K, the highest barrier in the O_2-dissociation mechanism is 1.27 eV, which is associated with H_2O formation.

In the OOH-dissociation mechanism H-addition is also somewhat inhibited by the inclusion of thermal contributions at 298 K. Thus, the overall energy and reaction barrier for OOH formation increase from 0.22 to 0.52 eV and from 0.94 to 1.09 eV, respectively. As was also the case in the O_2-dissociation mechanism, the dissociation step in the OOH-dissociation mechanism is left essentially unaffected (the barrier is lowered from 0.62 to 0.59 eV). Thus, again, the formation of water is now the step with the highest barrier (1.27 eV).

Finally, the inclusion of thermal contributions to yield free energies for the H_2O_2-dissociation pathway increases the second H-addition step needed to form H_2O_2 from 1.23 to 1.46 eV, so that it remains the step with the highest reaction barrier. As was seen in the other mechanisms, the O—O bond dissociation step (here H_2O_2 dissociation) only changes slightly as it is reduced from 0.43 to 0.31 eV.

1.6.8
Influence of an Electrode Potential

Thus far we have not considered the influence of the electrode potential. As mentioned in Section 1.3.4.3, the electrode potential can be modeled to a first order of approximation by multiplying the electrode potential by the number of electrons transferred at each electron-transfer step. In our ORR mechanisms an electron is transferred to the surface every time a H atom leaves the electrolyte solution phase and electrically attaches to the surface, either by adsorbing directly (as is the case within LH-type mechanisms) or by bonding to another molecule which is already adsorbed to the surface (as in the case in ER-type mechanisms).

Results at zero potential ($U = 0.0$ V vs. the reversible hydrogen electrode, RHE) and the ideal electrode potential ($U = 1.23$ V vs. the RHE) are summarized in Figure 1.24a and b, respectively. Here we have included both the LH- and ER-type mechanisms to illustrate how turning to an ER-type mechanism can be used to avoid the large H-addition barriers present in the LH-type mechanisms at zero potential.

At zero potential the overall ORR process is downhill -4.57 eV. The LH-type reaction barriers are rate limiting in all LH-type pathways and range from 1.09 to 1.19 to 1.28 to 1.46 eV for the formation of OOH, OH, H_2O and H_2O_2, respectively.

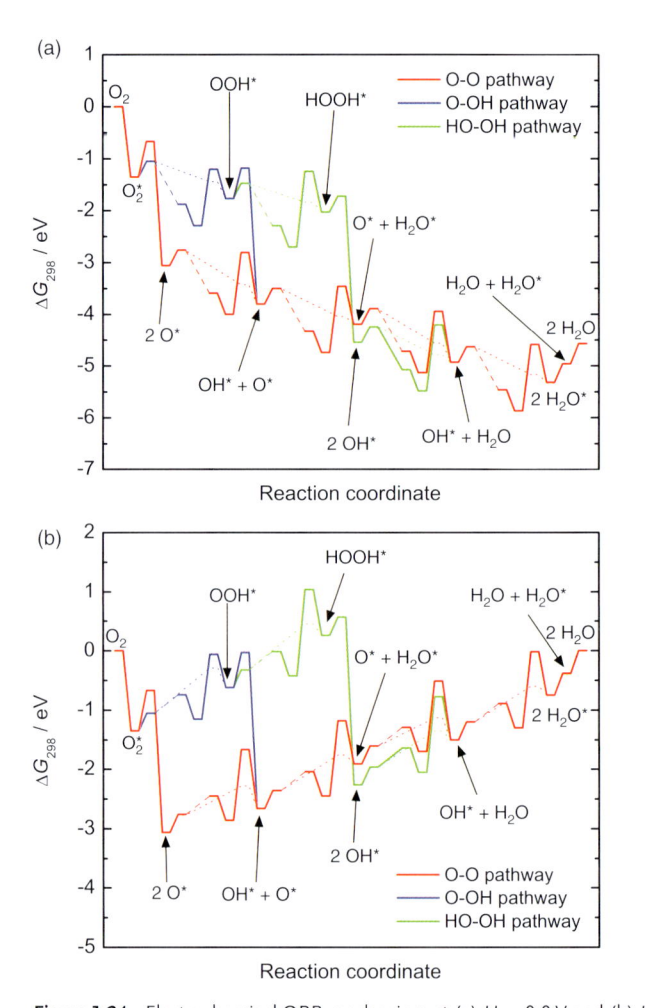

Figure 1.24 Electrochemical ORR mechanism at (a) $U = 0.0$ V and (b) $U = 1.14$ V vs. RHE (calculated equilibrium potential). The LH steps are denoted by dashed lines and the ER steps by dotted lines.

The corresponding reaction barriers for the addition of a proton taken directly from the electrolytic phase in the ER-type pathways are all much lower (approximated to be 0.3 eV for downhill reactions) [86]. Thus one would expect ER-type pathways to dominate at low potentials.

At the ideal potential, the overall ORR process is now energetically neutral ($\Delta G(298) = 0$). While the relative energies of intermediates in the LH-type mechanisms change, the barriers remain unaffected (only the reverse barriers are affected within our model) by the application of the electrode potential. In contrast, the barriers for the ER steps in the ER-type mechanisms increase significantly, since the barriers here are calculated relative to the products rather than relative to the reactants, as is the case for LH-type reaction steps. Thus the H-addition barriers for the formation of OOH, H_2O_2, OH and H_2O via ER-type processes are now 1.15, 1.16, 0.87 and 0.99 eV, respectively. As a result LH-type mechanisms are expected to be competitive with ER-type processes at similar potentials and among the ER-type mechanisms, the H_2O_2-dissociation pathway is expected to dominate as its highest barrier is 0.15 eV lower than the highest barrier for each of the other two mechanisms.

1.6.9
Modeling the Kinetic Rates

To better understand the phenomenological implications of these free energy landscapes, we take advantage of a simple kinetic model to describe the competition between the investigated ORR mechanisms. At the heart of this kinetic model is Eyring's canonical transition state theory (see Section 1.3.3.3 for more details) [24,87], which enables us to convert the reaction barriers acquired using DFT into kinetic rates. Thus the relationship between the DFT calculations and the kinetic model is a straightforward example of serial coupling. The kinetic model utilized here, takes two types of processes into consideration: those which end in surface O or OH (i.e., adsorption and dissociation processes), and those which remove OH from the surface (i.e., H_2O formation and desorption) or lead away from its formation (desorption of H_2O_2). Thus, the three ORR mechanisms investigated using DFT are broken down into the following six processes (shown in Figure 1.25) which we label as: O_2-dissociation, OOH-dissociation, HOOH-dissociation, HOOH desorption, and H_2O formation (starting from either O or OH).

The rate of each of these five processes is assumed to correspond to the rate of the slowest step in it (i.e., the step with the highest barrier), and the concentration of reactants leading up to each of these bottleneck barriers is assumed to be one for the sake of making a simple comparison between pathways. Thus, the rate constant for each of the six processes is found using Eyring's equation:

$$k(U) = \frac{k_B T}{h} \exp\left[\frac{-\Delta G_T(U)}{k_B T}\right] \quad (1.69)$$

where k_B is Boltzmann's constant, T is temperature, h Planck's constant, and $\Delta G_T(U)$ is the potential-dependent barrier.

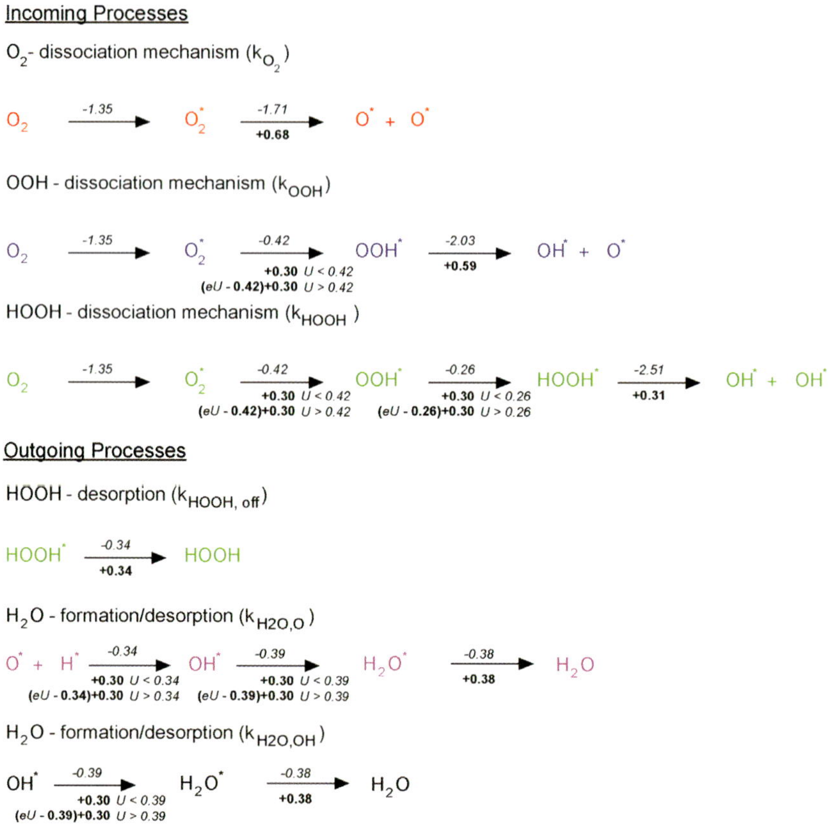

Incoming Processes

O_2- dissociation mechanism (k_{O_2})

$$O_2 \xrightarrow{-1.35} O_2^* \xrightarrow[+0.68]{-1.71} O^* + O^*$$

OOH - dissociation mechanism (k_{OOH})

$$O_2 \xrightarrow{-1.35} O_2^* \xrightarrow[\substack{+0.30\ U<0.42 \\ (eU-0.42)+0.30\ U>0.42}]{-0.42} OOH^* \xrightarrow[+0.59]{-2.03} OH^* + O^*$$

HOOH - dissociation mechanism (k_{HOOH})

$$O_2 \xrightarrow{-1.35} O_2^* \xrightarrow[\substack{+0.30\ U<0.42 \\ (eU-0.42)+0.30\ U>0.42}]{-0.42} OOH^* \xrightarrow[\substack{+0.30\ U<0.26 \\ (eU-0.26)+0.30\ U>0.26}]{-0.26} HOOH^* \xrightarrow[+0.31]{-2.51} OH^* + OH^*$$

Outgoing Processes

HOOH - desorption ($k_{HOOH,\ off}$)

$$HOOH^* \xrightarrow[+0.34]{-0.34} HOOH$$

H_2O - formation/desorption ($k_{H2O,O}$)

$$O^* + H^* \xrightarrow[\substack{+0.30\ U<0.34 \\ (eU-0.34)+0.30\ U>0.34}]{-0.34} OH^* \xrightarrow[\substack{+0.30\ U<0.39 \\ (eU-0.39)+0.30\ U>0.39}]{-0.39} H_2O^* \xrightarrow[+0.38]{-0.38} H_2O$$

H_2O - formation/desorption ($k_{H2O,OH}$)

$$OH^* \xrightarrow[\substack{+0.30\ U<0.39 \\ (eU-0.39)+0.30\ U>0.39}]{-0.39} H_2O^* \xrightarrow[+0.38]{-0.38} H_2O$$

Figure 1.25 Reaction schemes for single rate processes considered in kinetic rates analysis. Reaction energies (italics) and barriers (bold) are given in eV above and below the reaction arrows. The electrode-potential dependence for the reaction energies and barriers of CPET processes are given by including the appropriate correction as a function of the electrode potential, U, in the energy expression. The potential-dependent rate constants for these kinetic processes are plotted in Figure 1.26.

The incoming phase of the O_2-dissociation mechanism (k_{O_2}) includes no CPET processes, so that its rate is independent of the electrode potential. The highest barrier is for the O_2-dissociation step, 0.68 eV, making this the slowest incoming process at low potentials. As two O adatoms are the final products for this process, the outgoing process required to complete the O_2-dissociation mechanism is the formation of water from an O adatom. ($k_{H_2O,O}$).

The incoming phase of the OOH-dissociation mechanism (k_{OOH}) includes a single CPET process. At low potentials this CPET process is downhill and thus only has a small barrier of 0.30 eV, leaving the potential-independent dissociation step to be the rate-determining process with a barrier of 0.59 eV. However, once the electrode potential reaches $U = 0.42$ V the CPET barrier begins to increase with the increasing potential and overtakes the dissociation to become the rate-limiting step at $U = 0.71$ V. The overall rate for the incoming portion of the OOH-dissociation mechanism is now potential-dependent and is slowed down by further increases in potential. The result of the incoming portion of the OOH-dissociation is a mixture of O adatoms and adsorbed OH. To avoid a build-up of O adatoms on the surface, which could ultimately disable the cataylitic properties of the surface by blocking cataytic sites, the entire water formation pathway starting from an O adatom ($k_{H_2O,O}$) and not just the second portion starting from OH ($k_{H_2O,OH}$) is needed to make the OOH mechanism a viable catalytic pathway.

The incoming phase of the HOOH-dissociation mechanism is similar to that of the OOH-dissociation mechanism except that it incorporates an additional CPET step and has a significantly lower dissociation barrier (0.31 eV). This dissociation barrier is, nevertheless, still seen to be rate limiting at low potentials (although the difference between 0.30 and 0.31 is within the uncertainty of our methods). The second CPET step is less favorable than the first and so its potential dependence is triggered at $U = 0.26$ V and comes into play almost immediately (at $U = 0.27$ V) to make it the rate-limiting step. As OH is the sole product of the incoming portion of the HOOH-dissociation mechanism, only the shorter H_2O formation/desorption pathway is needed to arrive at the desired product. However, the desorption of H_2O_2 provides a second possible outgoing pathway ($k_{H_2O_2,off}$), leading to undesired products (H_2O_2). This desorption process is itself potential-independent, but does depend on the potential-dependent formation of HOOH for the formation of its necessary reactant. Any HOOH* which is formed can then either dissociate, with a barrier of 0.31 eV, or desorb, with a barrier of 0.34 eV. Because these barriers are similar, we anticipate that any utilization of the HOOH pathway will result in a mixture of desired H_2O and undesired H_2O_2 products. However, at potentials above 0.5 eV the slowdown in the CPET processes needed to remove OH from the system may result in OH overcrowding on the surface, which would encourage H_2O_2-desorption.

The two outgoing processes, which form H_2O from O and OH, respectively ($k_{H_2O,O}$ and $k_{H_2O,OH}$) consist of one or two CPET steps followed by H_2O desorption. At low potentials the potential-independent desorption barrier of 0.38 eV is rate limiting; however, the first CPET step in each process becomes rate limiting at $U = 0.43$ V and $U = 0.47$ eV, respectively. Thus for potentials greater than $U = 0.43$ V, H_2O formation starting from O is slower than starting from OH.

The potential-dependent rate constants for each of these six processes at 298 K are plotted in Figure 1.26. At low potentials the incoming rate for HOOH formation and dissociation (k_{HOOH}) is the fastest process. The outgoing processes relevant to the HOOH and OH formed ($k_{H_2O_2}$ and $k_{H_2O,OH}$ respectively) are one half and a whole order of magnitude slower than the incoming reaction sequence, suggesting that OH is likely to build up on and poison the surface at low potentials. At $U = 0.27$ V k_{HOOH} begins to decrease and drops below the outgoing processes, which form H_2O_2 (($k_{H_2O_2}$) and H_2O ((k_{H_2O}), at $U = 0.30$ V and $U = 0.34$ V, respectively. Until k_{HOOH} crosses k_{OOH} at $U = 0.55$ V, the HOOH-dissociation mechanism is the dominant incoming pathway and should result in a mixture of H_2O_2 and H_2O products. As the OOH-dissociation mechanism becomes the dominant incoming pathway the production of H_2O relative to that of H_2O_2 should increase, and the

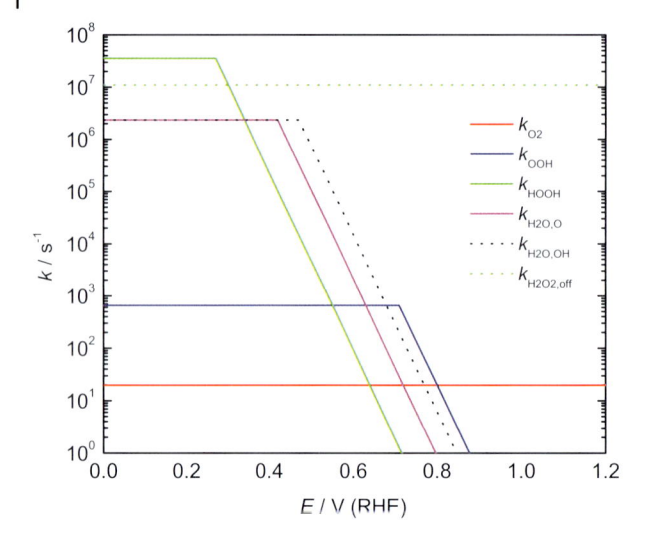

Figure 1.26 Potential-dependent rate constants corresponding to different ORR mechanisms. Rate constants k_{O_2}, k_{OOH}, and k_{HOOH} correspond to the smallest calculated rate constant for each mechanism ending with O—O, O—OH, and HO—OH dissociations, respectively. The rate constant k_{out} corresponds to the smallest calculated rate constant to create H_2O from O and/or OH intermediates. $k_{H_2O_2,off}$ corresponds to the rate constant for desorbing H_2O_2 from the Pt(111) electrode. The vertical axis is a measure of the relative magnitude of the rate constants. Solid lines represent ingoing mechanisms for species that adsorb onto the surface; dashed lines correspond to mechanisms that lead to the removal of surface-adsorbed species. The reaction pathways, reaction energies and barriers corresponding to each of the process rates are presented in Figure 1.25.

surface should remain relatively free from blockage (due to outgoing processes having faster rates than incoming processes) up to $U = 0.63$ V and $U = 0.68$ V, at which points O and OH, respectively, are expected to begin to accumulate on the surface. Nevertheless, the downward turn of k_{OOH} shortly thereafter may keep the incoming and outgoing rates similar enough to support efficient catalytic activity. At $U = 0.72$ V, $U = 0.77$ V and $U = 0.80$ V $k_{H_2O,O}$, $k_{H_2O,OH}$ and k_{OOH}, respectively, drop below k_{O_2}, leaving O_2-dissociation to be the dominant process for potentials above $U = 0.80$ V. Thus we anticipate the accumulation of oxygen on the surface at these high potentials.

1.6.10
Summary

To illustrate some of the approaches and methods presented in earlier sections, this section has summarized a computational study of the ORR on Pt(111). Three dissociation mechanisms were investigated with DFT, and both LH- and ER-type H-addition processes were considered within each mechanism. A Pt_{35} cluster was used to model the electrode surface. All reactions were first studied in a gas phase environment before the inclusion of an aqueous solvent by means of a Poisson–Boltzmann continuum model. The inclusion of solvent significantly changed the energetics of several reaction steps;

however, the same steps remained rate determining and these were hardly affected. Based on normal modes extracted from the Hessian of each structure, free energies at 298 K were calculated within the ideal gas approximation. The inclusion of these free energy corrections resulted in a scenario in which water formation became the highest barrier in two of the three mechanisms. The influence of an electrode potential showed that at high potentials LR-type reaction pathways are competitive with ER-type pathways, which are strongly preferred at low potentials. Finally, a kinetic rate model was developed and used to explain the roles of various incoming and outgoing reaction pathways and their influence on the overall catalytic behavior at electrode potentials ranging from $U=0.0\,\text{V}$ to $U=1.2\,\text{V}$.

1.7
Formic Acid Oxidation on Pt(111)

1.7.1
Introduction to Formic Acid Oxidation

For our second and final example we present modeling of a second major fundamental reaction in electrochemistry: formic acid oxidation. Once again, DFT provides the foundation, but a two-dimensionally infinite Pt(111) slab model is used in place of the Pt_{35} cluster used to study the ORR. The role of solvation effects is considered, however, an explicit solvation model is used. A kinetic rates model is extracted from DFT-characterized reaction pathways, and an electrode potential picture of the formic acid oxidation is compared to the experimental cyclic voltammogram (CV).

The electro-oxidation of formic acid (HCOOH) on Pt(111) and analogous transition metals has been the topic of numerous studies [65,108–137] due both to its fundamental role in electrochemistry and its importance for low-temperature fuel cell technologies. It has been studied under both gas phase (UHV) surface science and electrochemical (solvated) conditions. Gas phase HCOOH oxidation has been well characterized in surface science experiments [115–118], which reveal a single decomposition path which proceeds through a formate intermediate.

In contrast, there has been difficulty in characterizing the electrochemical oxidation of HCOOH experimentally; however, a dual-path mechanism, in which both so-called direct and indirect paths are utilized, is generally accepted [120–129]. The indirect pathway proceeds through a CO intermediate, a poisoning species which has been identified using *in situ* infrared reflection-adsorption spectroscopy (IRAS) [132,133]. In contrast, the reactive intermediate along the direct pathway is still being disputed, with some arguing for COH or CHO [134], some for COOH [119–120], others for HCOO [129–131], and still others for HCOOH [126–128]. Thus, as seen in Figure 1.27, there are at least three pathways to consider: the indirect pathway via CO, the direct pathway which begins with C—H activation, and a formate pathway (a second possibility within the direct path) which begins with O—H activation and thus proceeds via a formate intermediate.

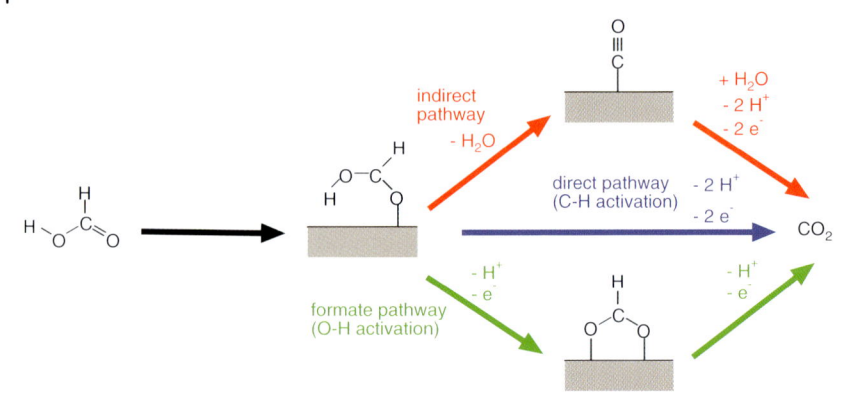

Figure 1.27 Schematic representation of the proposed mechanisms for the oxidation of electrochemical HCOOH.

1.7.2
Density Functional Theory Calculations

The direct and formate pathways, which make up the so-called direct path, have been the topic of recent theoretical studies [135–137] aimed at nailing down the relevant reaction intermediate(s) for the direct path. In this example we present modeling strategies used to characterize these reaction pathways and point to how the mechanistic insights thus obtained can be used to explain CV measurements.

The DFT calculations presented here were performed using PBE, a generalized, gradient approximation (GGA) exchange-correlation functional proposed by Perdew, Burke and Einzerhof [138], as implemented in the CASTEP code [111]. Vanderbilt-type, ultrasoft pseudopotentials [139] were used to describe the nuclei and core electrons, while band states occupied by valence electrons were described using plane-waves with an energy cut-off of 400 eV. The Pt electrode was modeled as a five layer, p(3 × 3) Pt(111) slab, allowing adsorbate coverages as low as 1/9 ML. A vacuum region of >13 Å was left to separate neighboring slabs in the supercell geometry. For the numerical integration in reciprocal space required for periodic systems a 2 × 2 Monkhorst-Pack k-grid was used following convergence tests. During geometry convergence the top three layers of the Pt(111) slab were allowed to relax, while the bottom two layers were fixed at their bulk positions. A transition state search procedure employing linear and quadratic synchronous transit (LST/QST) algorithms in conjunction with conjugate gradient methods was employed to locate transition states using CASTEP.

1.7.3
Gas Phase Reactions

In thoroughly investigating HCOOH oxidation on Pt(111) a variety of surface coverages and compositions as well as all reasonable, possible reaction pathways need to be considered. This has indeed been done [135–137]. Here, for the sake of

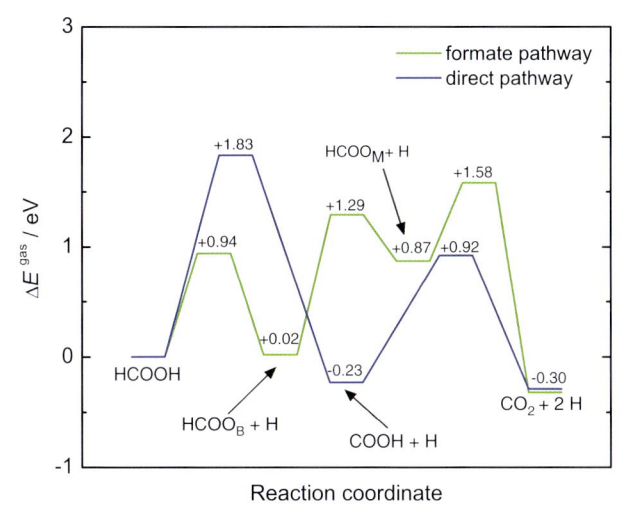

Figure 1.28 Reaction mechanisms for HCOOH oxidation without water. Here the potential profile at $\Theta_{HCOOH} = 1/9$ ML is illustrated.

illustration, we focus on two reaction pathways at low coverage (1/9 ML). Both reaction pathways begin with adsorbed HCOOH and end with CO_2 and two H adatoms. In the direct pathway, the C—H bond is cleaved first, resulting in a COOH intermediate which is further dehydrogenated (O—H bond dissociation) to form the final products (CO_2 and 2 H). In the formate pathway the O—H bond dissociates first to form a bi-dentate formate ($HCOO_B$) intermediate (both O atoms are bonded to the surface). One of the surface-O bonds then dissociates forming mono-dentate formate ($HCOO_M$), which then dissociates to yield CO_2 and H.

The energetics for these two pathways, based on DFT calculations are shown in Figure 1.28. The critical barrier for the direct pathway is the 1.83 eV needed to cleave the C—H bond in the first step. The two dehydrogenation barriers within the formate pathway are both less than 1 eV, making conversion of $HCOO_B$ into $HCOO_M$, with a barrier of 1.27 eV, the rate-determining step. Thus under low-coverage, high vacuum (i.e., surface science) conditions, we anticipate that the formate pathway is preferred. Indeed, on the basis of vibrational frequencies obtained in experimental UHV measurements, it has been proposed that a HCOO species is the active intermediate [117].

1.7.4
Explicit Solvation Model

In Section 1.6.6 we illustrated the use of an implicit solvation model to account for electrostatic interactions between the solvent and solvated reaction molecules in the ORR. Here, we illustrate the use of an explicit solvent model in which explicit, solvating water molecules are introduced into the model to account for hydrogen bonding between the reaction molecules and water molecules which are part of the electrochemical environment. In order to determine an appropriate number of water molecules to use in the explicit solvation model, along with their optimal locations and orientations, various configurations involving one, two and four water molecules per p(3 × 3) super cell were explored. Because solvating molecules are free to move around in an experiment, we anticipate that the lowest energy structure (for a given number of water molecules) is the most likely experimental structure for the number of water molecules. Similarly, among various possible solvation structures at a transition state, the most stable (i.e., the lowest energy) structure provides the barrier which is most relevant to the reaction rate. Among the solvation models tested, a water bilayer model consisting of four water molecules in an ice-like bilayer structure on the surface with HCOOH incorporated into the thus formed water network was found to be the most appropriate model using the critera discussed above and thus adopted [136]. The energy landscape for the direct and formate pathways using a water bilayer solvation model are presented in Figure 1.29. The most obvious effect of the solvating water molecules is the stabilization of H by formation of H_3O following O—H dissociation. This not only increases the overall exothermicity of the reaction from $\Delta E = -0.29$ eV to $\Delta E = -0.58$ eV, but reduces the O—H dissociation barrier by more than 50% from 0.94 to 0.29 eV and from 1.15 to 0.48 eV in the formate and direct pathways, respectively. The stabilizing influences of the hydrogen bonding within the water network also result in significant reductions to the barrier of the C—H dissociation step in the direct pathway (1.83 to 0.79 eV), and in the $HCOO_B$ to $HCOO_M$ transformation in the formate pathway (1.27 to 0.61 eV). In contrast, the C—H dissociation barrier in the formate pathway is only modestly reduced (0.71 to 0.58 eV). The net result is that the effective barriers for both pathways are considerably lower (i.e., the highest barriers for the formate

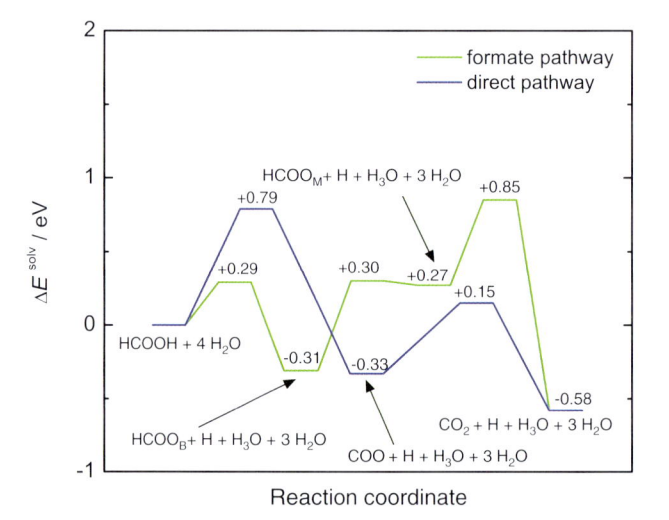

Figure 1.29 Potential energy diagram comparing the two lowest reaction pathways of the Langmuir–Hinshelwood HCOOH oxidation reaction mechanisms using the water bilayer solvation model.

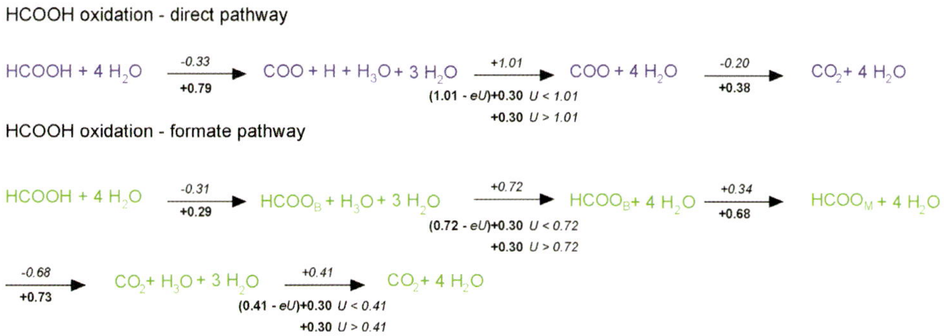

HCOOH oxidation - direct pathway

HCOOH + 4 H$_2$O $\xrightarrow[+0.79]{-0.33}$ COO + H + H$_3$O + 3 H$_2$O $\xrightarrow[\substack{(1.01 - eU)+0.30\ U<1.01 \\ +0.30\ U>1.01}]{+1.01}$ COO + 4 H$_2$O $\xrightarrow[+0.38]{-0.20}$ CO$_2$ + 4 H$_2$O

HCOOH oxidation - formate pathway

HCOOH + 4 H$_2$O $\xrightarrow[+0.29]{-0.31}$ HCOO$_B$ + H$_3$O + 3 H$_2$O $\xrightarrow[\substack{(0.72 - eU)+0.30\ U<0.72 \\ +0.30\ U>0.72}]{+0.72}$ HCOO$_B$ + 4 H$_2$O $\xrightarrow[+0.68]{+0.34}$ HCOO$_M$ + 4 H$_2$O

$\xrightarrow[+0.73]{-0.68}$ CO$_2$ + H$_3$O + 3 H$_2$O $\xrightarrow[\substack{(0.41 - eU)+0.30\ U<0.41 \\ +0.30\ U>0.41}]{+0.41}$ CO$_2$ + 4 H$_2$O

Figure 1.30 Reaction schemes for single rate processes considered in kinetic rates analysis. Reaction energies (italics) and barriers (bold) are given in eV above and below the reaction arrows. The electrode-potential dependence for the reaction energies and barriers of CPET processes are given by including the appropriate correction as a function of U in the energy expression.

and direct pathways are reduced from 1.27 to 0.61 eV and from 1.83 to 0.79 eV, respectively), in the presence of the water bilayer. Nevertheless, the preference for the formate pathway over the direct pathway remains unchanged, as is evident in Figure 1.30.

1.7.5
Eley–Rideal Mechanisms and the Electrode Potential

Thus far we have only considered dehydrogenation processes resulting in neutral H or hydronium adsorbates. These are so-called LH-type reactions whose energetic landscapes have no direct dependence on the electrode potential. Alternatively we might consider ER-type reactions in which the dissociating H atom gives up its electron to the electrode surface and is dissolved into the electrolyte as a solvated proton. When we further consider that this proton is in equilibrium with solvated protons at the counter electrode, which are involved in hydrogen evolution (and is indeed free to migrate and participate in this reaction itself), then we can view the whole as a coupled proton-electron-transfer (CPET) process: If we note that the standard states

$$H^+_{(aq)} + e^- \longrightarrow \frac{1}{2} H_{2(g)}$$

of $H^+_{(aq)}$ and $H_{2(g)}$ are at equilibrium at the hydrogen reference electrode (HRE), then we can calculate the energy of a solvated proton as a function of electrode potential and pH as follows:

$$E_{H^+_{(aq)}} = \frac{1}{2} E_{H_{2(g)}} - pH \cdot k_B T \ln(10) - U = \frac{1}{2} E_{H_{2(g)}} - k_B T \ln\left(\left[H^+_{aq}\right]\right) - U \quad (1.70)$$

Now that the electrode potential dependence of the reaction energy of a CPET process has been determined, a means for determining the barrier of a CPET process is needed. A CPET process consists of a series of proton transfers between water molecules which shuttle the proton from one electrode to the other. This barrier has been estimated to be 0.3 eV based on coupled cluster singlet doublet (CCSD) calculations [86]. The reaction barrier is then assumed to be 0.3 eV above the energetically higher endpoint (i.e., product or reactant state). Thus, in the uphill direction the reaction has a barrier which is 0.3 eV higher than the reaction energy, and the downhill direction a barrier of 0.3 eV.

To model ER-type reaction pathways we need to introduce ER-type reaction steps, which remove the activated H atoms by transferring them as protons into the electrolytic phase, after they have dissociated from HCOOH. The energetics for these CPET steps can then be computed as explained above. The energetic landscapes for the resulting ER-type pathways at 0 V and 1.0 V are shown in Figure 1.31a and b respectively. Along the direct pathway both CPET steps occur at the same stage of the pathway and can thus be combined into a single effective barrier. At zero

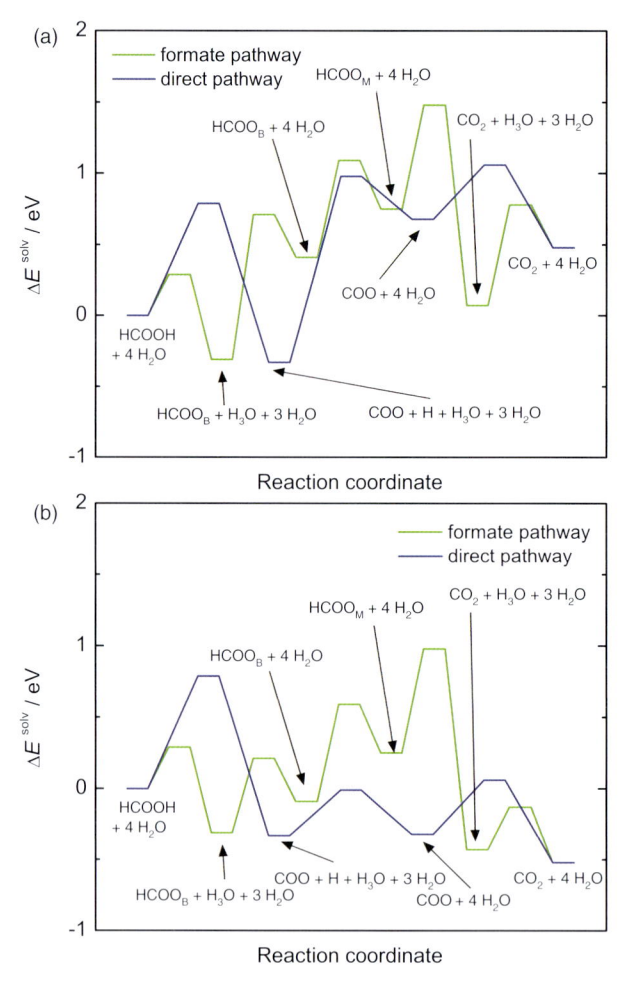

Figure 1.31 Comparison of Eley–Rideal HCOOH oxidation reaction mechanisms with the water bilayer solvation model (a) $U = 0.0$ V and (b) $U = 1.00$ V vs. RHE.

potential this barrier is large, as the reaction step is uphill, and is rate limiting. At $U = 1.0$ V the CPET steps are now downhill, so that the associated barrier is only 0.3 eV. Thus a different, potential-independent barrier (i.e., the barrier for the initial double deprotonation step) is rate limiting. In the case of the formate pathway, the added CPET steps take place immediately after each of the deprotonation steps. At zero potential the first of these has the largest barrier and is rate determining. However, since CPET steps accelerate as the potential increases, a potential-independent step (C—H bond dissociation) is rate limiting at 1.0 V. In the end the rate-limiting barrier for the direct pathway is always higher than that of the formate pathway, so that the formate pathway is always preferred.

1.7.6
Kinetic Rate Model of Formic Acid Oxidation

Finally, the DFT results for the energetics of the different competing reaction pathways were combined into a kinetic rate model using the method explained in Section 1.3.3.3. The role of co-adsorbed CO^* and OH^* was also considered [137]. From all of these possible pathways, the rate for the fastest pathway at the appropriate CO^* and OH^* coverage for electrode potentials between 0.0 and 1.2 V was considered to be a good approximation of the overall reaction rate. This potential-dependent rate is plotted in Figure 1.32 in conjunction with the experimental CV [118].

At low potentials (0.0 V $< U <$ 0.4 V) the formate pathway, limited by a potential-dependent CPET process, is the dominant mechanism. This is corroborated by the experimental observation of $HCOO_B$ at low potentials [127–131, 140]. Near 0.4 V the

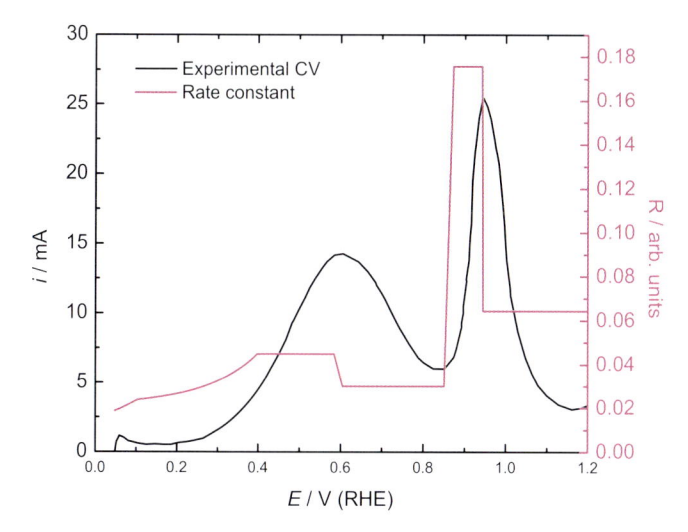

Figure 1.32 Comparison of the experimental CV [118] (taken at a potential sweep rate of 50 mV s^{-1}) and the calculated potential-dependent rate constants.

potential-independent formate pathway becomes dominant and is the primary pathway going into the so-called negative differential resistance (NDR) region. Around 0.6 V the surface concentration of CO^* increases dramatically and succeeds in tying up OH and thus disabling the direct pathway. This probably explains in part the low reactivity of the NDR region. The region of peak activity (0.85 V < U < 0.95) follows. Here the direct pathway, whose reactivity is enhanced by the presence of co-adsorbed CO^* and OH^*, dominates the reaction mechanism. Further increasing the electrode potential leads to the oxidative removal of CO^*, which suppresses the direct pathway and leaves the formate pathway as the primary mechanism above 0.95 V [137].

1.7.7
Summary

The study of HCOOH electro-oxidation on Pt(111) presented in this section has given the opportunity to show the use of slab and explicit solvent models in DFT calculations. The resulting solvent network plays a crucial role in (de)stabilizing important reaction steps, and the use of experimental results to determine the appropriate solvent model (i.e., OH^* and CO^* coverages) for various electrode potentials illustrates an important strategy in developing simulation models: that is, taking advantage of experiment hints. While it would have been possible to derive these coverages entirely from theory, it was more efficient to take advantage of experimental results already in place, and thus avoid additional errors the theoretical derivation might have introduced. The qualitative comparison between the theoretically derived rate constants and the experimental CV, as well as rough comparisons involving other points of contact between theory and experiment (e.g., the experimental observation of $COOH_B$ at low potentials as corroboration of the theoretically predicted mechanism) are typical examples of what must be done to compare results from theory and experiment and combine them to reach conclusions consistent with both.

1.8
Concluding Remarks

In this chapter much attention has been given to electronic structure theory and in particular to DFT. This is appropriate as these methods provide us with the most fundamental models of electrochemical systems (or parts thereof) available to us today, and because other levels within a multiscale modeling framework are typically built on this foundation of applied quantum mechanics. Furthermore, DFT has become a common tool in electrochemistry, such that the treatment of simple electrochemical systems is now a fairly routine application of DFT. The results of such DFT simulations can be readily and regularly translated into the language of experimental observables without the aid of intermediate models. Thus, DFT has often been able to stand on its own as a modeling tool in electrochemistry.

Nevertheless, there are many phenomena, which are not so easily described using only DFT models. One of these is growth processes, where the long-range structure and/or the order of events can be vital to the characterization of the overall process. Thus, kMC methods are a popular means of combining the individual adsorption and diffusion events, which can be modeled individually using DFT, into a coherent model of the entire growth process. Another example is the electrochemical double layer, a description of which requires the analysis of many more structures than routine applications of DFT are capable of tackling. Thus, molecular force fields are an ideal middleman *en route* from DFT to an experimentally meaningful description of the electrochemical double layer. While the quantitative capabilities of DFT increase, in terms of the number of atoms and geometries for which energies and densities can be routinely calculated, the size and complexity of the systems of interest to many electrochemists are increasing as well. At the same time the incentive for leveraging DFT's power by magnifying its results by placing it at the bottom of a multiscale modeling framework only increases as DFT calculations become more reliable and routine. Multiscale modeling in electrochemistry is still in its early stages and has a bright future ahead.

Acknowledgment

The authors would like to acknowledge Dr. John A. Keith and Dr. Wang Gao, who performed the studies presented in the ORR and formic acid oxidation examples. Financial support for the ORR and formic acid oxidation studies was provided by the DFG within the framework of the Emmy-Noether-Program, the Alexander von Humboldt foundation, and the European Union through the Marie-Curie Initial Training Network ELCAT. Computational resources for these studies were provided by the BWGrid. JEM gratefully acknowledges financial support from the Alexander von Humboldt foundation. DF gratefully acknowledges the DFG for financial support. The authors are also grateful to Dr. Attila Farkas for reading portions of the manuscript and providing helpful comments.

References

1 Ertl, G. (2008) Reactions at surfaces: From atoms to complexity (Nobel lecture). *Angew. Chem. Int. Ed.*, **47** (19), 3524–3535.
2 Kolb, D.M. (2011) Electrochemical surface science: past, present and future. *J. Solid State Electrochem.*, **15**, 1391–1399.
3 Groß, A. (2002) The virtual chemistry lab for reactions at surfaces: Is it possible? Will it be useful? *Surf. Sci.*, **500** (1–3), 347–367.
4 Besenbacher, F., Chorkendorff, I., Clausen, B.S., Hammer, B., Molenbroek, A.M., Nørskov, J.K., and Stensgaard, I. (1998) Design of a surface alloy catalyst for steam reforming. *Science*, **279** (5358), 1913–1915.
5 Lozovoi, A.Y., Alavi, A., Kohanoff, J., and Lynden-Bell, R.M. (2001) *Ab initio* simulation of charged slabs at constant chemical potential. *J. Chem. Phys.*, **115** (4), 1661–1669.

6 Taylor, C.D., Wasileski, S.A., Filhol, J.S., and Neurock, M. (2006) First principles reaction modeling of the electrochemical interface: Consideration and calculation of a tunable surface potential from atomic and electronic structure. *Phys. Rev. B*, **73**, 165 402.

7 Skulason, E., Karlberg, G.S., Rossmeisl, J., Bligaard, T., Greeley, J., Jonsson, H., and Nørskov, J.K. (2007) Density functional theory calculations for the hydrogen evolution reaction in an electrochemical double layer on the Pt(111) electrode. *Phys. Chem. Chem. Phys.*, **9**, 3241–3250.

8 Schmickler, W. (1999) Recent progress in theoretical electrochemistry. *Annu. Rep. Prog. Chem., Sect. C: Phys. Chem.*, **95**, 117–162.

9 Desai, S.K. and Neurock, M. (2003) First-principles study of the role of solvent in the dissociation of water over a Pt-Ru alloy. *Phys. Rev. B*, **68**, 075 420.

10 Nørskov, J.K., Bligaard, T., Logadottir, A., Kitchin, J.R., Chen, J.G., Pandelov, S., and Stimming, U. (2005) Trends in the exchange current for hydrogen evolution. *J. Electrochem. Soc.*, **152** (3), J23–J26.

11 Roudgar, A. and Groß, A. (2003) Local reactivity of metal overlayers: Density functional theory calculations of Pd on Au. *Phys. Rev. B*, **67**, 033 409.

12 Venkatachalam, S. and Jacob, T. (2009) Hydrogen adsorption on Pd-containing Au(111) bimetallic surfaces. *Phys. Chem. Chem. Phys.*, **11**, 3263–3270.

13 Kibler, L.A. (2008) Dependence of electrocatalytic activity on film thickness for the hydrogen evolution reaction of Pd overlayers on Au(111). *Electrochim. Acta*, **53** (23), 6824–6828.

14 Feynman, R.P., Leighton, R.B., and Sands, M. (1964) *The Feynman Lectures on Physics*, Addison Wesley.

15 Braatz, R.D., Seebauer, E.G., and Alkire, R.C. (2008) Multiscale modeling and design of electrochemical systems, in *Electrochemical Surface Modification Thin Films, Functionalization and Characterization*, Wiley-VCH Verlag GmbH, pp. 289–334.

16 Dirac, P.A.M. (1929) Quantum mechanics of many-electron systems. *Proc. Roy. Soc. London Ser. A*, **123** (792), 714–733.

17 Griffiths, D.J. (2004) *Introduction to Quantum Mechanics*, 2nd edn, Benjamin Cummings.

18 Cohen-Tannoudji, C., Diu, B., and Laloe, F. (1992) *Quantum Mechanics*, Wiley-Interscience.

19 McQuarrie, D.A. (2007) *Quantum Chemistry*, 2nd edn, University Science Books.

20 Atkins, P.W. and Friedman, R.S. (2010) *Molecular Quantum Mechanics*, 2nd edn, Oxford University Press.

21 Szabo, A. and Ostlund, N.S. (1996) *Modern Quantum Chemistry: Introduction to Advanced Electronic Structure Theory*, Dover Publications.

22 McQuarrie, D.A. (2000) *Statistical Mechanics*, University Science Books.

23 Hill, T.L. (1987) *An Introduction to Statistical Thermodynamics*, Dover Publications.

24 Steinfeld, J.I., Fransisco, J.S., and Hase, W.L. (1998) *Chemical Kinetics and Dynamics*, 2nd edn, Prentice Hall.

25 Cramer, C.J. (2004) *Essentials of Computational Chemistry: Theories and Models*, 2nd edn, John Wiley & Sons Ltd.

26 Jensen, F. (2007) *Introduction to Computational Chemistry*, 2nd edn, John Wiley & Sons, Ltd.

27 Koper, M.T.M. (2003) Ab initio quantum-chemical calculations in electrochemistry. *Mod. Aspects Electrochem.*, **36**, 51–130.

28 Born, M. and Oppenheimer, R. (1927) Zur quantentheorie der molekeln. *Ann. Phys-Berlin*, **84**, 457–484.

29 Hartree, D.R. (1928) The wave mechanics of an atom with a non-coulomb central field. Part i. theory and methods. *Math. Proc. Cambridge*, **24**, 89–110.

30 Hartree, D.R. (1928) The wave mechanics of an atom with a non-coulomb central field. Part ii. Some results and discussion. *Math. Proc. Cambridge*, **24**, 111–132.

31 Hartree, D.R. (1928) The wave mechanics of an atom with a non-coulomb central field. Part iii. Term values and intensities in series in optical spectra. *Math. Proc. Cambridge*, **24**, 426–437.

32 Slater, J.C. (1930) Atomic shielding constants. *Phys. Rev.*, **36**, 57–64.

33 Boys, S.F. (1950) Electronic wave functions. i. A general method of

calculation for the stationary states of any molecular system. *Proc. Roy. Soc. Lond. A*, **200**, 542–554.

34 Boys, S.F. (1950) Electronic wave functions. ii. A calculation for the ground state of the beryllium atom. *Proc. Roy. Soc. Lond. A*, **201**, 125–137.

35 Hohenberg, P. and Kohn, W. (1964) Inhomogeneous electron gas. *Phys. Rev.*, **136**, B864–B871.

36 Kohn, W. and Sham, L.J. (1965) Self-consistent equations including exchange and correlation effects. *Phys. Rev.*, **140**, A1133–A1138.

37 Tao, J.M., Perdew, J.P., Staroverov, V.N., and Scuseria, G.E. (2003) Climbing the density functional ladder: Nonempirical meta-generalized gradient approximation designed for molecules and solids. *Phys. Rev. Lett.*, **91** (14) 146401.

38 Perdew, J.P., Ruzsinszky, A., Tao, J.M., Staroverov, V.N., Scuseria, G.E., and Csonka, G.I. (2005) Prescription for the design and selection of density functional approximations: More constraint satisfaction with fewer fits. *J. Chem. Phys.*, **123** (6), 062201.

39 Bernhard Schlegel, H. (2002) Exploring potential energy surfaces for chemical reactions: An overview of some practical methods. *J. Comput. Chem.*, **24**, 1514–1527.

40 Henkelman, G., Johannesson, G., and Jonsson, H. (2002) Methods for finding saddle points and minimum energy paths, in *Theoretical Methods in Condensed Phase Chemistry*, Springer, pp. 269–302.

41 Sheppard, D., Terrel, R., and Henkelman, G. (2008) Optimization methods for finding minimum energy paths. *J. Chem. Phys.*, **128**, 106–134.

42 Henkelman, G. and Jonsson, H. (2000) Improved tangent estimate in the nudged elastic band method for finding minimum energy paths and saddle points. *J. Chem. Phys.*, **113**, 9978–9985.

43 Whitten, J.L. and Yang, H. (1996) Theory of chemisorption and reactions on metal surfaces. *Surf. Sci. Rep.*, **24**, 55–124.

44 van Santen, R.A. and Neurock, M. (1995) Concepts in theoretical heterogeneous catalytic reactivity. *Catal. Rev.*, **37**, 557–698.

45 Greeley, J., Nørskov, J.K., and Mavrikakis, M. (2002) Electronic structure and catalysis on metal surfaces. *Annu. Rev. Phys. Chem.*, **53**, 319–348.

46 Keith, J.A., Anton, J., Kaghazchi, P., and Jacob, T. (2011) *Modeling Catalytic Reactions on Surfaces with Density Functional Theory*, Wiley-VCH Verlag GmbH, pp. 1–38.

47 Groß, A. and Schnur, A. (2011) Computational chemistry applied to reactions in electrocatalysis, in *Catalysis in Electrochemistry*, John Wiley & Sons, Inc., pp. 165–196.

48 Rossmeisl, J., Nørskov, J.K., Taylor, C.D., Janik, M.J., and Neurock, M. (2006) Calculated phase diagrams for the electrochemical oxidation and reduction of water over Pt(111). *J. Phys. Chem. B*, **110**, 21833–21839.

49 Frenkel, D. and Smith, B. (2001) *Understanding Molecular Simulation: From Algorithms to Applications*, 2nd edn, Academic Press.

50 Benjamin, I. (1997) Molecular dynamic simulations in interfacial electrochemistry, in *Modern Aspects of Electrochemistry*, Plenum Press, pp. 115–179.

51 van Duin, A.C.T., Dasgupta, S., Lorant, F., and Goddard, W.A. (2001) ReaxFF: A reactive force field for hydrocarbons. *J. Phys. Chem. A*, **105** (41), 9396–9409.

52 Brenner, D.W. (1990) Empirical potential for hydrocarbons for use in simulating the chemical vapor depostion of diamond films. *Phys. Rev. B*, **42** (15), 9458–9471.

53 Tersoff, J. (1989) Modelling solid-state chemistry: Interatomic potentials for multicomponent systems. *Phys. Rev. B*, **39** (8), 5566–5568.

54 Car, R. and Parrinello, M. (1985) Unified approach for molecular dynamics and density-functional theory. *Phys. Rev. Lett.*, **55**, 2471–2474.

55 Helgaker, T., Uggerud, E., and Jensen, H.J.A. (1990) Integration of the classical equations of motion on ab initio molecular potential energy surfaces using gradients and hessians: application to translational energy release upon fragmentation. *Chem. Phys. Lett.*, **173**, 145–150.

56 Li, X., Tully, J.C., Schlegel, H.B., and Frisch, M.J. (2005) *Ab initio* ehrenfest dynamics. *J. Chem. Phys.*, **123** (8), 084 106.

57 Binder, K. and Heermann, D.W. (2010) *Monte Carlo Simulation in Statistical Physics*, Springer Berlin Heidelberg.

58 Field, M.J. (1999) *A Practical Introduction To The Simulation Of Molecular Systems*, Cambridge University Press.

59 Greiner, W. (1993) *Thermodynamik und Statistische Mechanik*, Verlag Harri Deutsch.

60 Lin, H. and Truhlar, D.G. (2007) QM/MM: What have we learned, where are we, and where do we go from here? *Theor. Chem. Acc.*, **117**, 185–199.

61 Zwanzig, R. (1965) Time-Correlation Functions and Transport Coefficients in Statistical Mechanics. *Annu. Rev. Phys. Chem.*, **16**, 67–102.

62 Ernst, M.H., Hauge, E.H., and van Leeuwen, J.M.J. (1971) Asymptotic time behavior of correlation functions. i. Kinetic terms. *Phys. Rev. A*, **4**, 2055–2065.

63 Ernst, M.H., Hauge, E.H., and Leeuwen, J.M.J. (1976) Asymptotic time behavior of correlation functions. iii. Local equilibrium and mode-coupling theory. *J. Stat. Phys.*, **15**, 23–58.

64 Haile, J.M. (1997) *Molecular Dynamics Simulation: Elementary Methods*, Wiley Professional Paperback Series, John Wiley & Sons.

65 Hartnig, C., Grimminger, J., and Spohr, E. (2007) Adsorption of formic acid on Pt(111) in the presence of water. *J. Electroanal. Chem.*, **607** (1–2), 133–139.

66 Goddard, W.A., III, Merinov, B., Van Duin, A., Jacob, T., Blanco, M., Molinero, V., Jang, S.S., and Jang, Y.H. (2006) Multi-paradigm multi-scale simulations for fuel cell catalysts and membranes. *Mol. Simul.*, **32** (3–4), 251–268.

67 Voter, A.F., Montalenti, F., and Germann, T.C. (2002) Extending the time scale in atomistic simulation of materials. *Annu. Rev. Mater. Res.*, **32**, 321–346.

68 Battaile, C.C. and Srolovitz, D.J. (2002) Kinetic Monte Carlo simulation of chemical vapor deposition. *Ann. Rev. Mater. Res.*, **32**, 297–319.

69 Chatterjee, A. and Vlachos, D.G. (2007) An overview of spatial microscopic and accelerated kinetic Monte Carlo methods. *J. Comput Aided. Mol. Des.*, **14**, 253–308.

70 Voter, A.F. (2007) Introduction to the kinetic Monte Carlo method, in *Radiation Effects in Solids*, Springer, pp. 1–23.

71 Henkelman, G. and Jonsson, H. (2001) Long time scale kinetic Monte Carlo simulations without lattice approximation and predefined event table. *J. Chem. Phys.*, **115**, 9657–9666.

72 Biehl, M. and Much, F. (2005) Off-lattice kMC simulations of Stranski-Krastanou-like growth, in *Quantum Dots: Fundamentals, Applications, and Frontiers*, Springer.

73 El-Mellouhi, F., Mousseau, N., and Lewis, N.J. (2008) Kinetic activation-relaxation technique: An off-lattice self-learning kinetic Monte Carlo algorithm. *Phys. Rev. B*, **78**, 153 202.

74 Kara, A., Truschin, O., Yildirim, H., and Rahman, T.S. (2009) Off-lattice, self-learning kinetic Monte Carlo: application to 2D cluster diffusion on the fcc(111) surface. *J. Phys-Condens. Mater.*, **21**, 084 213.

75 Kleiner, K., Comas-Vives, A., Naderian, M., Mueller, J.E., Fantauzzi, D., Mesgar, M., Keith, J.A., Anton, J., and Jacob, T. (2011) Multiscale modeling of Au-island ripening on Au(100). *Adv. Phys. Chem.*, 252591.

76 Trushin, O., Karim, A., Kara, A., and Rahman, T.S. (2005) Self-learning kinetic Monte Carlo method: Application to Cu(111). *Phys. Rev. B*, **72**, 115 401.

77 Metropolis, N., Rosenbluth, A.W., Rosenbluth, M.N., Teller, A.H., and Teller, E. (1953) Equation of state calculations by fast computing machines. *J. Chem. Phys.*, **21** (6), 1087–1092.

78 Fichthorn, K.A. and Weinberg, W.H. (1991) Theoretical foundations of dynamical Monte Carlo simulations. *J. Chem. Phys.*, **95** (2), 1090–1096.

79 Battaile, C.C., Srolovitz, D.J., and Butler, J.E. (1997) A kinetic Monte Carlo method for the atomic-scale simulation of chemical vapor deposition: Application to diamond. *J. Appl. Phys.*, **82**, 6293–6300.

80 Reuter, K. and Scheffler, M. (2006) First-principles kinetic Monte Carlo simulations for heterogeneous catalysis:

Application to the co oxidation at RuO$_2$(110). *Phys. Rev. B*, **73**, 045 433.

81 Oppelstrup, T., Bulatov, V.V., Donev, A., Kalos, M.H., Gilmer, G.H., and Sadigh, B. (2009) First-passage kinetic Monte Carlo method. *Phys. Rev. E*, **80**, 066 701.

82 Adzic, R. (1998) *Electrocatalysis*, Wiley-VCH Verlag GmbH, pp. 197–242.

83 Marković, N.M. and Ross, P.N. (1999) *Interfacial Electrochemistry: Theory, Experiments and Applications*, Marcel Dekker, pp. 821–841.

84 Marković, N.M., Schmidt, T.J., Stamenković, V., and Ross, P.N. (2001) Oxygen reduction reaction on Pt and Pt bimetallic surfaces: A selective review. *Fuel Cells*, **1** (2), 105–116.

85 Ross, P.N. (2003) *Handbook of Fuel Cells: Fundamentals, Technology, Applications*, Wiley-VCH, pp. 465–480.

86 Keith, J.A., Jerkiewicz, G., and Jacob, T. (2010) Theoretical investigations of the oxygen reduction reaction on Pt(111). *ChemPhysChem*, **11** (13), 2779–2794.

87 Keith, J.A. and Jacob, T. (2010) Theoretical studies of potential-dependent and competing mechanisms of the electrocatalytic oxygen reduction reaction on Pt(111). *Angew. Chem. Int. Ed.*, **49** (49), 9521–9525.

88 Becke, A.D. (1993) Density-functional thermochemistry. iii. The role of exact exchange. *J. Chem. Phys.*, **98** (7), 5648–5652.

89 Lee, C., Yang, W., and Parr, R.G. (1988) Development of the Colle-Salvetti correlation-energy formula into a functional of the electron density. *Phys. Rev. B*, **37**, 785–789.

90 Vosko, S.H., Wilk, L., and Nusair, M. (1980) Accurate spin-dependent electron liquid correlation energies for local spin density calculations: A critical analysis. *Can. J. Phys.*, **58** (8), 1200–1211.

91 Stephens, P.J., Devlin, F.J., Chabalowski, C.F., and Frisch, M.J. (1994) Ab initio calculation of vibrational absorption and circular dichroism spectra using density functional force fields. *J. Phys. Chem.*, **98** (45), 11 623–11 627.

92 Hay, P.J. and Wadt, W.R. (1985) Ab initio effective core potentials for molecular calculations. Potentials for the transition metal atoms Sc to Hg. *J. Chem. Phys.*, **82** (1), 270–283.

93 Boys, S.F. (1950) Electronic wave functions. i. A general method of calculation for the stationary states of any molecular system. *Proc. Roy. Soc. London Ser. A*, **200** (1063), 542–554.

94 Hehre, W.J., Ditchfield, R., and Pople, J.A. (1972) Self-consistent molecular orbital methods. xii. Further extensions of Gaussian-type basis sets for use in molecular orbital studies of organic molecules. *J. Chem. Phys.*, **56** (5), 2257–2261.

95 Frisch, M.J., Pople, J.A., and Binkley, J.S. (1984) Self-consistent molecular orbital methods 25. Supplementary functions for gaussian basis sets. *J. Chem. Phys.*, **80** (7), 3265–3269.

96 Francl, M.M., Pietro, W.J., Hehre, W.J., Binkley, J.S., Gordon, M.S., DeFrees, D.J., and Pople, J.A. (1982) Self-consistent molecular orbital methods. xxiii. a polarization-type basis set for second-row elements. *J. Chem. Phys.*, **77** (7), 3654–3665.

97 (2007) Jaguar, Schrödinger, LLC, New York, NY.

98 Mueller, J.E., van Duin, A.C.T., and Goddard, W.A. (2009) Structures, energetics, and reaction barriers for CHx bound to the nickel (111) surface. *J. Phys. Chem. C*, **113** (47), 20290–20306.

99 Jacob, T., Muller, R.P., and Goddard, W.A. (2003) Chemisorption of atomic oxygen on Pt(111) from DFT studies of Pt-clusters. *J. Phys. Chem. B*, **107** (35), 9465–9476.

100 Jacob, T. and Goddard, W.A. (2006) Water formation on Pt and Pt-based alloys: A theoretical description of a catalytic reaction. *ChemPhysChem.*, **7** (5), 992–1005.

101 Jacob, T. (2006) The mechanism of forming H$_2$O from H$_2$ and O$_2$ over a Pt catalyst via direct oxygen reduction. *Fuel Cells*, **6** (3–4), 159–181.

102 Gland, J.L., Sexton, B.A., and Fisher, G.B. (1980) Oxygen interactions with the Pt(111) surface. *Surf. Sci.*, **95** (2–3), 587–602.

103 Campbell, C.T., Ertl, G., Kuipers, H., and Segner, J. (1981) A molecular beam study of the adsorption and desorption of

oxygen from a Pt(111) surface. *Surf. Sci.*, **107** (1), 220–236.

104 Parker, D.H., Bartram, M.E., and Koel, B.E. (1989) Study of high coverages of atomic oxygen on the Pt(111) surface. *Surf. Sci.*, **217** (3), 489–510.

105 Lehwald, S., Ibach, H., and Steininger, H. (1982) Overtones and multiphonon processes in vibration spectra of adsorbed molecules. *Surf. Sci.*, **117** (1–3), 342–351.

106 Steininger, H., Lehwald, S., and Ibach, H. (1982) Adsorption of oxygen on Pt(111). *Surf. Sci.*, **123** (1), 1–17.

107 Nolan, P.D., Lutz, B.R., Tanaka, P.L., Davis, J.E., and Mullins, C.B. (1999) Molecularly chemisorbed intermediates to oxygen adsorption on Pt(111): A molecular beam and electron energy-loss spectroscopy study. *J. Chem. Phys.*, **111** (8), 3696–3704.

108 Parsons, R. and VanderNoot, T. (1988) The oxidation of small organic molecules: A survey of recent fuel cell related research. *J. Electroanal. Chem., Interfacial Electrochem.*, **257** (1–2), 9–45.

109 Neurock, M., Janik, M., and Wieckowski, A. (2009) A first principles comparison of the mechanism and site requirements for the electrocatalytic oxidation of methanol and formic acid over Pt. *Faraday Discuss*, **140**, 363–378.

110 Wang, H.F. and Liu, Z.P. (2009) Formic acid oxidation at Pt/H_2O interface from periodic DFT calculations integrated with a continuum solvation model. *J. Phys. Chem. C*, **113** (40), 17 502–17 508.

111 Segall, M.D., Lindan, P.J.D., Probert, M.J., Pickard, C.J., Hasnip, P.J., Clark, S.J., and Payne, M.C. (2002) First-principles simulation: ideas, illustrations and the CASTEP code. *J. Phys-Condens. Mater.*, **14** (11), 2717.

112 Abbas, N. and Madix, R.J. (1983) Surface reaction modification: The effect of structured overlayers of sulfur on the kinetics and mechanism of the decomposition of formic acid on pt(111). *Applic. Surf. Sci.*, **16** (3–4), 424–440.

113 Beden, B., Leger, J.M., and Lamy, C. (1992) *Modern Aspects of Electrochemistry*, Plenum.

114 Sun, S.G. (1998) *Electrocatalysis*, Wiley-VCH Verlag GmbH, p. 243.

115 Avery, N.R. (1982) Adsorption of formic acid on clean and oxygen covered Pt(111). *Applic. Surf. Sci.*, **11–12** (0), 774–783.

116 Avery, N.R. (1983) Reaction of HCOOH with a Pt(111)-O surface; identification of adsorbed monodentate formate. *Applic. Surf. Sci.*, **14** (2), 149–156.

117 Columbia, M.R., Crabtree, A.M., and Thiel, P.A. (1992) The temperature and coverage dependences of adsorbed formic acid and its conversion to formate on platinum(111). *J. Am. Chem. Soc.*, **114** (4), 1231–1237.

118 Columbia, M.R. and Thiel, P.A. (1990) The reaction of formic acid with clean and water-covered Pt(111). *Surf. Sci.*, **235** (1), 53–59.

119 Sun, S.G., Clavilier, J., and Bewick, A. (1988) The mechanism of electrocatalytic oxidation of formic acid on Pt(100) and Pt(111) in sulphuric acid solution: an EMIRS study. *J. Electroanal. Chem., Interfacial Electrochem.*, **240** (1–2), 147–159.

120 Capon, A. and Parsons, R. (1973) The oxidation of formic acid at noble metal electrodes Part iii. Intermediates and mechanism on platinum electrodes. *J. Electroanal. Chem., Interfacial Electrochem.*, **45** (2), 205–231.

121 Breiter, M.W. (1967) A study of intermediates adsorbed on platinized-platinum during the steady-state oxidation of methanol, formic acid and formaldehyde. *J. Electroanal. Chem., Interfacial Electrochem.*, **14** (4), 407–413.

122 Capon, A. and Parson, R. (1973) The oxidation of formic acid at noble metal electrodes: I. Review of previous work. *J. Electroanal. Chem., Interfacial Electrochem.*, **44** (1), 1–7.

123 Kizhakevariam, N. and Weaver, M.J. (1994) Structure and reactivity of bimetallic electrochemical interfaces: infrared spectroscopic studies of carbon monoxide adsorption and formic acid electrooxidation on antimony-modified Pt(100) and Pt(111). *Surf. Sci.*, **310** (1–3), 183–197.

124 Rice, C., Ha, S., Masel, R.I., and Wieckowski, A. (2003) Catalysts for direct formic acid fuel cells. *J. Power Sources*, **115** (2), 229–235.

125 Chen, Y.X., Ye, S., Heinen, M., Jusys, Z., Osawa, M., and Behm, R.J. (2006) Application of in-situ attenuated total reflection-Fourier transform infrared spectroscopy for the understanding of complex reaction mechanism and kinetics: Formic acid oxidation on a Pt film electrode at elevated temperatures. *J. Phys. Chem. B*, **110** (19), 9534–9544.

126 Chen, Y.X., Heinen, M., Jusys, Z., and Behm, R.J. (2006) Kinetics and mechanism of the electrooxidation of formic acid – spectroelectrochemical studies in a flow cell. *Angew. Chem. Int. Ed.*, **45** (6), 981–985.

127 Chen, Y.X., Heinen, M., Jusys, Z., and Behm, R.J. (2006) Bridge-bonded formate: Active intermediate or spectator species in formic acid oxidation on a Pt film electrode? *Langmuir*, **22** (25), 10 399–10 408.

128 Chen, Y.X., Heinen, M., Jusys, Z., and Behm, R.J. (2007) Kinetic isotope effects in complex reaction networks: Formic acid electro-oxidation. *ChemPhysChem*, **8** (3), 380–385.

129 Samjeské, G. and Osawa, M. (2005) Current oscillations during formic acid oxidation on a Pt electrode: Insight into the mechanism by time-resolved IR spectroscopy. *Angew. Chem. Int. Ed.*, **44** (35), 5694–5698.

130 Samjeské, G., Miki, A., Ye, S., Yamakata, A., Mukouyama, Y., Okamoto, H., and Osawa, M. (2005) Potential oscillations in galvanostatic electrooxidation of formic acid on platinum: A time-resolved surface-enhanced infrared study. *J. Phys. Chem. B*, **109** (49), 23509–23516.

131 Mukouyama, Y., Kikuchi, M., Samjeské, G., Osawa, M., and Okamoto, H. (2006) Potential oscillations in galvanostatic electrooxidation of formic acid on platinum: A mathematical modeling and simulation. *J. Phys. Chem. B*, **110** (24), 11912–11917.

132 Kunimatsu, K. and Kita, H. (1987) Infrared spectroscopic study of methanol and formic acid absorbates on a platinum electrode: Part ii. role of the linear co(a) derived from methanol and formic acid in the electrocatalytic oxidation of ch3oh and hcooh. *J. Electroanal. Chem., Interfacial Electrochem.*, **218** (1–2), 155–172.

133 Corrigan, D.S. and Weaver, M.J. (1988) Mechanisms of formic acid, methanol, and carbon monoxide electrooxidation at platinum as examined by single potential alteration infrared spectroscopy. *J. Electroanal. Chem., Interfacial Electrochem.*, **241** (1–2), 143–162.

134 Wilhelm, S., Vielstich, W., Buschmann, H.W., and Iwasita, T. (1987) Direct proof of the hydrogen in the methanol adsorbate at platinum — An ECTDMS study. *J. Electroanal. Chem., Interfacial Electrochem.*, **229** (1–2), 377–384.

135 Gao, W., Keith, J.A., Anton, J., and Jacob, T. (2010) Oxidation of formic acid on the Pt(111) surface in the gas phase. *Dalton Trans.*, **39**, 8450–8456.

136 Gao, W., Keith, J.A., Anton, J., and Jacob, T. (2010) Theoretical elucidation of the competitive electro-oxidation mechanisms of formic acid on Pt(111). *J. Am. Chem. Soc.*, **132** (51), 18 377–18 385.

137 Gao, W., Mueller, J.E., Jiang, Q., and Jacob, T. (2012) The role of co-adsorbed CO and OH in the electrooxidation of formic acid on Pt(111). *Angew. Chem. Int. Ed.*, **51** (37), 9448–9452.

138 Perdew, J.P., Burke, K., and Ernzerhof, M. (1996) Generalized gradient approximation made simple. *Phys. Rev. Lett.*, **77**, 3865–3868.

139 Vanderbilt, D. (1990) Soft self-consistent pseudopotentials in a generalized eigenvalue formalism. *Phys. Rev. B*, **41**, 7892–7895.

140 Grozovski, V., Vidal-Iglesias, F.J., Herrero, E., and Feliu, J.M. (2011) Adsorption of formate and its role as intermediate in formic acid oxidation on platinum electrodes. *ChemPhysChem*, **12** (9), 1641–1644.

2
Statistical Mechanics and Kinetic Modeling of Electrochemical Reactions on Single-Crystal Electrodes Using the Lattice-Gas Approximation

Marc T.M. Koper

2.1
Introduction

Theory, modeling and simulation are playing an increasingly important role in surface electrochemistry and electrocatalysis [1,2]. Especially, the use of density functional theory calculations for the computation of the energetics of surface adsorption and surface reactions has become very popular in recent years [3,4]. Such calculations typically focus on small systems of at best a hundred atoms or so, and compute the energetics of elementary adsorption or reaction steps at zero temperature. Proper comparison of such calculations to experiment still requires a statistical averaging over all different adsorption and reaction possibilities, that is, due account needs to be taken of configurational entropy.

The link from the energetics of elementary processes to the properties of a real macroscopic system at some realistic temperature is provided by statistical mechanics [5,6]. If the aim is to model a dynamic phenomenon in macroscopic chemical kinetics, one usually resorts to kinetic modeling. However, kinetic modeling often implicitly makes certain statistical-mechanical approximations the accuracy of which is not so easy to estimate and, therefore, typically not confirmed.

The aim of this chapter is to give an introduction to the statistical-mechanical description of electrochemical surface reactions, with the specific aim to elucidate the effect of approximations made in its derivation, and to assess the appropriateness of such approximations by explicit comparison to experimental results. For such a comparison to be meaningful, the experiments must involve well-defined single-crystalline surfaces.

The chapter starts with the introduction of an important model in the description of surface reactions, namely the lattice-gas model. After having discussed the necessary input parameters for such a model, which can for instance be obtained from modern first-principles density functional theory (DFT) calculations, we will summarize the main statistical mechanical equations to calculate the static and dynamic properties of a macroscopic system. The solution of these equations can be obtained by using certain approximations, which ideally lead to analytical solutions

---

Electrocatalysis: Theoretical Foundations and Model Experiments, First Edition.
Edited by Richard C. Alkire, Ludwig A. Kibler, Dieter M. Kolb, and Jacek Lipkowski.
© 2013 Wiley-VCH Verlag GmbH & Co. KGaA. Published 2013 by Wiley-VCH Verlag GmbH & Co. KGaA.

or equations that are easily solved numerically, or the exact solution can be generated numerically, by using so-called Monte Carlo simulation methods. Finally, we will discuss the lattice-gas modeling of a number of selected model reactions, including both adsorption (or electrosorption) reactions and catalytic reactions on well-defined surfaces. This comparison will give us insight into the proper description and interpretation of electrochemical data on single-crystal surfaces, and will point out the potential limitations and pitfalls of the application of statistical-mechanical models and kinetic modeling to less well-defined systems.

2.2
Lattice-Gas Modeling of Electrochemical Surface Reactions

A lattice-gas model assumes that certain properties of a physical or chemical system can be projected accurately onto a lattice of discrete points. Properties of solids are often well described using the lattice-gas model, the Ising model for ferromagnetism being a classical example [5,6]. The lattice-gas model is also very popular in the description of heterogeneous catalytic reactions, in which the reactive surface is modeled as a two-dimensional lattice of adsorption and reaction sites. Typically, the lattice has the same symmetry as the experimental surface, the reason why lattice-gas modeling is particularly attractive (if not uniquely applicable) to the modeling of reactions on single-crystal surfaces. It is possible to include various different adsorption sites, steps and defects, and even surface reconstructions, into the lattice-gas model, though at the cost of increasing modeling complexity. Another advantage of the lattice-gas model is the availability of analytical solutions of some of its properties, as well as the relative ease with which the solutions of a lattice-gas model can be generated numerically, by for instance using (kinetic) Monte Carlo simulations (see below).

When modeling an electrochemical surface reaction using the lattice-gas model, one needs to consider a number of ingredients. The first choice to be made is the symmetry of the lattice. Figure 2.1 shows a typical lattice of points modeling a (111) surface of an fcc metal (such as platinum), also incorporating the existence of step sites which break the overall symmetry of the lattice. A similar figure could have been generated for a (100) surface. In many cases, the model is simplified to only a single reaction site with no step or defect sites, such that the model has perfect hexagonal symmetry (or, in the case of a (100) surface, square symmetry).

Once the lattice has been specified, one needs to decide whether to model a static, time-independent system, or a kinetic, time-dependent system. In the former case, we only need to consider the interaction energies; in the latter case, we also need to specify activation barriers and reaction rate constants. The types of reactions included in the model are either adsorption or desorption steps (which may be electrosorption processes if they involve charge transfer), diffusion steps, and reaction steps between species adsorbed on the lattice.

In the static description, one of the input parameters that we need for an adsorption/desorption step of a species A is its interaction energy with the lattice site i, $\Delta E_{ads}^i(A)$. Note that this energy may depend on the specific lattice site onto

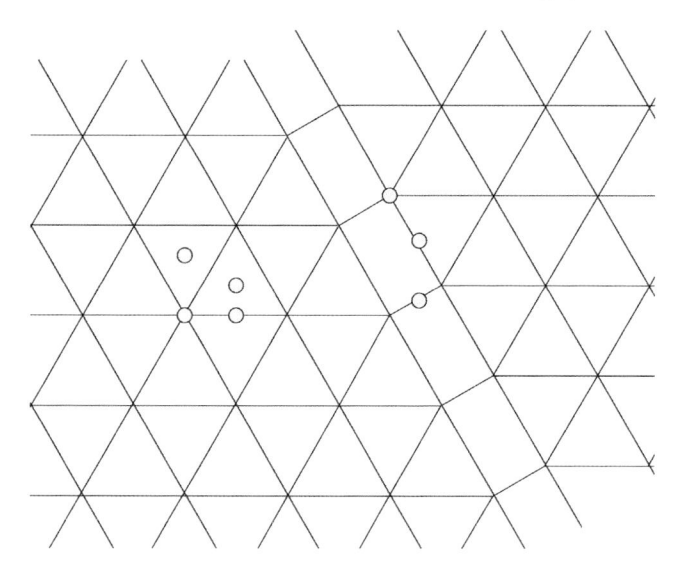

Figure 2.1 Section of a lattice representation of a stepped Pt(111) surface indicating some of the lattice sites that could be considered in the model.

which the species adsorbs. Also note that this energy is partially a free energy, in the sense that it includes changes in entropy that A experiences as it adsorbs from the solution onto the surface. However, configurational entropy related to the final adsorption state of an ensemble of A species is not yet accounted for. That free energy will be denoted by the symbol ΔG and its calculation is the subject of statistical mechanics. The second set of input parameters that we need for determining the likelihood of the presence of species A on lattice site i, is its interaction with other species on the lattice. If a species B is adsorbed on site j this interaction energy will be denoted as $\Delta E_{int}^{ij}(AB)$. Both energies, that is, $\Delta E_{ads}^{i}(A)$ and $\Delta E_{int}^{ij}(AB)$, can be obtained from first-principles quantum-chemical calculations, such as those based on DFT. Reviews on the application of such techniques to electrochemical interfaces can be found elsewhere in the literature [3,4,7]. However, they may also be obtained from fitting an adsorption-interaction model to experimental data. Strategies for doing so in a systematic way have been summarized in review papers by Rikvold et al. [8,9].

In the static description, reaction events such as diffusion and reaction are not considered. However, in a dynamic description, such events may play a decisive role in the kinetic behavior of the system. In formulating their rates, all reactions that are considered reversible, must satisfy the principle of *detailed balance* (or *microscopic reversibility*) [5], even if the corresponding rates are slow. This ensures that the correct equilibrium expressions are satisfied when the overall system is considered at thermodynamic equilibrium. For an adsorption/desorption process:

$$A + *_i \underset{k_{des}}{\overset{k_{ads}}{\rightleftarrows}} A_{ads} \tag{2.1}$$

where $*_i$ denotes the empty lattice site i, one typically writes the rate constants for adsorption and desorption as:

$$k_{ads} = k_{ads}^0 \exp\left(\frac{-\alpha \Delta E_{ads}^i(A)}{k_B T}\right) \qquad (2.2)$$

$$k_{des} = k_{des}^0 \exp\left(\frac{(1-\alpha)\Delta E_{ads}^i(A)}{k_B T}\right) \qquad (2.3)$$

where the sub- or superscripts "A" or "i" on the rate constants have been omitted in order to avoid a needlessly complex notation. In Eqs. (2.2) and (2.3), α is a so-called Brønsted coefficient, expressing how a change in adsorption energy would affect the activation energy for adsorption, and k_B and T have their usual meaning. The ratio of the adsorption and desorption rate constants is the equilibrium constant:

$$K = \frac{k_{ads}}{k_{des}} = \frac{k_{ads}^0}{k_{des}^0} \exp\left(\frac{-\Delta E_{ads}^i(A)}{k_B T}\right) \qquad (2.4)$$

where $\Delta E_{ads}^i(A)$ is defined as a negative quantity if adsorption is favorable. The introduction of the Brønsted coefficient ensures that detailed balance is always satisfied. Since adsorption is often an activationless process, one often assumes α is close to 0. For reactions for which the potential energy surface is more symmetric (e.g., diffusion, see below), α is typically chosen close to 0.5. In principle, the real value of α needs to be calculated from first principles for each reaction.

Diffusion on a lattice gas is usually restricted to neighboring or nearby sites. Just as adsorption and desorption, diffusion rate constants must satisfy detailed balance. If diffusion of species A is from site i to j, the energy of A changes by:

$$\Delta E_{dif}^{ij}(A) = \Delta E_{ads}^j(A) + \sum_k \Delta E_{int}^{jk}(A_j B_k) - \Delta E_{ads}^i(A) - \sum_l \Delta E_{int}^{il}(A_i B_l) \qquad (2.5)$$

The concomitant rate constants for diffusion are

$$k_{dif}^{i \to j}(A) = k_{dif}^0 \exp\left(\frac{-\alpha \Delta E_{dif}^{ij}(A)}{k_B T}\right) \qquad (2.6)$$

$$k_{dif}^{j \to i}(A) = k_{dif}^0 \exp\left(\frac{(1-\alpha)\Delta E_{dif}^{ij}(A)}{k_B T}\right) \qquad (2.7)$$

For diffusion, barriers tend to be symmetric, and α is typically chosen equal to 0.5. Note that fast diffusion is typically observed when the adsorption energy of a certain species is not strongly dependent on the adsorption site. When a species has a strong preference for a certain adsorption site, diffusion is slow. One often assumes that strongly adsorbed species show correspondingly slow diffusion, but this assumption is only valid if the strong adsorption applies to a single surface site or coordination. More generally, it is the surface corrugation potential that determines the diffusion rate, not the adsorption energy.

For surface reactions between surface-adsorbed species, one can derive similar expressions, with rate constants that follow from transition state theory and that must obey detailed balance if the reaction is modeled as reversible. Activation barriers for such reactions can be obtained from DFT calculations (though one should not forget that such calculations practically always apply to zero temperature, and their relevance to electrochemistry remains somewhat difficult to assess). Pre-exponential factors are far more difficult to estimate from first principles, and typically their values are chosen within a reasonable range such that the correspondence to experiment is not too sensitively dependent on their actual values.

For all surface reactions that involve electron or charge transfer, one must consider the effect of the electrode potential. In order to clearly distinguish between the energy E and electrode potential, we will use symbol ϕ for the electrode potential. However, in some of the figures reproduced from the literature, E stands for electrode potential. First, adsorption energies of species depend on the electrode potential [1]:

$$\Delta E^i_{ads}(A) = \Delta E^{i0}_{ads}(A) - e_0 \gamma_A (\phi - \phi^0) + k_B T \ln c_A \tag{2.8}$$

where γ_A is the so-called electrosorption valency of A, and c_A is the activity of A in solution. At equilibrium, Eq. (2.8) is equal to the chemical potential μ_A of species A in solution. The electrosorption valency is a thermodynamically measurable quantity, and is the amount of charge flowing through the external circuit as one species of A adsorbs or desorbs. In many practical cases, the electrosorption valency is taken as a constant equal to the valency of the ion adsorbing or equal to the number of electrons that are transferred in the electrosorption reaction, but in principle γ_A need not be an integer, and can depend on the electrode potential and the coverage or the local environment (see Section 2.5 for a well-documented example). Note that by inserting Eq. (2.8) into Eqs. (2.2) and (2.3), we obtain the usual Butler–Volmer (BV) equation for the potential dependence of the rate of electron transfer, where α is now the BV transfer coefficient [1]. The BV equation is also used to express the potential dependence of surface reactions involving electron transfer. Surface reactions that do not involve interfacial charge transfer are normally assumed to be independent of electrode potential.

2.3
Statistical Mechanics and Approximations

2.3.1
Static System

For the modeling of a static lattice gas system, one usually collects all energetic interactions into a lattice gas Hamiltonian. For instance, if the model has two adsorbates, A and B, the Hamiltonian reads as:

$$H(A, B) = \sum_i \left(\Delta E^i_{ads}(A) c^i_A + \Delta E^i_{ads}(B) c^i_B \right) + \sum_{\langle ij \rangle} \Delta E^{ij}_{int}(AB) c^i_A c^j_B \tag{2.9}$$

where c_A^i and c_B^i are either 1 or 0 depending on whether A and B are adsorbed on site i or not, and the sums $\sum_i$ and $\sum_{\langle ij \rangle}$ run over all sites and pairs of sites on the lattice, respectively. Note that this entails an important assumption, namely that the interaction energies are pairwise additive. This assumption is very common, and it is normally a reasonably good approximation (see for instance ref. 10 for a good study into this issue).

All thermodynamic quantities follow from the so-called partition function. For a canonical system (i.e., with fixed number of particles n) with a discrete set of states (with index s labeling the states) and corresponding energies, such as a lattice gas, the partition function is:

$$Z_n = \sum_s \exp(-E_s/k_B T) \tag{2.10}$$

where E_s is the energy of state s, as calculated from the Hamiltonian, and $\exp(-E_s/k_B T)$ is the Boltzmann factor corresponding to state s. For a grand-canonical ensemble (i.e., with a fixed chemical potential μ, such that the number n can fluctuate), the grand canonical partition function is:

$$\Xi_\mu = \sum_{n=0}^N Z_n \exp\left(\frac{n\mu}{k_B T}\right) \tag{2.11}$$

where N is the total number of sites on the lattice. The natural logarithm of the partition function is related to the free energy relevant to the ensemble studied, and from the free energy and the Maxwell relations one derives expressions for the various observables [5,6].

The partition function can be expressed exactly for a system of non-interacting adsorbates on the lattice: it is simply the number permutations of n adsorbates on a lattice with N sites [5,6]. For a single adsorbate A, the total occupancy or coverage as a function of $\mu_A = \Delta E_{ads}^i(A)$, is given by the well-known Langmuir isotherm:

$$\theta = \frac{N_A}{N} = \frac{1}{1 + \exp(\mu_A/k_B T)} \quad \text{or} \quad \frac{\theta}{1-\theta} = \exp(\mu_A/k_B T) \tag{2.12}$$

In the case of lateral interactions between adsorbates A, there is no exact expression for the partition function. One useful and very common approximation is the so-called mean-field approximation (MFA) [5,6]. The MFA assumes that all adsorbates on the lattice are randomly mixed and uncorrelated. As a result, the *average* interaction energy an adsorbate feels from the other adsorbates is proportional to the coverage θ. If we restrict ourselves to nearest-neighbor interactions ε_{nn}, this interaction energy is $z\varepsilon_{nn}\theta$, with z the number of nearest-neighbor sites, which is then added to μ_A:

$$\frac{\theta}{1-\theta} = \exp\left(\frac{\mu_A + z\varepsilon_{nn}\theta}{k_B T}\right) = K \exp(g\theta) \tag{2.13}$$

This equation is known as the Frumkin isotherm in the electrochemistry literature. The key consequence of the MFA is that all adsorbates on the surface can be described by their average coverages. The MFA is, in general, a rather strong assumption, and care must be exercised when a detailed interpretation of experimental data is made on the basis of such an approximation.

The quasi-chemical approximation (QCA) gives an exact treatment of the correlations between nearest neighbors, but treats all the other interactions (including those of the nearest neighbors with their nearest neighbors) in the MFA. The expression for the coverage in the QCA is [6]:

$$\frac{\theta}{1-\theta}\left(\frac{(1-\theta)}{\theta}\frac{(\beta-1+2\theta)}{(\beta+1-2\theta)}\right)^{z/2} = \exp\left(\frac{\mu_A + z\varepsilon_{nn}/2}{k_B T}\right) \quad (2.14)$$

with $\beta = \{1 - 4[1 - \exp(\varepsilon_{nn}/k_B T)]\theta(1-\theta)\}^{1/2}$. The QCA works quite well for a square lattice, as will be seen below, because on such a surface nearest neighbors do not share other nearest neighbors.

There are a few specific models for which exact analytical solutions are known. One is the one-dimensional lattice gas, which can serve, as we will see later, as a model for the adsorption on step sites. If the interaction between two neighboring sites on the one-dimensional lattice is ε_{nn}, the coverage of the one-dimensional lattice is given by the following expression [11]:

$$\frac{\beta - 1 + 2\theta}{\beta + 1 - 2\theta} = \exp\left(\frac{\mu_A + z\varepsilon_{nn}\theta}{k_B T}\right) = K \exp(g\theta) \quad (2.15)$$

with $\beta = \exp(1 - 4\theta(1-\theta)(1-\exp(g)))^{1/2}$, where $z = 2$ and $g = 2\ \varepsilon_{nn}$.

Two-dimensional models with (partially) exact solutions are the so-called hard-square model and the hard-hexagon model. The hard-square model is the lattice gas equivalent of the famous two-dimensional Ising model for ferromagnetism. In the hard-square model, all nearest-neighbor interactions on a square lattice are infinitely repulsive, that is, $\varepsilon_{nn} = \infty$. The hard-square model does not have a known analytical solution for its isotherm, but many of its properties and so-called critical behavior are known exactly [12]. At $\mu_A/k_B T \approx 1.334$, corresponding to a critical coverage $\theta_c \approx 0.368$, the hard-square model predicts an order–disorder second-order phase transition: below $\mu_A/k_B T \approx 1.334$ the adsorbates are distributed randomly on the lattice, above $\mu_A/k_B T \approx 1.334$ the adsorbates start forming a c(2 × 2) overlayer (see Figure 2.2). Note that the MFA isotherm predicts no adsorption at all for the hard-square model, whereas the QCA correctly predicts the saturation coverage of 0.5 for large μ_A. A simple mean-field model that does predict a saturation coverage of 0.5 is the "sublattice model", in which one assumes that only one of the c(2 × 2) lattices is available for adsorption, and adsorption on that sublattice follows the Langmuir isotherm:

$$\frac{2\theta}{1 - 2\theta} = \exp(\mu_A/k_B T) \quad (2.16)$$

Figure 2.2 compares the solution of the hard-square model obtained by Monte Carlo simulations, as compared to the QCA and the sublattice MFA [13]. Only the exact MC solution predicts the phase transition.

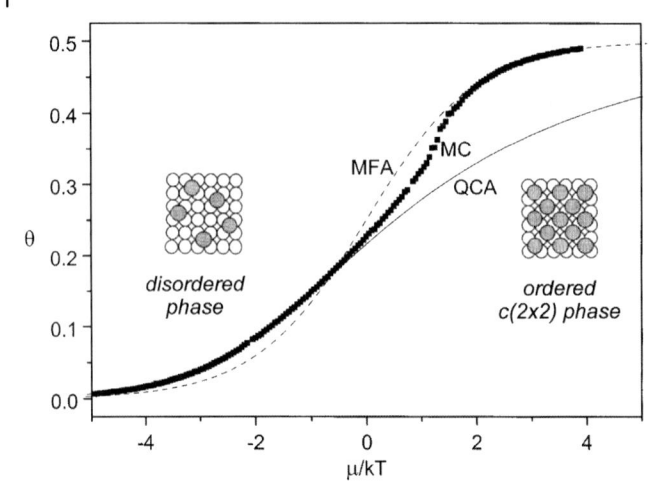

Figure 2.2 Solutions for the hard-square model in the MFA, Eq. (2.16), QCA, Eq. (2.14), and the exact solution obtained by MC simulations.

The only two-dimensional model with a fully analytical solution is the so-called hard-hexagon model [14], for which nearest-neighbor interactions on a hexagonal lattice are infinitely repulsive. The expression for the isotherm is too complicated to be repeated here, but Figure 2.3a shows the result and its derivative, together with an isotherm for the situation in which also the next-nearest neighbor is excluded from adsorption

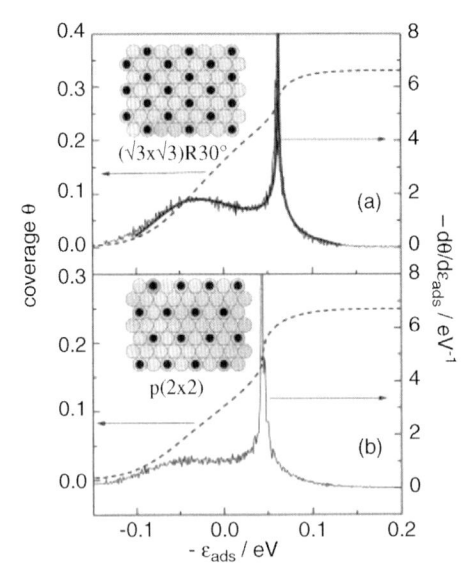

Figure 2.3 Isotherms of their derivatives for (a) the hard-hexagon model (the wiggly curve is the MC simulation, the bold line is Baxter's exact solution), and (b) a hard-hexagon model with also the next-nearest neighbor site being excluded from adsorption.

(Figure 2.3b) [15]. For both interaction models, disorder–order phase transitions are predicted, to $(\sqrt{3} \times \sqrt{3})R30°$ and $p(2 \times 2)$ structures, respectively. Note that the derivative exhibits a sharp peak (discontinuity) exactly at the phase transition.

The relevance of showing the derivative lies in the following observation. Equations (Equations (2.12)–(2.16)) give expressions for the coverage of adsorbate A as a function of the electrochemical potential μ_A. Such a curve is called an isotherm. Cyclic voltammetry in fact measures the derivative of the isotherm, if the electrosorption reaction is in equilibrium. This follows from:

$$j = \gamma \Gamma_n F \frac{d\theta}{dt} = \gamma \Gamma_n F \frac{d\theta}{dE}\frac{dE}{dt} = \gamma \Gamma_n F v \frac{d\theta}{dE} \tag{2.17}$$

where γ is the above-mentioned electrosorption valency, Γ_n is the total number of surface sites per surface area, and v the sweep rate of the voltammetric scan. Note that the integral of the current therefore gives the isotherm, provided, of course, that γ is constant.

2.3.2
Dynamical System

For the modeling of a dynamical system, one usually first formulates a so-called Master equation. The Master equation is based on the assumption that the processes taking place on the lattice are Markovian. A dynamical process is Markovian if its next state only depends on its current state and not on preceding states. A Markovian process is therefore random and memoryless. The result of a Markovian process is, in discrete time, a Markov chain. From the assumption that a process is Markovian, one derives the Chapman–Kolmogorov equation [16]. The Chapman–Kolmogorov equation is an equation for the conditional probability p to reach state x_3 from state x_1 via state x_2 by integrating over all intermediate states x_2:

$$p(x_3; t_3|x_1; t_1) = \int dx_2 p(x_3; t_3|x_2; t_2) p(x_2; t_2|x_1; t_1) \tag{2.18}$$

The differential version of the Chapman–Kolmogorov equation gives the Master equation for the probability P to find the system in state x_3:

$$\frac{\partial P(x_3, t)}{\partial t} = \int dx_2 w(x_3|x_1) P(x_2, t) - \int dx_2 w(x_2|w_3) P(x_3, t) \tag{2.19}$$

or in discrete form:

$$\frac{\partial P_k}{\partial t} = \sum_{l \neq k} w_{kl} P_l - \sum_{l \neq k} w_{lk} P_k \tag{2.20}$$

where w are transition probabilities (i.e., rate constants). Again, these transition probabilities must satisfy detailed balance:

$$\frac{w_{kl}}{w_{lk}} = \frac{P_{eq,k}}{P_{eq,l}} = \exp(-(U_k - U_l)/k_B T) \tag{2.21}$$

As with static systems, approximate solutions to the Master equation can be generated by making certain assumptions regarding the (lack of) correlations between the adsorbate configurations on the lattice. The simplest assumption is that of a complete lack of correlations, that is, random mixing, which is equivalent to the MFA. In the literature, this is often referred to as "kinetic modeling", in which all adsorbates are represented by their average coverages. Such an assumption forms the basis of the kinetic derivation of the Frumkin isotherm. The adsorption rate in the MFA of species A is given by:

$$v_{ads}(\phi, \theta) = k^0_{ads}(\phi) c_A (1 - \theta) \exp(-\alpha g \theta) \tag{2.22}$$

and the corresponding desorption rate is given by:

$$v_{des}(\phi, \theta) = k^0_{des}(\phi) \theta \exp((1 - \alpha) g \theta) \tag{2.23}$$

Equating Eqs. (2.22) and (2.23)) yields Eq. (2.13).

It is possible to take into account correlations between surface adsorbates in a dynamic approach, but these models have a relatively low popularity, as they are mathematically much more involved than the MFA and still do not give an accurate description of the real correlations. Zhdanov [17] has formulated a "dynamic QCA", and Nieminen and Jansen [18] have introduced a so-called "cluster approximation" to treat correlations between adsorbates under reactive conditions. The latter approximation also allows the incorporation of surface diffusion in the formalism, something which is not possible within the MFA and QCA. For an application to electrochemical surface reactions, the interested reader is referred to ref. 19.

In the dynamic description, the relation to voltammetry is given by Faraday's law:

$$j = F \sum_i n_i v_i \tag{2.24}$$

where the summation is over all reactions i in which charge is transferred, with n_i the charge number, being positive for oxidation reactions and negative for reduction reactions.

2.4
Monte Carlo Simulations

Computer simulations are extremely useful to generate exact solutions to statistical-mechanical models. One such simulation method that is uniquely applicable to approximating the partition function of a lattice-gas hamiltonian is the Monte Carlo method [5]. Such methods are based on the use of a random number generator. The most popular algorithm is an algorithm devised by Metropolis. Basically, the Metropolis algorithm samples the phase space of the system by focusing on those states which have a high Boltzmann weight in Eq. (2.10), that is, which contribute significantly to the partition function. Such a tactic is called "importance sampling". In brief, the Metropolis algorithm starts from a configuration, makes a random change, and computes the probability p of the change in configuration from their respective Boltzmann factors. By generating a random number between 0 and 1, the change is accepted if the random number is smaller than the probability p. By repeating this procedure, the method samples configurations according to their Boltzmann weight, and the average of these configurations can be used to generate thermodynamic properties.

A variation on the Metropolis algorithm can also be used to solve the Master equation of a dynamical system. A popular method is the so-called first reaction method (FRM) [20]. The FRM makes a list of all possible changes and their corresponding rates or transition probabilities, and assigns a prospective execution time Δt to each event, drawing it randomly from the corresponding Poissonian distribution. Next, the event with the smallest Δt is selected and executed, and the clock is advanced by Δt, followed by an update of the above list in accordance with the new configuration. A simulation method for solving the Master equation, such as the FRM, is called a kinetic Monte Carlo (KMC) simulation method.

2.5
Applications to Electrosorption, Electrodeposition and Electrocatalysis

In this section, I will illustrate the application of lattice-gas models to a number of electrochemical systems, including electrocatalytic reactions and other systems of relevance to electrocatalysis. This will shed some light on the applicability of some of the assumptions very commonly made in the modeling of electrochemical processes on surfaces less well-defined than single crystals.

2.5.1
Electrosorption and Electrodeposition

A good starting point for the application of lattice-gas modeling is the blank voltammetry of Pt(111), which is probably the most frequently measured single-crystal voltammogram in surface electrochemistry. Figure 2.4 shows the voltammograms of Pt(111) in sulfuric acid, perchloric acid, and sodium hydroxide. We can distinguish three regions in these voltammograms. Between 0.05 and 0.35 V, hydrogen is adsorbed onto Pt(111) through the reactions:

$$H_3O^+ + * + e^- \rightleftharpoons H_{ads} + H_2O \qquad (2.25)$$

in acidic media, and

$$H_2O + * + e^- \rightleftharpoons H_{ads} + HO^- \qquad (2.26)$$

in alkaline media ("*" denoting a free adsorption site on the surface). On the pH-corrected reversible hydrogen electrode (RHE) scale, these reactions will take place at the same potentials. Between 0.6 and 0.9 V, hydroxyl is adsorbed onto Pt(111) through the reactions:

$$H_2O + * \rightleftharpoons OH_{ads} + H^+ + e^- \qquad (2.27)$$

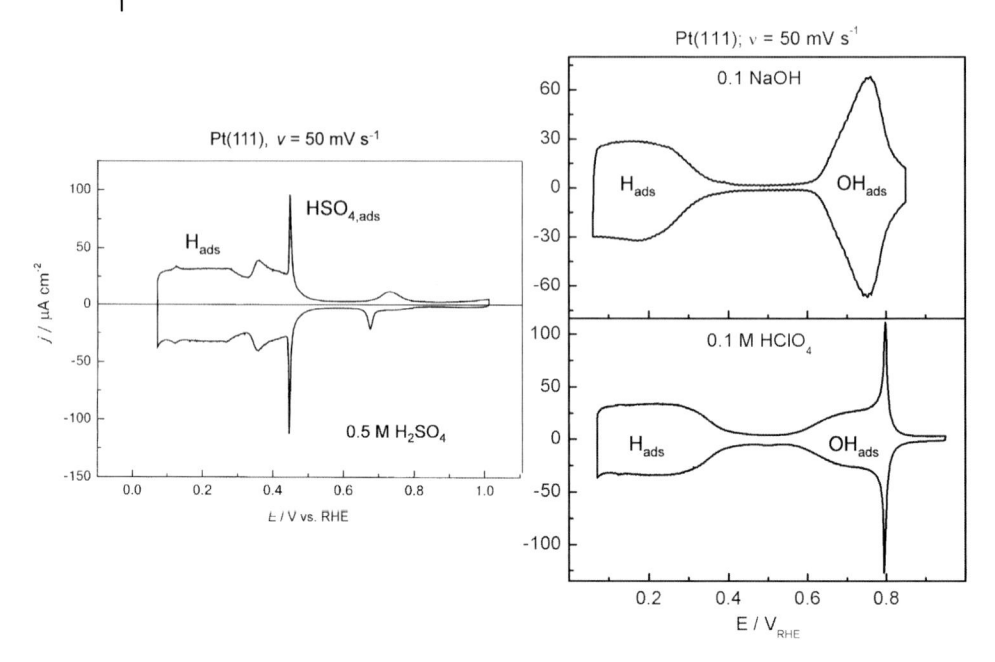

Figure 2.4 Cyclic voltammetry of a Pt(111) in electrode in three different electrolyte solutions.

acidic media, and

$$OH^- + * \rightleftharpoons OH_{ads} + e^- \tag{2.28}$$

in alkaline media. Finally, between 0.3 and 0.5 V in 0.5 M H_2SO_4, anion adsorption takes place, presumably bisulfate adsorption [21,23]:

$$HSO_4^- + * \rightleftharpoons HSO_{4,ads}^- \tag{2.29}$$

Note that both OH adsorption in perchloric acid and bisulfate adsorption in sulfuric acid display a sharp peak in their voltammetric signal. Such signals have been referred to as a "butterfly" by Clavilier, the father of single-crystal electrochemistry [22]. Based on the general results derived in Section 2.3, a butterfly presumably signals the occurrence of a disorder–order phase transition in the adlayer. The nature of the irreversible feature at about 0.7 V in sulfuric acid is not completely understood. It has been ascribed to a reordering of the (bi)sulfate adlayer [23] but also to adsorption on OH inside the domain walls (defects) of the (bi)sulfate adlayer [24].

The underpotential deposition of hydrogen (H-upd) between 0.05 and 0.3 V on a Pt(111) electrode surface is perhaps the best example of a system that may be modeled quantitatively using a lattice-gas approach, and for which the resulting numbers for the adsorption energy of hydrogen and lateral interaction energy between adsorbed hydrogen atoms can be given a clear physical meaning. By analyzing experimental data, Jerkiewicz concluded that the H-upd on Pt(111) is well fitted using a Frumkin isotherm [25]. The binding energy extracted by Jerkiewicz was −0.21 eV (vs. H in vacuum) and the lateral interaction energy, when quoted as a nearest-neighbor interaction on a hexagonal surface, was 0.047 eV. Koper and Lukkien [15] showed that for such a value of the lateral interaction, there is only a small difference between the mean-field predictions and the Monte Carlo simulations. Later, Karlberg et al. [26] calculated the binding energy of hydrogen and the lateral interaction energy by first-principles DFT, and came to the same conclusions. Their DFT-computed lateral interaction was somewhat smaller than the experimental one, however, so that an even better agreement between mean-field and MC was to be expected. In conclusion, H-upd on Pt(111) stands out as one of the very few examples of a quantitatively meaningful application of the Frumkin isotherm in electrochemistry, supported by experiments on well-defined electrodes, density functional theory calculations, and the comparison of the mean-field prediction with exact MC simulations [27]. The applicability of the mean-field Frumkin isotherm is due to the relatively weak interactions between the adsorbed H atoms.

A good example of the much stronger lateral interactions involved in anion adsorption is the well-studied case of bromide on a Ag(100) electrode [13,28–30]. Figure 2.5 shows the coverage of bromide as measured using chronocoulometry by Wandlowski et al. [29] and a fit by a lattice-gas model solved using Monte Carlo simulations [13]. The key feature of the lattice-gas model is the inclusion of hard nearest-neighbor interactions (i.e., the hard-square model) together with long-range repulsive interactions beyond nearest neighbors:

$$\varepsilon_{ij} = \varepsilon_{nnn} \frac{a^3}{r_{ij}^3} \, \text{eV} \tag{2.30}$$

where r_{ij} is the distance between site i and j and a is the distance between two lattice sites (2.889 Å for silver), and $\varepsilon_{nnn} = 0.078$ in the fit of Figure 2.5. Without such long-range interactions, no good fit of the experimental isotherm could be obtained.

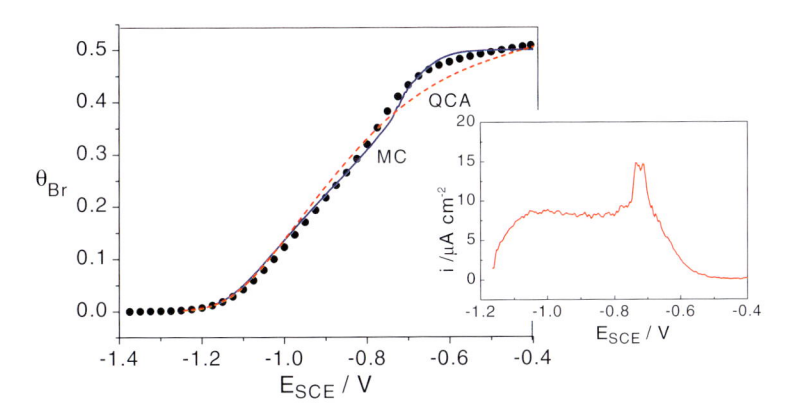

Figure 2.5 Comparison between the experimental potential-dependent bromide coverage on an Ag(100) electrode [29] and the MC simulations (solid blue line) and the QCA isotherm (dashed line) using the interaction potential of Eq. (2.30) [13]. The inset shows the corresponding positive-going voltammogram calculated from the MC results.

The functional form represented by Eq. (2.30) can be derived from a model assuming electrostatic interactions between the adsorbed bromide species [13].

The electrosorption of chloride on Ag(100) can also be modeled using a lattice-gas approach, using a similar interaction potential as that used for Br on Ag(100), but a good fit to experimental data is obtained only if it is assumed that the interaction energy ε_{nnn} and the electrosorption valency γ are not constants but depend on the coverage [31,32]. This was modeled by the charge on the adsorbed chloride q being a function of the overall coverage, $q = -(1 + \gamma_0 + \gamma_1 \theta)e_0$. As a result, the electrosorption valency is modeled as:

$$\gamma = \gamma_0 + \gamma_1 \theta \tag{2.31}$$

and the interaction in Eq. (2.30) as:

$$\varepsilon_{nnn} = A(1 + \gamma_0 + \gamma_1 \theta)^2 \tag{2.32}$$

The model assumes that as the coverage of chloride on the surface increases, the chloride adsorbates discharge and their electrostatic repulsive interaction becomes weaker. This phenomenon is called mutual depolarization [33]. The model gives a very good fit of the experimental data. Abou-Hamad et al. [32] ascribe the difference between chloride and bromide to the higher electronegativity of chloride.

The adsorption of bromide on an Au(100) electrode displays an incommensurate phase transition and can, therefore, not be modeled using a lattice-gas approach. Mitchell and Koper constructed an off-lattice model with an interaction potential based on DFT calculations to obtain a reasonable agreement with experiment [34]. Nevertheless, such simulations are far more computationally intensive than lattice-gas simulations.

So far, there have been no successful and realistic models suggested for the bisulfate and OH butterflies on Pt(111) (Figure 2.4). The adsorption of bisulfate on Pt(111) leads to a $\sqrt{3} \times \sqrt{7}$ structure [35], meaning that the interaction potential between the bisulfate species on the surface does not have the same symmetry as the underlying lattice. Hermse et al. [36] have studied a number of models for the formation of such an adlayer. The OH butterfly is still somewhat controversial because the ordering of the OH adlayer suggested by the voltammetric feature has not been confirmed by spectroscopic techniques, such as STM or surface X-ray scattering. In general, the exact shape of the butterfly (i.e., the shape of the voltammetric feature) is determined by the detailed interaction potential between the adsorbed species, rather than by the structure of the adsorbate overlayer formed on the lattice [37].

On a Pt(100) electrode, bromide adsorption also leads to a c(2 × 2) overlayer [38] but, in contrast to Ag(100), it competes with the H-upd reaction, Eq. (2.25). This leads to some interesting new phenomena, which have been modeled in some detail by Garcia-Araez et al. [39]. On Pt(100), bromide adsorption is accompanied by a single sharp peak, that can be explained by a model that assumes the displacement of H-upd by bromide. Figure 2.6 shows a fit of the experimental isotherm with a Monte Carlo simulation. The simulation is based on a lattice-gas model in which the bromide adsorbates experience hard nearest-neighbor interactions (hard-square

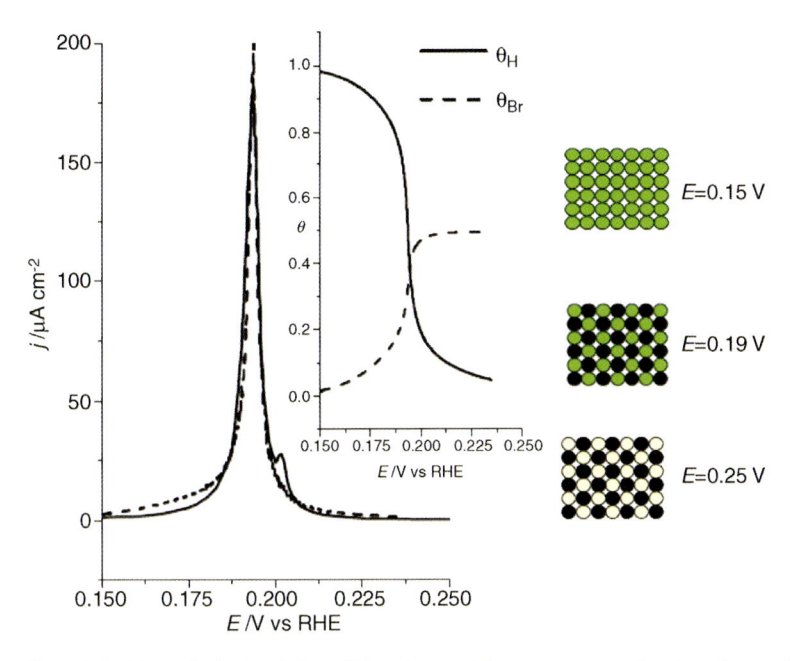

Figure 2.6 Monte Carlo simulation of H-upd and bromide adsorption on a Pt(100) electrode. Solid line is the experimental voltammogram; dashed is the MC simulation using the model explained in the text. Inset shows the H-upd and Br coverages as a function of potential. Snapshots on the right show typical surface configurations at three different potentials. Green circle is hydrogen, black is bromine, and white is empty site [39].

model), leading to a c(2 × 2) bromine structure at high potentials, and nearest-neighbor interaction between two H-upd and between H and Br. All these interactions are *repulsive*, whereas the sharp peak could in fact be fitted by a Frumkin isotherm assuming strong *attractive* interactions. Garcia-Araez et al. [39] showed that this attractive interaction is effective, in the sense that this "effective" interaction energy for competing adsorbates may be negative (i.e., attractive) if the interparticle repulsion exceeds the sum of the intraparticle repulsions. Such a situation is typically satisfied if a smaller adsorbate (i.e., H-upd) competes with a bigger adsorbate (i.e., Br). However, the good fit by the Frumkin isotherm is misleading. If the interaction parameters estimated from the approximate Frumkin isotherm are used as input for the exact Monte Carlo simulation, a completely different voltammogram is obtained. The situation may be improved considerably by considering a sublattice adsorption for bromide, for which the analytical Frumkin-type isotherm matches the MC results closely [40]. Still, the example illustrates nicely how deceptive a good fit by some analytical isotherm expression can be, if no proper care is exercised in checking the result against exact results.

Another voltammetric feature in which a displacement reaction has been suggested to play a key role, is the sharp peak observed in the voltammetry of stepped single-crystal Pt surfaces with (111) terraces and (111) or (100) type steps [41]. Steps

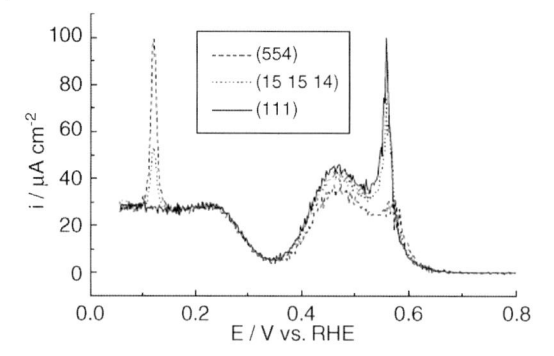

Figure 2.7 Monte Carlo simulation of the H-upd region of Pt(111) and two stepped Pt surfaces, including H-upd adsorption on terrace and steps, and anion adsorption on terraces.

give rise to a sharp peak in the hydrogen region: for instance, (111) step sites lead to a peak at 0.125 V in the experimental cyclic voltammetry as modeled using MC simulations in Figure 2.7 [42]. In fact, the shape of the peak can be modeled exactly and analytically using Eq. (2.15) [43]. Remarkably, the sharpness of the peak suggests strongly attractive interactions. If only H-upd would adsorb on the step sites at this potential, the nature of these strongly attractive interactions would be difficult to understand, especially considering that H-upd on Pt(111) exhibits repulsive interactions. On the other hand, Van der Niet et al. [41] have suggested that the sharpness can be explained if it is caused by a displacement reaction:

$$H_{ads} + H_2O \rightleftharpoons OH_{ads} + 2H^+ + 2e^- \qquad (2.33)$$

Since a large adsorbate (OH) replaces a small adsorbate (H) in this reaction, it satisfies the conditions that give rise to an effective attractive interaction. Such an explanation would also explain a few other anomalies associated with the voltammetry of stepped Pt electrodes, as discussed in detail in ref. 41. Another important effect illustrated in Figure 2.7 is the dependence of the shape of a butterfly peak on increasing step density. In the MC simulation of Figure 2.7, there is anion adsorption on the (111) terrace modeled by the hard-hexagon model [42]. With the introduction of step sites, the sharpness of the peak associated with the order–disorder transition quickly disappears. For the Pt(554) surface (terrace of 10 atoms wide), the peak is almost gone. This observation agrees with the experimental observations, and confirms the experimental rule of thumb that the quality of a (111) surface can be judged from the sharpness of its butterfly peak in sulfuric or perchloric acid (Figure 2.4).

Electrodeposition has been studied quite extensively using lattice-gas modeling. One of first studies using lattice-gas Monte Carlo simulations for a detailed comparison to single-crystal experiments was the underpotential deposition of Cu^{2+} onto an Au(111) surface in sulfuric acid [8,9,44]. Figure 2.8a shows the interaction model proposed by Zhang et al. [45] for the adsorption of Cu and sulfate on the Au(111) surface, as well as for their co-adsorption. As can be seen, not only nearest-neighbor interactions are included, as even three-body interactions are

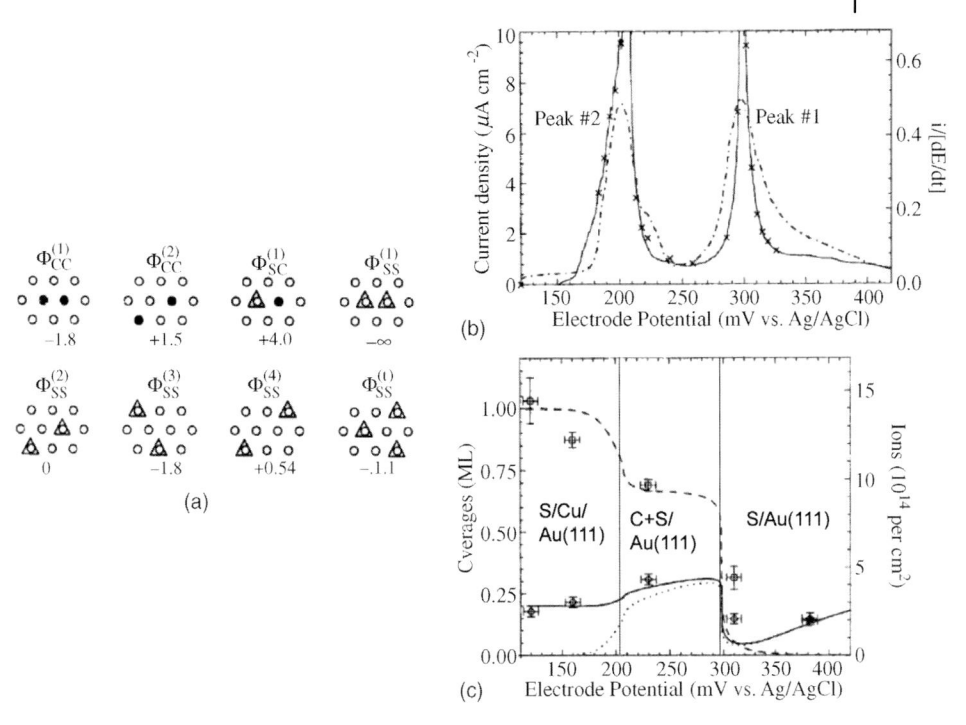

Figure 2.8 (a) Interaction model for Cu (C,●) and sulfate (S,△) adsorption on an Au(111) electrode. Interaction energies are in kJ mol^{-1}, positive interaction energies are attractive, negative interaction energies are repulsive. (b) Voltammetry calculated using the model in (a) using MC simulations. (c) Coverages of Cu and sulfate on the Au(111) surface as a function of potential, as calculated from the MC simulations [9,45].

introduced to obtain a good agreement with experiment. Figure 2.8b shows the calculated voltammetry from MC simulations, in comparison to the experimental voltammetry, as well as the corresponding coverages of Cu and sulfate on the surface (Figure 2.8c). Going from right (more positive potential) to left (less positive potential), the Au(111) is initially covered with sulfate, slowly desorbing with more negative potential. Peak #1 corresponds to the underpotential deposition of Cu, which also induces renewed sulfate adsorption. Between peak #1 and peak #2, there is a surface with a 2/3 ML (monolayer equivalent) of Cu and a 1/3 ML of sulfate. In peak #2, the sulfate desorbs from the Au(111) surface, a full monolayer of Cu is formed, and sulfate adsorbs on top of this Cu monolayer. In fact, one can argue that it is the strong attractive interaction between Cu and sulfate, labeled $\Phi_{SC}^{(1)}$ in Figure 2.8, that effectively causes the underpotential deposition of Cu on Au(111) as it stabilizes the Cu-sulfate adlayer on Au(111).

There have been many statistical-mechanical models and associated Monte Carlo simulations proposed for the various stages of electrodeposition, the growth of surface layers, and island formation. The interested reader is referred to a selective sampling of the literature [46–53].

2.5.2
Electrocatalysis

There are remarkably few detailed statistical-mechanical modeling studies of electrochemical surface reactions in close comparison to experimental data on well-defined surfaces. The only reaction that has been studied in some detail is the oxidative stripping of carbon monoxide. This reaction has been studied extensively on single-crystal surfaces (mainly Pt), and various kinetic modeling and kinetic Monte Carlo simulation studies have been performed [54–57].

All models of the oxidation of adsorbed carbon monoxide (CO) are based on the commonly accepted Langmuir–Hinshelwood-type mechanism:

$$H_2O + * \rightleftharpoons OH_{ads} + H^+ + e^- \tag{2.34}$$

$$CO_{ads} + OH_{ads} \rightarrow CO_2 + H^+ + e^- + 2* \tag{2.35}$$

In these models, the back reaction to Eq. (2.35) is neglected. Such a model assumes that CO and OH compete for the same sites. The formation of OH_{ads} is assumed to be in equilibrium, and the rate of CO oxidation is rate determining. The kinetics of this reaction has been considered in two different limits. In the first limit, one assumes that the adsorbates on the surface are all perfectly mixed. This is equivalent to assuming that the surface diffusion of adsorbed CO is very fast. Under these assumptions, the rate of CO oxidation can be calculated using the mean-field approximation:

$$\frac{d\theta_{CO}}{dt} = k(\phi)\theta(1-\theta) \tag{2.36}$$

For a chronoamperometric experiment, in which the electrode potential ϕ is stepped from a value at which the CO adlayer is stable to a more positive value at which the reaction is "switched on", this yields the following expression for the current transient j as a function of time [54]:

$$j(t) = 2F\Gamma_m k(\phi) C \frac{\exp(-k(\phi)t)}{[1 + C\exp(-k(\phi)t)]^2} \tag{2.37}$$

where $C = \ln[\theta_{in}/(1-\theta_{in})]$ with θ_{in} the coverage of CO at $t=0$. In the second limit, one considers a model in which CO and OH do not mix at all. This is equivalent to assuming that surface diffusion of adsorbed CO is absent. Surface diffusion of OH is irrelevant as reaction Eq. (2.34) is in equilibrium. In this case, the transient can be calculated from a nucleation-and-growth (N&G) model. If the nucleation of OH into the holes or "nucleation sites" of the initial CO adlayer is instantaneous, that is, all parts of the surface not covered by CO are instantaneously occupied by OH, the current transient is given by [1,54]:

$$j(t) = 2F\Gamma_m M_0 k^2 t \exp\left[-\pi M_0 k^2 t^2\right] \tag{2.38}$$

where M_0 is the number of nucleation sites in the adlayer and $k = k(\phi)$. If the formation of OH in the nucleation sites is slow, the transient is modeled by a progressive N&G transient [1,54]:

$$j(t) = 2F\Gamma_m M_0 k_{OH} k^2 t^2 \exp\left[-\pi M_0 k_{OH} k^2 t^3 / 3\right] \tag{2.39}$$

where k_{OH} is the rate constant of formation of OH through reaction 2.34. Note that the assumption that the OH formation reaction is in equilibrium, implies that the instantaneous N&G mechanism would be the most appropriate.

However, from experiments on stepped Pt electrodes in acidic media one concludes that step sites are actually the active sites for CO oxidation [58], as those sites adsorb OH at lower potentials than terrace sites (if the interpretation of ref. 41 as discussed in Section 2.5.1 is correct, the OH adsorption on steps takes place already in the "H-upd" region). Therefore, neither of the above models appears to be directly applicable. However, Lebedeva et al. [58] have shown that current transients on stepped Pt surfaces (consisting of (111) terraces and (110) steps) can be fitted well with the MFA transient Eq. (2.37), and that the rate constant k as extracted from the fit is proportional to the step density. If the reaction takes place at steps, and CO needs to reach those steps by slow diffusion, one would expect a diffusional tailing in the transients for long times on surfaces with a low step density [58,59]. However, such a diffusional tailing was never observed in the experiments on single-crystal electrodes, justifying the conclusion that CO diffusion on (111) terraces must be too fast to measure. Detailed kinetic modeling simulations of CO oxidation on Pt(111) electrodes have also been reported by Angelucci et al., assuming the MFA [60].

Various authors have observed transients with a pronounced tailing during the chronoamperometric oxidation of a CO adlayer on non-single-crystalline Pt electrodes, and have ascribed such tailing to slow diffusion [61,62]. We have shown that on a polycrystalline Pt electrode in alkaline media asymmetric transients with apparent tailing may be observed, whereas under the same conditions single-crystal electrodes give much more symmetric transients [63]. This suggests that transients on a polycrystalline surface may reflect contributions from different facets that do not interact, and that therefore the interpretation of such a transient using a model assuming only a single adsorption site and perfect mixing may not always be adequate.

The crossover from N&G-type kinetics, with no CO surface diffusion, to MFA-type kinetics, with infinitely fast CO surface diffusion, has been studied by MC simulations. Figure 2.9 compares the stripping voltammetry on a homogeneous square lattice gas for three values of the CO diffusion coefficient ($D = 0, 100\ \text{s}^{-1}$ and ∞, i.e., mean-field) together with snapshots of the surface for the two values of the diffusion coefficient. It is clearly seen that fast diffusion leads to a sharper voltammetric peak and a lower peak potential, due to the overall faster reaction kinetics. Similar KMC simulations were performed for model PtRu alloy surfaces [64], where Ru surface atoms provide the active sites for OH adsorption, yielding good agreement with experiment [65] only in the limit of high CO surface diffusion. Because of the importance of steps and defects in the CO oxidation on Pt single crystals, Housmans

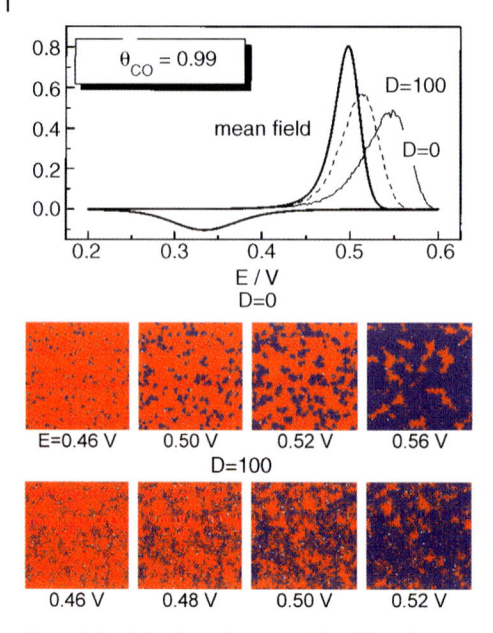

Figure 2.9 Stripping voltammetry for CO oxidation on square model surface for different values of the CO surface diffusion coefficient [54].

et al. [66] considered a similar model for CO oxidation that included the role of steps explicitly. In the limit of high CO diffusion coefficient, the transients simulated by KMC simulations agree well with Eq. (2.37). In the limit of vanishing diffusion rate, the chronoamperometric transients exhibit two features: an initially decaying current corresponding to the oxidation of CO at and near the steps (which indeed can be shown analytically to lead to an exponentially decaying current due to the one-dimensional nature of a "step lattice" [69]), and a peak current due to the oxidation of CO on a terrace, which has initiated at the steps. Such transients were indeed observed for CO oxidation on single-crystalline platinum electrodes in alkaline media, as shown for Pt(111) in Figure 2.10 [67]. Stepped Pt electrodes show similar transients. The initial decay in the transient is ascribed to CO oxidation near steps and defects, and the following peaked current transient is ascribed to CO oxidation on the (111) terraces. Remarkably, this part of the transient is fitted well by an MFA model (Eq. (2.37)) and is not fitted well by an N&G model, as also shown in Figure 2.9. Garcia and Koper have ascribed this effect to the adsorption of carbonate (i.e., the product of CO oxidation in alkaline media) at the step sites, blocking access to the step sites after oxidation has occurred there. Carbonate adsorbs less strongly on the (111) terrace and therefore it does not interfere with the CO oxidation dynamics on the terrace, leading to the expected MFA-type transient. On the other hand, Herrero et al. [68] have measured transients on stepped Pt electrodes and obtained a better fit with models that assume an N&G-type mechanism for the terrace oxidation, and hence argued that the mobility of CO on Pt(111) terraces is

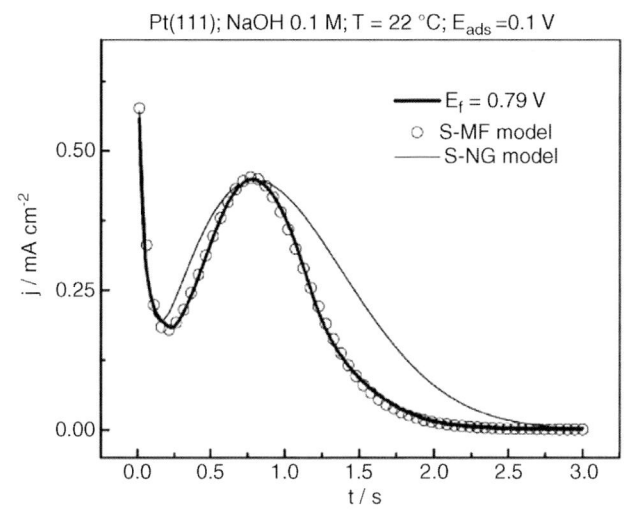

Figure 2.10 Transient of CO oxidation on a Pt(111) electrode in 0.1 M NaOH. Two fits to the experimental are shown, one assuming the MFA approximation for oxidation on the terrace ("S-MF"), and one assuming the N&G model for oxidation on the terrace ("S-NG") [67].

lower in alkaline media compared to acidic media. It illustrates how sensitive such theory–experiment conclusions can be to the details of the experiments, even if they involve well-defined electrodes. Finally, effects of anions that interfere with the mobility of CO on electrode surfaces were observed by Housmans et al. [69] in their study of CO stripping on stepped rhodium electrodes. Sulfate adsorbs strongly on rhodium so that it lowers the mobility of CO on the terrace, yielding the same two-stage chronoamperometric transients that were found in the KMC simulations referred to above.

Finally, the role of lateral interactions on the kinetics of electrocatalytic reactions on single-crystal surfaces has not received much detailed attention. Koper [70] considered a simple model for the stripping of an adsorbate assuming first-order kinetics and incorporating lateral interactions in the MFA:

$$\frac{d\theta}{dt} = -k\theta \exp(g\theta)$$

In the limit of strongly repulsive interactions, this equation predicts a hyperbolic transient, that is, $j(t) \propto 1/t$. Kinetic Monte Carlo simulations showed that the MFA prediction of a hyperbolic transient was close to the exact behavior. The model was used to interpret experimental transients of the reductive stripping of nitric oxide on a Pt(111) electrode, which indeed appear to be hyperbolic [71]. From the fit, an estimate could be made for the lateral interaction between the NO adsorbates, giving a reasonable value of 0.051 eV. To the best of my knowledge, this remains the only example in surface electrocatalysis where a detailed comparison between a kinetic model, Monte Carlo simulations, and single-crystal experiments has yielded insight into the energetic interactions between reactants.

2.6
Conclusions

This chapter has given a brief overview of the theory and applications of the modeling of electrochemical and electrocatalytic processes on single-crystal electrode surfaces using statistical-mechanical modeling employing the lattice-gas approximation. Such a model often provides the conceptual and theoretical basis for kinetic modeling. The results reviewed here have clearly shown that the adsorption of ions and other species on single-crystal electrodes is associated with strong lateral interactions between such species. These lateral interactions are almost never modeled adequately using the mean-field approximation (better known as the Frumkin isotherm in the electrochemical literature), the adsorption of H on Pt(111) and Pt(100) being perhaps two of the very few well-documented exceptions. For other species, such as bromide, chloride, sulfate, Cu, and hydroxide, mean-field models essentially fail to give an accurate description. This is not the same as saying that mean-field models cannot give good fits of the experimental data. In fact, mean-field models may give very good fits, but unfortunately the meaning of the fitting parameters is doubtful and should always be considered very carefully. However, MC simulations will always give the exact solution for the suggested interaction model, and this chapter has shown how the detailed comparison between experiments and MC simulations can yield very insightful information. For the modeling of dynamic and catalytic processes on single-crystal electrodes, one may assume that similar conclusions would apply, though a detailed well-documented undisputed example of a careful comparison between experiment, kinetic modeling, and kinetic MC simulations is still missing. This has to do in part with the high complexity of such processes, such as the influence of steps and defects, unknown lateral interaction, and the importance of reactant mixing and surface diffusion, as has been discussed in some detail in the chapter. Therefore, as a general conclusion, Monte Carlo simulations will remain an essential ingredient of any complete lattice-gas modeling of atomic processes on single-crystal surfaces.

Acknowledgments

I gratefully acknowledge the award of a Long-Term Fellowship of the Japanese Society for the Promotion of Science (JSPS), No. L-11527, which supported my stay at Hokkaido University.

References

1 Schmickler, W. and Santos, E. (2010) *Interfacial Electrochemistry*, 2nd edn, Springer, Heidelberg.

2 Koper, M.T.M. (ed.) (2008) *Fuel Cell Catalysis, A Surface Science Approach*, Wiley & Sons Inc., Hoboken (NJ).

3 Rossmeisl, J., Greeley, J., and Karlberg, G.S. (2008) *Fuel Cell Catalysis, A Surface Science Approach* (ed. M.T.M. Koper), Wiley & Sons Inc., Hoboken (NJ), p. 57.
4 Jacob, T. (2008) *Fuel Cell Catalysis, A Surface Science Approach* (ed. M.T.M. Koper), Wiley & Sons Inc, Hoboken (NJ), p. 129.
5 Chandler, D. (1987) *Introduction to Modern Statistical Mechanics*, Oxford University Press.
6 Spanjaard, M. and Desjonqueres, D. (1996) *Concepts in Surface Physics*, Springer, Heidelberg.
7 Koper, M.T.M. (2003) *Modern Aspects of Electrochemistry*, vol. **36** (ed. C.G. Vayenas), p. 51.
8 Rikvold, P.A., Zhang, J., Sung, Y.-E., and Wieckowski, A. (1996) *Electrochim. Acta*, **41**, 2175.
9 Rikvold, P.A., Brown, G., Novotny, M.A., and Wieckowski, A. (1998) *Colloids Surf. A: Physicochem. Eng. Aspects*, **134**, 3.
10 Curulla Ferré, D., van Bavel, A.P., and Niemantsverdriet, J.W. (2005) *ChemPhysChem*, **6**, 473.
11 Hill, T.L. (1960) *An Introduction to Statistical Thermodynamics*, Addison-Wesley, pp. 235–241.
12 Pearce, P.A. and Seaton, K.A. (1988) *J. Stat. Phys.*, **53**, 1061.
13 Koper, M.T.M. (1998) *J. Electroanal. Chem.*, **450**, 189.
14 Baxter, R.J. (1980) *J. Phys. A*, **13**, L61.
15 Koper, M.T.M. and Lukkien, J.J. (2000) *J. Electroanal. Chem.*, **485**, 161.
16 van Kampen, N.G. (1981) *Stochastic Processes in Physics and Chemistry*, Elsevier, Amsterdam.
17 Zhdanov, V.P. (1981) *Surf. Sci.*, **111**, 63.
18 Nieminen, R.M. and Jansen, A.P.J. (1997) *Appl. Catal. A: Gen.*, **160**, 99.
19 Koper, M.T.M., Jansen, A.P.J., and Lukkien, J.J. (1999) *Electrochim. Acta*, **45**, 645.
20 Lukkien, J.J., Segers, J.P.L., Hilbers, P.A.J., Gelten, R.J., and Jansen, A.P.J. (1998) *Phys. Rev. E*, **58**, 2598.
21 Jinnouchi, R., Hatanaka, T., Morimoto, Y., and Osawa, M. (2012) *Phys. Chem. Chem. Phys.*, **14**, 3208.
22 (a) Clavilier, J., Faure, R., Guinet, G., and Durand, R. (1980) *J. Electroanal. Chem.*, **107**, 205; (b) Clavilier, J. (1980) *J. Electroanal. Chem.*, **107**, 211.
23 Garcia-Araez, N., Climent, V., Rodriguez, P., and Feliu, J.M. (2008) *Electrochim. Acta*, **53**, 6793.
24 Saravanan, C., Koper, M.T.M., Markovic, N.M., Head-Gordon, M., and Ross, P.N. (2002) *Phys. Chem. Chem. Phys.*, **4**, 2660.
25 Jerkiewicz, G. (1998) *Prog. Surf. Sci.*, **57**, 137.
26 Karlberg, G.S., Jaramillo, T.F., Skulason, E., Rossmeisl, J., Bligaard, T., and Nørskov, J.K. (2007) *Phys. Rev. Lett.*, **99**, 126101.
27 Koper, M.T.M. (2008) *Faraday Discuss.*, **140**, 11.
28 Ocko, B.M., Wang, J.X., and Wandlowski, T. (1997) *Phys. Rev. Lett.*, **79**, 1511.
29 Wandlowski, Th., Wang, J.X., and Ocko, B.M. (2001) *J. Electroanal. Chem.*, **500**, 418.
30 Mitchell, S.J., Brown, G., and Rikvold, P.A. (2001) *Surf. Sci.*, **471**, 125.
31 Abou Hamad, I., Wandlowski, Th., Brown, G., and Rikvold, P.A. (2003) *J. Electroanal. Chem.*, **554–555**, 211.
32 Abou Hamad, I., Mitchell, S.J., Wandlowski, Th., Rikvold, P.A., and Brown, G. (2005) *Electrochim. Acta*, **50**, 5518.
33 Koper, M.T.M. (1998) *Surf. Sci.*, **395**, L196.
34 Mitchell, S.J. and Koper, M.T.M. (2004) *Surf. Sci.*, **563**, 169.
35 Funtikov, A.M., Stimming, U., and Vogel, R. (1997) *J. Electroanal. Chem.*, **428**, 147.
36 Hermse, C.G.M., van Bavel, A.P., Koper, M.T.M., Lukkien, J.J., van Santen, R.A., and Jansen, A.P.J. (2004) *Surf. Sci.*, **572**, 247.
37 Koper, M.T.M. and Lukkien, J.J. (2002) *Surf. Sci.*, **498**, 105.
38 Garcia-Araez, N., Climent, V., Herrero, E., and Feliu, J.M. (2004) *Surf. Sci.*, **560**, 269.
39 Garcia-Araez, N., Lukkien, J.J., Koper, M.T.M., and Feliu, J.M. (2006) *J. Electroanal. Chem.*, **588**, 1.
40 Garcia-Araez, N. and Koper, M.T.M. (2010) *Phys. Chem. Chem. Phys.*, **12**, 143.
41 van der Niet, M.J.T.C., Garcia-Araez, N., Hernandez, J., Feliu, J.M., and Koper, M.T.M. (2013) *Catal. Today*, **202**, 105.
42 Koper, M.T.M., Lukkien, J.J., Lebedeva, N.P., Feliu, J.M., and van Santen, R.A. (2001) *Surf. Sci.*, **478**, L339.
43 Schouten, K.J.P., van der Niet, M.J.T.C., and Koper, M.T.M. (2010) *Phys. Chem. Chem. Phys.*, **12**, 15217.
44 Blum, L., Huckaby, D.A., and Legault, M. (1996) *Electrochim. Acta*, **41**, 2207.

45 Zhang, J., Sung, Y.-S., Rikvold, P.A., and Wieckowski, A. (1996) *J. Chem. Phys.*, **104**, 5699.

46 Giminez, M.C., Del Popolo, M.G., and Leiva, E.P.M. (1999) *Electrochim. Acta*, **45**, 699.

47 Berthier, F., Legrand, B., Braems, I., Creuze, J., and Tetot, R. (2004) *J. Electroanal. Chem.*, **573**, 377.

48 Saedi, A. (2006) *J. Electroanal. Chem.*, **588**, 267.

49 Frank, S. and Rikvold, P.A. (2006) *Surf. Sci.*, **600**, 2470.

50 Luque, N.B. and Leiva, E.P.M. (2005) *Electrochim. Acta*, **50**, 3161.

51 Pötting, K., Luque, N.B., Quaino, P.M., Ibach, H., and Schmickler, W. (2009) *Electrochim. Acta*, **54**, 4494.

52 Luque, N.B., Pötting, K., Ibach, H., and Schmickler, W. (2010) *Electrochim. Acta*, **55**, 5411.

53 Giminez, M.C., Ramirez-Pastor, A.J., and Leiva, E.P.M. (2010) *J. Chem. Phys.*, **132**, 184703.

54 Koper, M.T.M., Jansen, A.P.J., van Santen, R.A., Lukkien, J.J., and Hilbers, P.A.J. (1998) *J. Chem. Phys.*, **109**, 6051.

55 Petukhov, A.V., Akemann, W., Friedrich, K.A., and Stimming, U. (1998) *Surf. Sci.*, **402–404**, 182.

56 Orts, J.M., Louis, E., Sander, L.M., Feliu, J.M., Aldaz, A., and Clavilier, J. (1998) *Surf. Sci.*, **416**, 371.

57 Korzeniewski, C. and Kardash, D. (2001) *J. Phys. Chem. B*, **105**, 8663.

58 Lebedeva, N.P., Koper, M.T.M., Feliu, J.M., and van Santen, R.A. (2002) *J. Phys. Chem. B*, **106**, 12938.

59 Koper, M.T.M., Lebedeva, N.P., and Hermse, C.G.M. (2002) *Faraday Discuss.*, **121**, 301.

60 Angelucci, C.A., Herrero, E., and Feliu, J.M. (2010) *J. Phys. Chem. B*, **114**, 14154.

61 Andreaus, B., Maillard, F., Kocylo, J., Savinova, E.R., and Eikerling, M. (2006) *J. Phys. Chem. B*, **110**, 21028.

62 Kucernak, A. and Offer, G.J. (2008) *Phys. Chem. Chem. Phys.*, **10**, 3699.

63 Gisbert, R., García, G., and Koper, M.T.M. (2011) *Electrochim. Acta*, **56**, 2443.

64 Koper, M.T.M., Lukkien, J.J., Jansen, A.P.J., and van Santen, R.A. (1999) *J. Phys. Chem. B*, **103**, 5522.

65 Gasteiger, H.A., Markovic, N.M., Ross, P.N.Jr., and Cairns, E.J. (1994) *J. Phys. Chem.*, **98**, 617.

66 Housmans, T.H.M., Hermse, C.G.M., and Koper, M.T.M. (2007) *J. Electroanal. Chem.*, **607**, 69.

67 Garcia, G. and Koper, M.T.M. (2009) *Phys. Chem. Chem. Phys.*, **11**, 11437.

68 Herrero, E., Chen, Q.-S., Hernandez, J., Sun, S.-G., and Feliu, J.M. (2011) *Phys. Chem. Chem. Phys.*, **13**, 16762.

69 Housmans, T.H.M. and Koper, M.T.M. (2005) *Electrochem. Commun.*, **7**, 581.

70 Koper, M.T.M. (2003) *Z. Phys. Chem.*, **217**, 547.

71 Rosca, V. and Koper, M.T.M. (2005) *Surf. Sci.*, **584**, 258.

3
Single Molecular Electrochemistry within an STM
Richard J. Nichols and Simon J. Higgins

3.1
Introduction

Performing and analyzing measurements at the single molecule limit captivates the imagination of scientists from many different subject areas, spanning across chemistry, physics and biology. Single molecule responses may be individually analyzed or they may be analyzed over a number of individual single molecule events to get a statistical view of the single molecule behavior. There have been large advances in single-molecule studies in recent years which have often been driven by technological advances in measurement techniques and instrumentation. However, the recording of electrical signals from single ion channels in biological cells dates back to over several decades ago since the development of techniques in electrophysiology used to monitor ion channels with the aid of patch clamps. There is now a growing collection of techniques which address the single molecule level; these include single molecule fluorescence, single molecule force spectroscopy with an AFM and detection of single molecules traveling through nanopores. These methods have found strong applications in the study of bio-macromolecules such as, for example, proteins or DNA. In the field of electrochemistry, the technique of scanning tunneling microscopy (STM) has opened up possibilities for single molecule electrical measurements under electrochemical control [1], while scanning electrochemical microscopy (SECM) has facilitated molecular detection and the control of redox electrochemistry at the single molecule level [2–4].

Figure 3.1 compares STM-based single molecule measurements with SECM-based ones. In the STM single molecule determination the electrochemically active molecule resides between the STM tip and the substrate. Typically, the redox molecule is tethered to the substrate, for instance in a self-assembled monomolecular layer, but it can be attached to the STM tip as well. Molecules can also be attached to both tip and susbtrate, as is apparent in the recognition tunneling of DNA bases using an STM [5,6]. If the molecule is not attached to the tip, the tip is still kept close to the molecule (typically < 1 nm) so electron tunneling across the short gap can occur. By contrast, the SECM techniques works by detecting

3 Single Molecular Electrochemistry within an STM

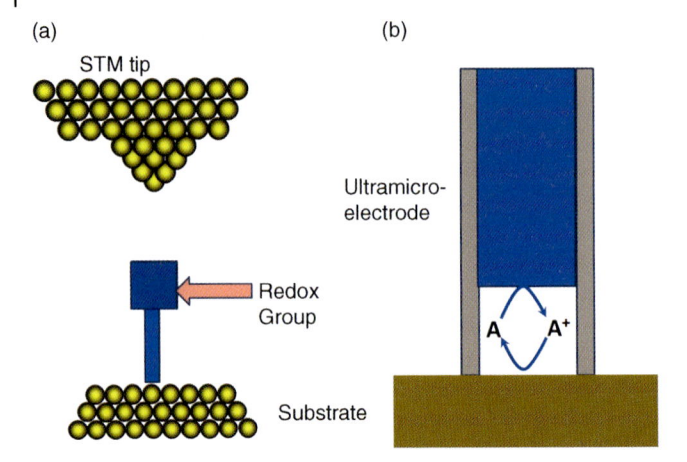

Figure 3.1 A comparison of the STM (a) and SECM (b) techniques, both of which can reach the level of studying electron transfer involving single molecules. In the STM technique the electrochemically addressable group resides between the STM tip and the substrate and tunneling current flow between tip and substrate can be monitored as a function of the electrochemical potential and the redox state. In the SECM technique a redox-active analyte in solution repeatedly and rapidly shuttles between the two electrodes by diffusion and is oxidized at one and reduced at the other. In favorable circumstance just a single molecule can be monitored.

electrochemical current flow from a redox molecule in solution to an ultra-microelectrode, which is placed in close proximity to a substrate surface [2–4]. However, in the SECM set-up the ultra-microelectrode is significantly further from the substrate so electron tunneling from the ultra-microelectrode to the substrate is not a possibility. The ultra-microelectrode is slightly recessed in its sheath and a small number of redox molecules (or even single ones) are trapped in the electrolyte in the cavity. The electrochemically active molecule repeatedly and rapidly shuttles between the two electrodes by diffusion and is oxidized at one and reduced at the other. In favorable circumstances current flow to or from a single redox active molecule in solution trapped between the ultra-microelectrode and the substrate can be monitored due to the current amplification resulting from the repeated shuttling between the electrodes. For discussion of "single molecule electrochemistry" with an SECM see, for example, references [2–4]. Conceptually related to this is redox cycling in nano-fluidic channels, where sensitivity to single redox molecules has been reported [7].

This chapter focuses on single molecule electrical measurements under electrochemical conditions. The majority of such measurements have been conducted with the aid of an *in situ* STM. This chapter is structured in the following manner. First, experimental methods suitable for determining the electrical properties of molecules in an electrochemical configuration are described (Section 3.2). An overview is then given of electron transfer mechanisms for single molecules in an STM nanogap configuration (Section 3.3). Section 3.4 then reviews single molecule electrical measurements made using an electrochemical STM. For the systems discussed, reference is made to the implications of these studies for the electron transfer mechanisms across the single molecular bridges. Section 3.5 concludes the chapter with an overview and outlook.

3.2
Experimental Methods for Single Molecule Electrical Measurements in Electrochemical Environments

A number of techniques have been developed for measuring the electrical properties of single molecules. These techniques have employed the STM [1,8,9], the conducting AFM [10] or mechanically controlled break junctions (MCBJs) [11–16]. The commonality between these techniques is that they can trap single molecules between metallic contacts for electrical characterization. Of these techniques, STM methods have been most commonly employed for the measurement of single molecule electrical properties in electrochemical environments with electrochemical potential control. However, MCBJs have also been adapted for *in situ* electrochemical measurements and measurements in liquid environments [15,17,18]. It is important to distinguish the two general modes of STM characterization of molecular electrical properties. In the first type, which is a scanning tunneling spectroscopy (STS) configuration, the STM is held above a self-assembled molecular monolayer. The tip is, however, retained within tunneling distance of the molecular monolayer. Alongside the electrochemical potential, important parameters are the set-point current (I_0) and the bias voltage (V_{bias}) which determines the tunneling resistance and hence the proximity of the STM tip to the surface. The tip-to-surface distance can be estimated if the decay characteristics of the complete junction are known. The scanning tunneling spectroscopy mode is illustrated in Figure 3.2a. In the STS mode, tunneling current versus electrochemical potential or tunneling current versus the tip bias voltage (V_{bias}) can be recorded. Before sweeping the electrochemical potential or the tip bias, the STM tip is stabilized at a given height above the surface. The feedback loop is then disabled and the electrochemical or tip potential swept rapidly and returned to its original value and then the feedback loop is re-engaged. It is important that there is little drift in the STM tip position, particularly its height, during the potentiodynamic sweep.

The other STM mode is the "single molecule conductance" mode, in which a metal|molecule|metal junction is formed, with the molecule tethered between the substrate surface and the STM tip (Figure 3.2b). Such junctions may be formed from a low or higher coverage molecular adlayer on the substrate; the target molecule may also be dissolved in the electrolyte solution, but this is less commonly the case. There are two variants of the "single molecule conductance modes" both of which use an STM. These are the *in situ* break junction technique [8] and the $I(s)$ technique [1] (and the related $I(t)$ technique [9]). In the *in situ* break junction (BJ) method the metal STM tip is driven a certain distance into the metal substrate forming a metal-to-metal contact between the two [8]. The tip is then rapidly retracted, which first results

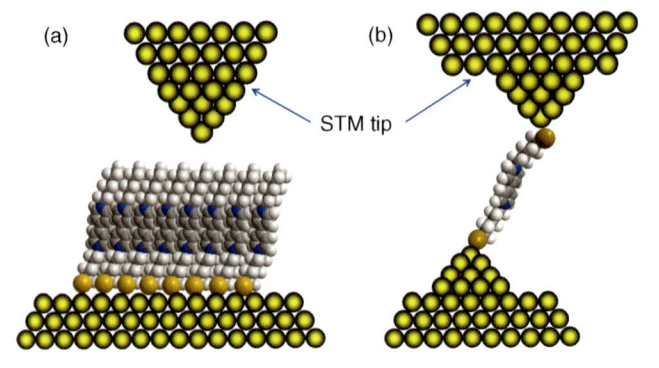

Figure 3.2 Schematic illustration of (a) scanning tunneling spectroscopy characterization of a molecular self-assembled monolayer in which the tip is held above the monolayer and (b) a single molecular junction formed in an STM determination of single molecule conductance.

in a bridge of metal atoms that then narrows down to a single metal atom diameter. This single atom diameter metallic bridge exhibits a conductance of $\sim G_0$ (77 µS) for clean gold. As the STM tip is retracted and the metallic junction is pulled apart, jumps in conductance corresponding to G_0 and multiples thereof can be readily seen in this *in situ* STM BJ experiment with gold contacts. However, these large current jumps are not usually the focus in single molecule conductance experiments. Instead, these large current jumps are followed at slightly greater distances during the tip retraction by smaller current jumps. These smaller current steps correspond to the cleavage of molecular bridges which have been formed following the opening of the metallic break junction.

Like the *in situ* BJ technique the $I(s)$ technique also uses an STM. The term $I(s)$ refers to the recording of tunneling current (I) as the tip is retracted (distance, s) [1]. Unlike the *in situ* BJ method, in this technique there is no metallic contact between the STM tip and the metal substrate. Instead the tip is approached to given set-point values (the set-point current, I_0, and the bias voltage V_{bias}) which bring the tip close to the substrate surface. Molecular bridges form stochastically in the narrow nanogap between tip and substrate (Figure 3.3b). The tip is rapidly retracted from the set-point position and the tunneling current monitored as the molecule is pulled-up in the junction (Figure 3.3b–d). A typical junction conductance curve for a single $I(s)$ trace is shown in the plot in Figure 3.3. The upper curve shows the conductance of a molecular bridge as it is pulled up in the junction. Notably, a plateau is observed, which ends with a rapid fall in junction conductance as the gold–molecule–gold junction cleaves. The lower curve in this figure shows the junction conductance versus distance response for an empty STM nanogap (i.e., in the absence of a molecular bridge). In this case the current decay is much more rapid and no current plateau is observed.

Another method for determining single molecule conductance is the $I(t)$ method [9]. In this method the tip is positioned at a constant distance from the substrate surface by first fixing the STM set-point conditions (the bias voltage and the set-point

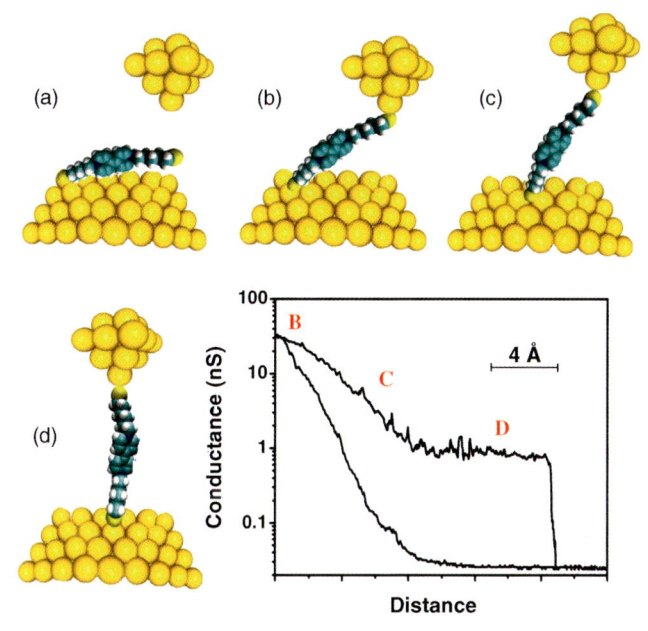

Figure 3.3 A schematic illustration of the $I(s)$ method for determining single molecule conductance. In this method the STM tip is brought close to a substrate onto which the target molecule is adsorbed (a). This initial approach is controlled by the set-point parameters (the set-point current, I_0, and the bias voltage V_{bias}). Molecular bridges can stochastically form between tip and substrate at these close proximities. The tip is then rapidly retracted (b–d) while the tunneling current (or junction conductance) is recorded. The plot shows two representative junction conductance versus distance curves. The lower one is in the absence of molecular bridges, while the upper one, which features a current plateau, is characteristic of a metal|molecule|metal bridge being formed. This upper curve is for the molecular target 6-[1′-(6-mercapto-hexyl)-[4,4′] bipyridinium]-hexane-1-thiol dication, 6V6, discussed in Section 3.4.2. From Nichols et al., *Phys. Chem. Chem. Phys.*, 2010, **12**, 2801–2815 [19] – reproduced by permission of the PCCP owner societies. http://pubs.rsc.org.ezproxy.liv.ac.uk/en/content/articlelanding/2010/cp/b922000c.

current value). The feedback loop of the STM is then temporarily disabled and current recorded as a function of time (hence the term "$I(t)$ method" is used to describe the technique). Figure 3.4 shows an example of a resulting $I(t)$ trace with switching between a lower current value and a higher one. The higher current corresponds to the formation of a metal|molecule|metal junction and hence higher tunneling current flow.

3.3
Electron Transfer Mechanisms

The fundamental aspects of electron transfer are increasingly well understood although there are many new challenges associated with understanding and

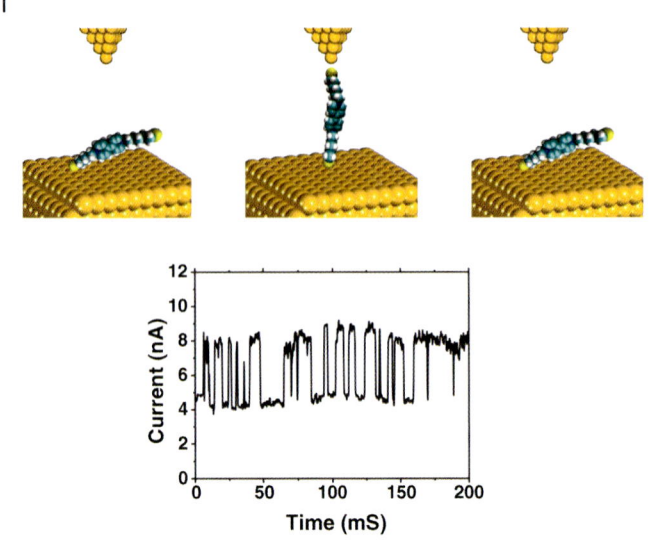

Figure 3.4 A schematic illustration of the $I(t)$ method. At constant tip-to-substrate separation, where molecules can bridge the gap, metal|molecule|metal junctions are stochastically formed and broken giving rise to a train of current pulses. These can be analyzed to obtain single molecule conductance. From Nichols et al., *Phys. Chem. Chem. Phys.*, 2010, **12**, 2801–2815 [19] – reproduced by permission of the PCCP owner societies. http://pubs.rsc.org.ezproxy.liv.ac.uk/en/content/articlelanding/2010/cp/b922000c.

modeling in greater detail, for instance, electron transfer in complex condensed matter surroundings such as those presented by the electrode–electrolyte environment, electron transfer though single molecule junctions, electron transfer through macromolecular and biomolecular systems and long-range electron transfer. Electron transfer through molecular bridges manifests itself in many different fields of science. Examples include electron transfer through proteins (which is important in photosynthetic and respiratory processes of cells), molecular electronics junctions, charge transfer in organic photocells and charge transfer at electrochemical interfaces. Three different platforms for studying charge transport across molecular bridges are illustrated in Figure 3.5. In the electrochemical platform, rates of electron transport to/from the redox group and the electrode can be measured by standard electrochemical kinetics methods or by temperature jump methods. In the two-terminal single molecule electronics platform, current across the junction is measured as a function of bias voltage and single molecule conductance can be obtained. On the other hand, in the donor–bridge–acceptor platform, photoexcitation is used to drive charge transport across the bridge from donor to acceptor and rates for these processes are measured using photophysics methods. The donor–bridge–acceptor platform has been very widely used to study electron transfer across molecular bridges, including synthetic "molecular wires" as well as DNA or proteins. The donor is photoexcited to provide the driving force for charge transfer through the bridge to an acceptor and the rates of charge transfer are measured

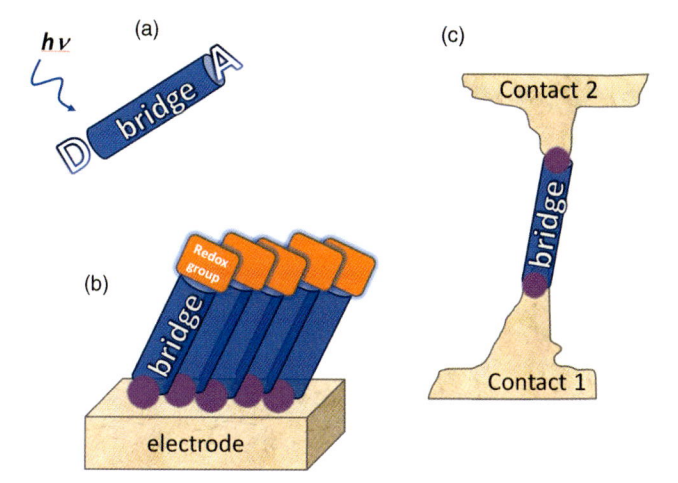

Figure 3.5 Three different platforms for studying charge transport across molecular bridges. (a) The photoexcitation of a molecule bridge with charge transfer between donor (D) and acceptor (A) moiety. (b) The electrochemical method for studying charge transfer between an electrode surface, through a molecular bridge and to a tethered redox moiety. (c) A single molecular junction in which the molecular bridge is attached to two metal contacts at each end through head groups.

using a variety of photophysics approaches. Semiclassical methods, and in particular the Marcus theory, have provided a cornerstone for interpreting many experimental measurements of electron transfer from donor to acceptor. When the donor–acceptor interaction is weak, the electron transfer is non-adiabatic (or diabatic) and electronic surfaces have to be crossed over in passing from donor to acceptor. As a consequence the transition state must be reached many times and transmission factors are low ($\ll 1$). The rate of electron transfer (k_{et}) according to the Marcus theory in the non-adiabatic limit is given by [20]:

$$k_{ET} = \frac{2\pi}{h}|H_{el}|^2 \frac{1}{\sqrt{4\pi\lambda k_B T}} \exp\left(\frac{-(\lambda + \Delta G_{ET})^2}{4\lambda k_B T}\right) \qquad (3.1)$$

Alongside the temperature, this equation introduces three other important terms which influence the rate of electron transfer: the electronic coupling (H_{el}), the total free energy change for the electron transfer (ΔG_{ET}) which provides the driving force for the electron transfer (ET) process and the total reorganization energy (both inner and outer-sphere), λ. When $-\Delta G_{ET} = \lambda$ the rate is only limited by the donor–acceptor coupling and the ET reaction is activationless. The reorganization energy reflects changes in the solvation and structure when the electron transfers from the donor to the acceptor. On the other hand, when the coupling strength is large, occurring for instance when the donor and acceptor are in intimate contact, the transfer is adiabatic. In this limit, transmission factors are close to 1, and ET occurs whenever the transition state configuration is reached. These concepts of coupling strength, driving force for the ET process, and reorganization energy are also important in models for charge transport across redox-active molecules within an electrochemical STM nanogap discussed later.

Although the three platforms for studying charge transport across molecular bridges illustrated in Figure 3.5 share commonalities, there are also important differences. The first is that different physical quantities are measured. In the donor–acceptor platform, rates of photoinduced charge transfer and also rates of other photophysical processes are measured. In the single molecule electronics platform, current–voltage response of the molecular junctions or molecular conductance is measured. On the other hand, in the electrochemical platform, electrochemical rate constants are obtained. Concerning the latter two, there has been some debate in the literature concerning the relationship between measured electrochemical rate constants and single molecule conductance values [21].

A common concern between all three platforms illustrated in Figure 3.5 is an understanding of how the electron (or hole) interacts with the molecular bridge as it is transferred across it. In this respect two limits can be identified for ET through the bridge molecule between the donor and acceptor, or by analogy also between two electrodes in metal–molecule–metal junctions. When the bridge state lies well above the energy of the donor (or Fermi levels of the electrodes in metal–molecule–metal junctions), superexchange tunneling is likely to be dominant. In this case charge (electron or hole) does not reside on the bridge, but is instead coherently transferred across. When the bridge states are closer in energy to the donor (or the Fermi levels of the enclosing electrodes in molecular junctions), population of the bridge states during charge transfer then becomes possible, leading to incoherent hopping mechanisms. Hopping is characterized by shallow distance decay and it becomes the most likely transport mechanism for charge transfer over longer distances. The following discusses some of the important models for electron transfer across a molecular bridge which is tethered between two metal electrodes.

3.3.1
Tunneling

The most prominent mechanisms of charge transport for relatively short molecules with large HOMO–LUMO gaps are direct tunneling or superexchange. For such systems it is likely (but not necessarily the case) that both the HOMO and LUMO are well separated from the Fermi levels of the metal contacts, making hopping or diffusive transport through these levels less likely at low or moderate bias voltage. Figure 3.6 illustrates electron tunneling through an arbitrary one-dimensional tunnel barrier between two metal electrodes, with the barrier height ($\phi(x)$) varying along the coordinate x between the left and right electrodes. The basic tunneling models replace the barrier profile with a simpler barrier shape, for instance a rectangular barrier at zero bias or a trapezoidal barrier at low to moderate bias voltage.

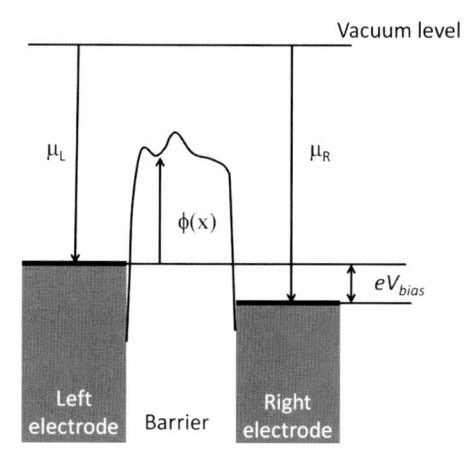

Figure 3.6 Energy levels for electron tunneling through an arbitrary tunneling barrier between two metal electrodes, with Fermi levels shown as solid bold horizontal lines and $\phi(x)$ the barrier height along the one-dimensional barrier.

Classic examples of tunneling transport in molecule junctions are the well-studied alkanethiol or alkanedithiol molecular systems, which have large HOMO–LUMO in the region of ~8 to 10 eV [22]. In these systems the HOMO lies closer to the Fermi level of the electrodes and transport is by hole tunneling through the barrier formed with the relevant HOMO on the bridge ($\phi = E_F - E_{HOMO}$). For these systems tunneling has been shown to be the transport mechanism over bias voltage ranges at least up to ± 1 V. This has been demonstrated by electrical measurements on self-assembled monolayers (SAMs) sandwiched between metal contacts in larger area junctions [23–25], as well in single molecule junctions [8,10]. The transport mechanism in other organic molecule-containing electrical junctions is generally less clear-cut. For instance, in certain π-conjugated molecular systems a transition from tunneling at shorter lengths, to hopping at longer lengths is observed [26,27]. Direct tunneling mechanisms typically exhibit no temperature dependence. Indeed, this is observed for alkane(di)thiols, where HOMO and LUMO levels are far removed from the Fermi energies of the metal electrodes. However, temperature dependence can also be apparent for phase coherent tunneling mechanisms, if molecular orbitals lie closer to Fermi energies [28]. The temperature dependence then arises from the thermal broadening of the Fermi energies [27,28].

From simple barrier tunneling models the decay of conductance with length of the molecular bridge (L) is given by:

$$G = G_c e^{-\beta L} \qquad (3.2)$$

Where G is the conductance of the junction, L is the length of the molecular bridge, G_c is the contact conductance (for $L=0$), and β is the decay constant for tunneling with units of reciprocal length (typically Å^{-1} or nm^{-1}). This equation can also be expressed in terms of the repeat units of the molecular bridge (N) for instance

$(CH_2)_N$ in alkane(di)thiol, in which case β and L are replaced by β_N and N, respectively. Alkanedithiols possess β around 0.9 Å^{-1} (9 nm^{-1}) ($\beta_N = \sim 1.1$) at low bias, and therefore the conductance falls off very rapidly with the length of the bridge. For instance, increasing the length of the chain by one CH_2 group causes the conductance to drop to about a third of its previous value. This decay is smaller than that in vacuum where, for instance, a decay constant of 3 Å^{-1} causes the current to fall 20-fold for a 1 Å increase in distance. Nevertheless, the decay of the tunneling current with distance is large for large band gap molecules, like alkane(di)thiols, when compared to conjugated molecular bridges with substantially smaller decay (β) factors.

There are a number of analytical models for considering electron tunneling across organic molecular bridges. The most widely employed model is that of Simmons, which describes tunneling using a one-dimensional barrier [29]. At zero bias the barrier is considered to be rectangular with a height of Φ_0. Simmons showed that for bias voltages $V < \Phi_0/e$ the effective barrier for tunneling from the left electrode is lowered to $\Phi_0 - eV/2$. The Simmons model has been widely used to understand and fit the voltage (V) dependence of the change density (J) through molecular layers and single molecules:

$$J = \frac{e}{4\pi^2 h d^2}\left\{\left(\Phi_B - \frac{eV}{2}\right)\exp\left[-\frac{2(2m)^{1/2}}{\hbar}a\left(\Phi_B - \frac{eV}{2}\right)^{1/2}d\right] - \left(\Phi_B + \frac{eV}{2}\right)\right.$$
$$\left. \times \exp\left[-\frac{2(2m)^{1/2}}{\hbar}a\left(\Phi_B + \frac{eV}{2}\right)^{1/2}d\right]\right\} \quad (3.3)$$

m and e represent the mass and the charge of an electron, d denotes the width of the tunneling barrier, Φ_B and α are the fit parameters where Φ_B is the effective barrier height of the tunneling junction and α is related to the effective mass of the tunneling electron. This model has been successfully employed to fit and understand tunneling current flow through both molecular monolayers and single molecules [25,30–33]. For higher bias voltages ($V > \Phi_0/e$) the barrier becomes triangular and may then be represented by Fowler Nordheim tunneling (or field emission) models.

Coherent tunneling through metal–molecule–metal junctions can be understood by considering the junction conductance represented in the Landauer formula. For conduction through a single channel the conductance G is given by:

$$G = \frac{2e^2}{h} \cdot \Gamma_L \cdot \Gamma_B \cdot \Gamma_R \quad (3.4)$$

where e is the electron charge and h is Plank's constant and Γ_L, Γ_B and Γ_R are the transmission coefficient of the left contact, the molecular bridge and the right contact, respectively. These terms can be combined together to give the transmission coefficient $T(E)$ for electrons of energy E passing through the whole junction yielding $G = \frac{2e^2}{h}T(E)$. The transmission coefficient $T(E)$ can, for example, be determined from *ab initio* DFT calculations combined with, for instance, non-equilibrium Green's function transport methods. For coherent tunneling the conductance G can be calculated by integrating the electron transmission coefficient $T(E)$ with energy E:

$$G(V, T) = \frac{I(V, T)}{V}$$

$$= \frac{2e}{h} \int_{-\infty}^{\infty} dE\, T(E) \frac{f(E - E_F - eV/2, T) - f(E - E_F + eV/2, T)}{V} \quad (3.5)$$

where $f(x, T) = (\exp x/k_B T + 1)^{-1}$ is the Fermi function, k_B is Boltzmann's constant, e the electron charge and E_F the Fermi energy. There are now a good number of literature examples of good agreement between measured values and conductance computed by these *ab initio* methods, at least when issues related to the inadequacies of mean-field theories based on DFT have been taken into account [28,34–38].

3.3.2
Resonant Tunneling

In resonant tunneling through a molecule placed between two electrodes, a molecular state is brought into resonance through the bias voltage or electrochemical potential, leading to an enhanced current flow through the junction. Any residence of the electron (or hole) transferring between two electrodes on the molecule is, however, only fleeting and the electron transfers without environmental or molecular relaxation. The situation when the electron resides sufficiently long on the molecule for vibrational relaxation to occur is discussed in the following section on "hopping."

Resonant tunneling is commonly encountered in the scanning tunneling spectroscopy configuration encountered in UHV STM. Here, the bias voltage between STM tip and substrate is scanned with tunneling resonances occurring as higher transmission channels mediated through adsorbates states open up.

Figure 3.7 illustrates resonant electron transfer through a redox center placed between two electrodes in an electrochemical environment. Such resonant electron transfer in an electrochemical STM configuration differs from the two-terminal ultralow temperature UHV STM configuration discussed above in that the energy levels of the redox active group can be tuned with the electrochemical potential (although gating can be also achieved under UHV conditions if an appropriately placed third terminal (gate) is introduced). In addition the redox center interacts with its dielectric solvent environment. The oxidized and reduced states of the redox-active group have broad energy distributions from variations in interactions between the redox group and its immediate environment. Gerischer described the distribution of oxidized and reduced state energies as assuming a Gaussian distribution, with the following distribution (D_{ox}) for the oxidized state [39]:

$$D_{ox}(\varepsilon, \eta) = c \left[\frac{\pi}{4\lambda kT} \right]^{1/2} \exp\left[-\frac{(\lambda - \varepsilon + e_0 \eta)^2}{4\lambda kT} \right] \quad (3.6)$$

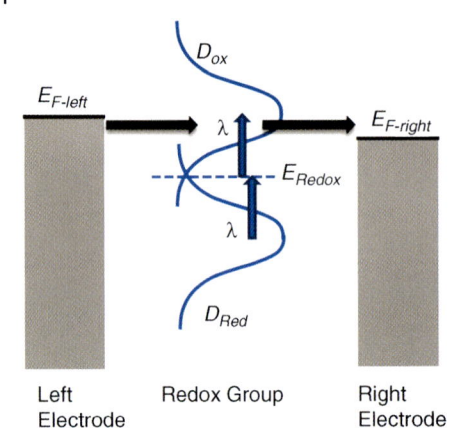

Figure 3.7 An illustration of resonant electron transfer through a redox center placed between two electrodes in an electrochemical environment. D_{ox} (and D_{red}) are the distributions of oxidized (and reduced) state energies using Gerischer's definitions. λ is the total reorganization energy, $E_{F\text{-left}}$ and $E_{F\text{-right}}$ refer to the Fermi levels of the substrate and tip, respectively, and E_{Redox} is the redox potential of the ox/red couple.

In this equation λ is the total reorganization energy (assumed here to be the same for the oxidized and reduced states), T is the temperature, k is Boltzmann's constant, ε is the energy of the empty (oxidized state) in solution. The distribution of reduced states has a similar form, with ε now being the energy of the occupied (reduced state) in solution.

$$D_{red}(\varepsilon, \eta) = c \left[\frac{\pi}{4\lambda kT}\right]^{1/2} \exp\left[-\frac{(\lambda + \varepsilon - e_0\eta)^2}{4\lambda kT}\right] \quad (3.7)$$

The distribution D_{ox} has a maximum at $\varepsilon = \lambda + e_0\eta$, while D_{red} has a maximum at $\varepsilon = -\lambda + e_0\eta$, so both are shifted by the same amount ($e_0\eta$) on application of an overpotential.

Schmickler et al. have described resonant tunneling through a redox-active adsorbate in an STM nanogap configuration by treating the redox process on the level of the Marcus theory and assuming step-like Fermi functions of the metal STM tip and substrate [40,41]. When the STM tip is positively biased, electrons tunnel through the molecular state to the tip, to give a tunneling current described by the following equation:

$$i \propto \int_0^r d\varepsilon . \rho_S(\varepsilon)\rho_S(\varepsilon) T_r(\varepsilon) D_{ox}(\varepsilon - e_0\eta + V) \quad (3.8)$$

In this equation T_r is the resonant tunneling probability and D_{ox} is the density of empty electronic states on the redox system as described above. ρ_s and ρ_t are the density of electronic states on the tip and substrate, respectively. Assuming that densities of the electronic states and the tunneling probabilities are independent of electronic energies ε they are taken outside the integral to leave the simple expression [40]:

$$i \propto \int_0^r D_{ox}(\varepsilon - e_0\eta + V).d\varepsilon \qquad (3.9)$$

The resonant tunneling current is simply the integral of D_{ox} in the energy range between the Fermi level of the substrate and the Fermi level of the tip ("the Fermi window"). A maximum in the resonant tunneling current through the oxidized state should occur when the density of empty electronic states on the redox system passes through the middle of the window between the Fermi energies. D_{ox} in the Gerischer picture described above has a Gaussian form and its maximum is shifted from the redox potential by the reorganization energy, λ. Therefore the electrode potential at which the maximum (ϕ_{max}) occurs is shifted away from the equilibrium potential (ϕ_0) by $\phi_{max} - \phi_0 - \lambda - V/2$. V is the tip to sample bias voltage, half of which is assumed to drop at the redox site.

This resonance tunneling model of Schmickler et al. has been used to describe tunneling currents which flow through Fe(III)-protoporphyrin adsorbed on HOPG in aqueous electrolyte as the electrochemical potential is tuned [40]. This is described in Section 3.4.1.

3.3.3
Hopping Models

When electronic levels within a bridge molecule lie close to resonance, these levels can become temporarily populated during charge transfer across the bridge. In the two-electrode configuration with the molecular bridge spanning the two electrodes, the levels can be brought close to or into resonance with the bias voltage. In the gated three-electrode set-up, resonance can be achieved with both the bias voltage and the gate voltage. In electrochemical STM the gating is controlled through the electrochemical potential. The temporary population of bridge sites during the electron or hole transfer process has given rise to the term "hopping". Hopping mechanisms have often been invoked to describe charge transfer across relatively long molecular bridges, such as DNA strands, where tunneling mechanisms would be inefficient over the relatively long distances.

Hopping or sequential electron transfer mechanisms have also been used to successfully describe a number of active redox molecular targets in the *in situ* STM configuration, with the molecule located in the nanogap between the substrate surface and the STM tip [42–46]. These systems have generally involved a single redox center tethered through chemisorption to the substrate surface. In some cases the molecule is also attached to the STM tip for the measurement, while in other cases there is a tunneling gap between the STM tip and the surface-immobilized redox-active molecule. Sequential two-step mechanisms have been the most successful for modeling charge transfer in this latter configuration [47–51]. These mechanisms involve two single electron/hole transfer processes, first from one electrode to the redox group, which is then followed by transfer of the electron/hole in the sequential process to the other electrode. Clearly these processes are electrochemical electron transfer processes with rates which can be modeled by electron transfer theory at defined electrochemical interfaces. In addition the dynamics of the redox group prior to transfer of the charge to the second electrode play an important role.

The rate constants for electron transfer to and from the redox group in the nanogap and the enclosing electrodes can be obtained from the analogous electrochemical electron transfer rates. In the following example, electrons are transferred from the substrate to the oxidized form of the molecule in the gap and then to the positively polarized STM tip, at a bias voltage of V_{bias}. $k^{r/o}$ represents the electron transfer rate from the substrate to the oxidized molecule, while $k^{o/r}$ represents the sequential electron transfer from the reduced molecule to the tip [52–55].

$$k^{o/r} \approx 8\kappa_{tip}\rho_{tip}k_B T \frac{\omega_{eff}}{2\pi} \exp\left[-\frac{[\lambda - e\eta - \gamma eV_{bias}]^2}{4\lambda k_B T}\right]$$

$$k^{r/o} \simeq 8\kappa_{substr}\rho_{substr}k_B T \frac{\omega_{eff}}{2\pi} \exp\left[-\frac{[\lambda + e\eta - (1-\gamma)eV_{bias}]^2}{4\lambda k_B T}\right]$$

(3.10)

These equations represent the rate constants in the case of weak interactions between the redox group and the electrodes (weak coupling limit). λ is the reorganization energy, k_B is Boltzmann's constant, T the temperature, κ the electronic transmission coefficient for interfacial electron transfer and ρ the metallic electron density. The subscripts "tip" and "substr" refer to the metallic tip and substrate, respectively. γ is the fraction of the bias voltage experienced at the site of the redox center, while η is the overpotential [54,55]. In the weak coupling limit the steady-state tunneling current can be found from these electrochemical rate constants using the following equation [54]:

$$i_{tunn}^{weak} = e\frac{k^{o/r}k^{r/o}}{k^{o/r} + k^{r/o}}$$

(3.11)

A schematic diagram illustrating the energy levels involved in the sequential two-step electron transfer, with a positive tip bias is shown in Figure 3.8. The energy levels involved are the Fermi levels of the tip (ε_{tip}) and substrate ($\varepsilon_{substrate}$) and the energy levels corresponding to the oxidized (ε_{ox}) and reduced (ε_{red}) redox group in the STM nanogap. The bias voltage is also shown (eV_{bias}). The bias voltage can be seen as a probing energy, with analogies to optical spectroscopy, with a narrower bias voltage ("window") increasing "tunneling spectroscopic resolution" of electronic levels on the redox molecule as the electrochemical potential is scanned.

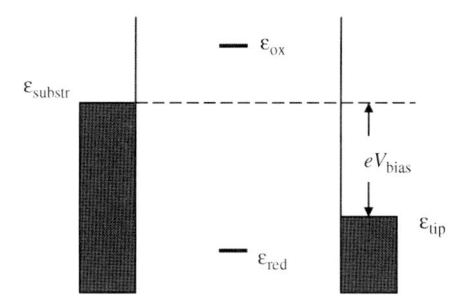

Figure 3.8 An illustration of the energy levels involved in the sequential two-step electron transfer across an STM nanogap containing a redox-active center. The energy levels involved are the Fermi levels of the tip (ε_{tip}) and substrate ($\varepsilon_{substrate}$) and the energy levels corresponding to the oxidized (ε_{ox}) and reduced (ε_{red}) redox group in the gap. The bias voltage is also shown (eV_{bias}), with positive tip bias in this illustration.

The initially oxidized redox group is reduced following the first electron transfer and then subsequently oxidized following electron transfer to the positively polarized tip electrode. In this two-step process the steady state tunneling current is given, in the limit of strong electronic interactions between the molecular redox group and the electrodes, by [54,55]:

$$i_{tunn}^{strong} = 2en_{o/r} \frac{k^{o/r} k^{r/o}}{k^{o/r} + k^{r/o}} \quad (3.12)$$

In this limit of strong electronic interaction the rate constants are given by [54]:

$$k^{o/r} \approx \frac{\omega_{eff}}{2\pi} \exp\left[-\frac{[\lambda - e\eta - \gamma eV_{bias}]^2}{4\lambda k_B T}\right]$$

$$k^{r/o} \approx \frac{\omega_{eff}}{2\pi} \exp\left[-\frac{[\lambda + e\eta - (1-\gamma)eV_{bias}]^2}{4\lambda k_B T}\right] \quad (3.13)$$

The term $n_{o/r}$ in Eq. (3.12) has important consequences for charge transfer across the junction. Prior to the first electron transfer onto the redox group, there is a pre-organization of the molecule which is facilitated by environmental fluctuations. Following electron transfer from the substrate, the vibrationally excited molecule relaxes and an electron is transferred onto the tip. However, while the molecule relaxes through the energy window between the Fermi level of the substrate and tip ("the window") it is possible for a large number of electron transfer cycles to occur, with successive reduction and re-oxidation of the redox site shuttling electrons across the molecular junctions. The term $n_{o/r}$ is the number of electrons transferred per initial environmental pre-organization event. This can take on values from 1 to several orders of magnitude. Equation (3.13) gives an expression for the enhanced tunneling current resulting from the two-step electron transfer process. This equation assumes that there is a symmetrical electronic contact $(\kappa_{substr}\rho_{substr} = \kappa_{tip}\rho_{tip})$

$$i_{tunn}^{symm} = \frac{1}{2} e n_{o/r} \frac{\omega_{eff}}{2\pi} \exp\left(-\frac{\lambda - eV_{bias}}{4k_B T}\right) \left\{ \cosh\left[\frac{\left(\frac{1}{2}-\gamma\right)eV_{bias} - e\eta}{2k_B T}\right]\right\}^{-1} \quad (3.13)$$

γ is the fraction of the bias voltage at the site of the redox center [54]. This equation also takes the electrochemical potential to be fully experienced at the site of the redox center; if this is not the case, another form of this equation has been derived in which only a fraction (ξ) of the electrochemical potential is experienced at the redox site. This model is known as the Kuznetsov Ulstrup (KU) model for sequential two-step electron transfer across the junction.

Two limits for Eq. (3.13) can be distinguished, the "weak" and "strong" coupling limits. In the weak coupling limit, the electron transfer process occurs in two consecutive one-electron processes with the redox center fully vibrationally relaxing before the second electron transfer occurs. This is the diabatic (nonadiabatic) limit, where the electron coupling between the redox center and the enclosing electrodes is sufficiently weak. $n_{o/r}$ in Eq. (3.13) in this limit is unity and the resulting current enhancement (i_{tunn}^{symm}) is relatively small.

The other limit is achieved when coupling between the redox group and the two enclosing electrodes is strong (adiabatic limit, see e.g., ref. 49). Electron transfer is fast so now only partial vibrational relaxation occurs prior to transfer of the electron onto the second contact. Due to environmental fluctuations in the initial pre-organization step, the oxidized state (ε_{ox} in Figure 3.8) is brought close to the Fermi level of the substrate and electron transfer occurs [49]. This now occupied state partially vibrationally relaxes, but before passing the Fermi level of the other contact (the tip) electron transfer occurs. The now re-oxidized redox group relaxes back towards the Fermi level of the substrate, accepting an electron before it fully relaxes or passes the Fermi level of the substrate. Many such cycles can occur per initial environmental pre-organization event, resulting in a large number of electron transfer events per initial excitation. This can lead to a large tunneling current amplification (large values of i_{tunn}^{symm}) for two-step electron transfer mediated through the redox center in the *in situ* electrochemical STM configuration. In this strong electronic coupling limit $n_{o/r}$ takes on the following form [54]:

$$n_{o/r} = \frac{eV_{bias}}{\Delta\varepsilon}; \Delta\varepsilon = \frac{1}{\kappa_{substr}\rho_{substr}} + \frac{1}{\kappa_{tip}\rho_{tip}} \ll k_B T \quad (3.14)$$

Where κ_{substr} and κ_{tip} are the electronic transmission coefficients for electron transfer between the substrate and redox group and the redox group and tip, respectively. ρ_{substr} and ρ_{tip} are the electronic level densities of substrate and tip, respectively.

3.4 Single Molecule Electrochemical Studies with an STM

The Kuznetsov Ulstrup model for two-step electron transfer across molecular junctions has been used to model a number of systems involving molecules in STM nanogap configurations and illustrative examples are discussed in Sections 3.4.4 and 3.4.5.

3.4.1 Adsorbed Iron Complexes

The first use of electrochemical STM to explicitly study electron transfer at the single molecule level at the electrode–electrolyte interface was by Tao [56]. A mixture of two porphyrins was adsorbed on high oriented pyrolytic graphite (HOPG). Both porphyrins consisted of the same porphyrin body, but differed in that one contained an iron atom in the central cavity ("FePP") while in the other this was absent (the free base form, protoporphyrin IX or "H_2PP"). These two molecules are illustrated in Scheme 3.1. FePP is electroactive at readily accessible electrode potentials in aqueous electrolyte, showing a defined redox wave at −0.48 V (versus SCE) for the adsorbed molecule. This corresponds to the one-electron reduction of the Fe(III)-porphyrin to the Fe(II) form. The free base analog shows no redox activity in this region. Figure 3.9a–e shows a series of STM images recorded in steps between −0.15 and −0.65 V (i.e., moving through the Fe(III)PP/Fe(II)PP redox potential) [56]. The central image in this series is recorded at a potential close to the peak electrochemical reduction current. The central spot in the image in Figure 3.9a

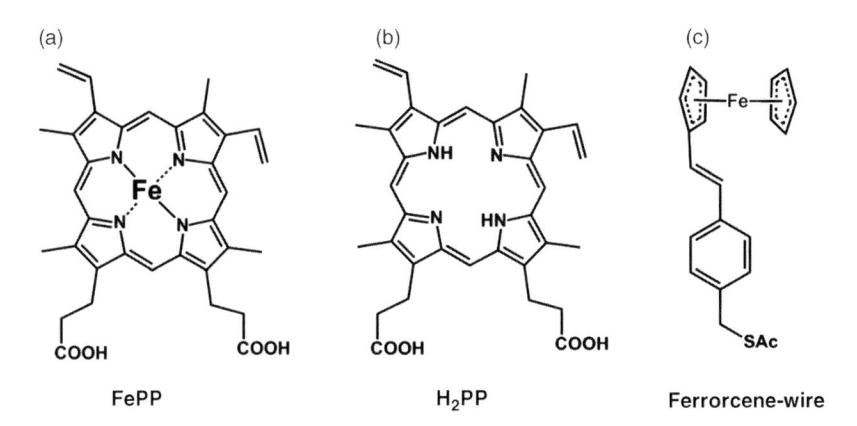

Scheme 3.1 The porphyrin structures used by Tao in reference [56] and the ferrocene wire featuring in reference [57]. (a) Fe(III)-protoporphyrin IX ("FePP"), (b) protoporphyrin IX ("H_2PP") and (c) ferrocene linked to an OPV bridge and a CH_2-thioactetate terminus used in the STM study of Davis et al. [57].

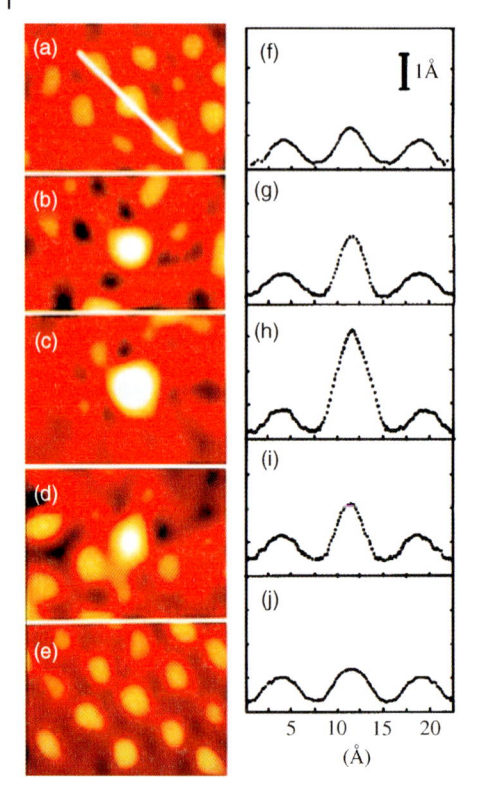

Figure 3.9 *In situ* electrochemical STM images of FePP (central) located within a monolayer matrix of the free base porphyrin anologue (H$_2$PP). The electrochemical potentials of the substrate were −0.15, −0.30, −0.42, −0.55 and −0.65 V, respectively for the sequence (a)–(e). Panels (f)–(j) show the corresponding apparent height cross-sections through the Fe(III) center (white line in Panel (A)). Reprinted with permission from N. J. Tao, *Phys. Rev. Lett.*, **76**, 4066–4069, 1996. Copyright 1996 by the American Physical Society [56].

is a FePP molecule surrounded by H$_2$PP which shows up with similar contrast. However, when the electrode potential is adjusted toward the reversible Fe(III)PP/Fe(II)PP potential the contrast difference between FePP and H$_2$PP molecules changes markedly, with the former "lighting" up. This contrast difference then decreases as the potential is increased away from the reversible potential (Figure 3.9d and e). By taking height cross-sections through the center of the FePP molecule, the increase in apparent height of the iron center could be monitored (Figure 3.9f – j). The apparent height increase has been attributed to the electrochemical alignment of the Fermi levels of the enclosing electrodes (the HOPG electrode and the metal STM tip) with the LUMO of the FePP adsorbate. The system is electrochemically moved into resonance, which is manifested in an increase in tunneling current flow mediated by the iron center of the FePP adsorbate.

3.4 Single Molecule Electrochemical Studies with an STM

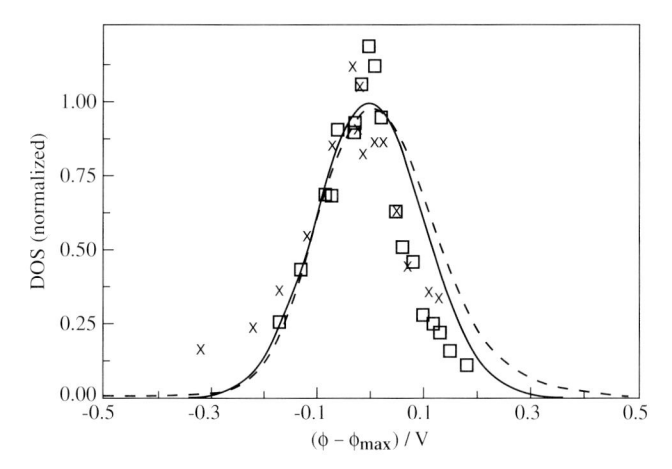

Figure 3.10 A comparison of experimentally derived data for FePP (crosses and squares) and simulations (full and dashed lines) based on the resonant tunneling model of Schmickler. Normalized apparent density of states (DOS) is plotted against electrochemical potential $(\phi - \phi_{max})$. The DOS is inferred from the relative height of FePP compared to surrounding PP molecules as measured with the STM. ϕ_{max} is the electrochemical potential where the STM measured height (and hence apparent DOS) is maximal. The squares correspond to experimental data measured with a tungsten STM tip, while the crosses refer to data measured with a Pt-Ir tip. The full line is a simulation for a semi-classical model, while the dashed line is from a quantum mechanical model. Reprinted from *Electrochim. Acta*, Vol. 42, Schmickler, W.; Tao, N. J., Measuring the inverted region of an electron transfer reaction with a scanning tunneling microscope, p. 2809, Copyright 1997, with permission from Elsevier [40].

The resonant tunneling model of Schmickler *et al.* described in Section 3.3.2 has been used to model enhanced tunneling currents which flow through Fe(III)-protoporphyrin adsorbed on HOPG in aqueous electrolyte as the electrochemical potential is tuned through the reversible one-electron reduction of Fe(III) to Fe(II). A plot of the normalized apparent density of states, inferred from the relative height of FePP compared to surrounding H_2PP molecules as measured with the STM, versus electrochemical potential is shown in Figure 3.10 [40]. To fit this to the resonant tunneling model, the relative heights measured at constant current have been converted to the current that would have been measured at constant height. This requires assumptions to be made regarding the decay constant (κ in $I \propto \exp{-\kappa z}$, where z is the height of the STM tip). Using this model and assumptions the data in Figure 3.10 could be fitted to the roughly Gaussian density of states and reorganization energies in the order of a few tenths of an eV were estimated. Although the fitting to this resonant tunneling model is not necessarily unique, the importance of this was to demonstrate that STM can be used at the single molecule level to investigate electron transfer reactions of redox-active species attached to electrode surfaces.

Other inorganic transition metal complexes with iron centers have been studied more recently, notably ferrocene tethered to electrode surfaces [57–59]. Bis-cyclopentadienyl iron(II), $(C_5H_5)_2Fe$ (ferrocene), in non-aqueous electrolyte solutions exhibits a reversible and well-defined one-electron oxidation process to the ferricenium cation. The electrochemical kinetics of this complex and multifarious derivatives has been widely studied so it is a natural choice for *in situ* electrochemical studies. In a conceptually similar experiment to the iron protoporphyrin studies of Tao described above, Davis et al. used *in situ* STM to study apparent height changes of a ferrocene-terminated molecular wire (Scheme 3.1) as it is electrochemically oxidized [57]. For these STM measurements the ferrocene wire was assembled into defects of a decanethiol self-assembled monolayer and the relative apparent height of the ferrocene wire with respect to the electrochemically inactive decanethiol matrix was measured. From changes in the apparent relative height of the ferrocene with electrode potential a large conductance enhancement at resonance of 3000 was estimated. This large gating factor was attributed to strong molecule–electrode electronic coupling of the ferrocene group attached to the underlying electrode with a conjugated oligophenylenevinylene (OPV) bridge with a terminal –CH_2–S linker (Scheme 3.1).

3.4.2
Viologens

The first electrochemical single molecule conductance measurements were performed on viologen derivatives [1], and this has also been the subject of subsequent studies [51,60–64]. These undergo reversible one-electron reduction from the dication V^{2+} form to the cation radical $V^{+\bullet}$. In the study by Haiss et al. the "viologen" molecular target 6-[1'-(6-mercapto-hexyl)-[4,4']bipyridinium]-hexane-1-thiol iodide was employed [1,64]. This is abbreviated to "6V6," with the "6" representing the HS–$(CH_2)_6$- chains on each side of the bipyridinium (viologen) group (R1 = R2 = HS–$(CH_2)_6$- in Scheme 3.2). The thiol anchoring groups assist binding to the gold surface and gold tip. 6V6 forms a low coverage phase on an Au(111) surface which has been characterized by surface infrared spectroscopy and STM [64]. When the STM tip is brought close to this phase, molecular bridges can form. Evidence for interaction between the STM and viologen adlayers which leads to formation of molecular bridges can be seen in STM imaging of the flat-lying phase. If the STM tip

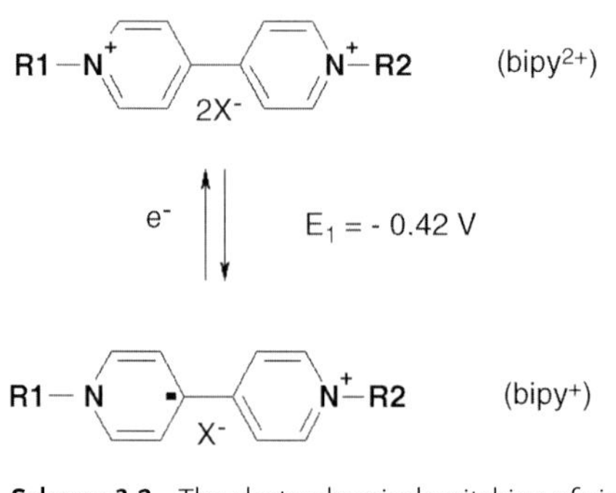

Scheme 3.2 The electrochemical switching of viologen molecular wires between the dication (2+) state and the cation radical (+$^\bullet$) state.

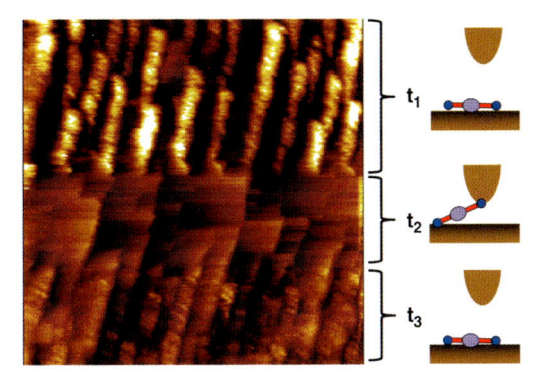

Figure 3.11 An STM image of the low coverage flat-lying phase of the viologen 10V10 on Au(111). This image was recorded in a line-by-line manner from top to bottom. At the start of the scan (top part of the image) the tip was held relatively far from the substrate. At the end of the period marked "t_1" the tunneling resistance was reduced by 1–2 orders of magnitude which brought it closer to the surface, leading to the adlayer structure being wiped aside. At the end of the period "t_2" the tunneling resistance was increased to its original value resulting in the adlayer structure reappearing. Adapted with permission from Haiss, W. et al., Langmuir **2004**, *20*, 7694. Copyright 2004 American Chemical Society [64].

is moved in close to the phase while scanning, the viologen adlayer is wiped away in the scanned area, as can be seen in Figure 3.11 for 10V10 [64]. In this figure, the image was recorded in a line-by-line manner from top to bottom with an initial relatively large tip to surface distance (high tunneling resistance). At the end of the period marked "t_1" in the figure, the tunneling resistance was decreased by 1–2 orders of magnitude during the imaging, which brought the tip closer to the viologen adlayer. The adlayer was then "wiped away" but gradually restored after the tip was subsequently moved back to its original height after period t_2 in the figure [64]. This is evidence of the interaction between the STM tip and the viologen adlayer which leads to the formation of molecular wires between the substrate and tip.

In the $I(s)$ determination of molecular conductance the tip is rapidly retracted from set-point conditions in which it is in close proximity to the viologen-covered substrate surface [1]. An example of two typical types of current–distance traces obtained during such current (I) versus distance (s) retraction scans for 6V6 is shown in Figure 3.12 [1]. In the first type marked "1" in Figure 3.12 a fast exponential decay of the current over a very short distance is observed. This is characteristic of the exponential decay of the tunneling current for an "empty" nanogap between the STM tip and the substrate. However, the other response ("2" in Figure 3.12) is more interesting. A plateau is formed which extends in this case to just over 2 nm, then decays to the zero tunneling current baseline at about 3 nm. The decay of the tunneling current upon retracting the tip is associated with the stretching and eventual cleavage of the gold–molecule–gold junction. Statistical analysis places the average value of the distance $s_{1/2}$ ($s_{1/2}$ is the distance for $I = I_w/2$, where I_w is the plateau current) at about (2.4 ± 0.6) nm; this matches well with the sulfur-to-sulfur distance of 2.4 nm for 6V6 with the alkyl chains in their extended all-trans

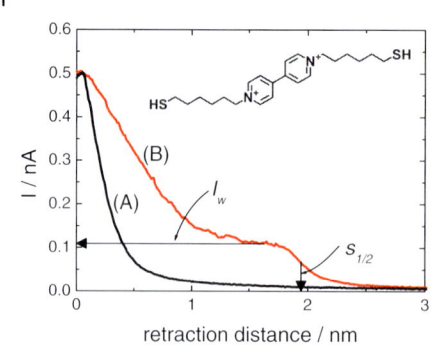

Figure 3.12 Current–distance decay curves for an adsorbate-free Au(111) surface (a) and for 6V6 on Au(111) (b). Adapted with permission from Haiss, W. et al., Langmuir **2004**, 20, 7694. Copyright 2004 American Chemical Society [64].

conformation [1]. Statistical analysis of the current distance curves or the current plateau (I_w values) as described in the methods section is used to find the single molecule conductance.

We now turn to the electrochemical dependence of the single molecule conductance of 6V6 as described by Haiss et al. [1] Figure 3.13 shows the single molecule

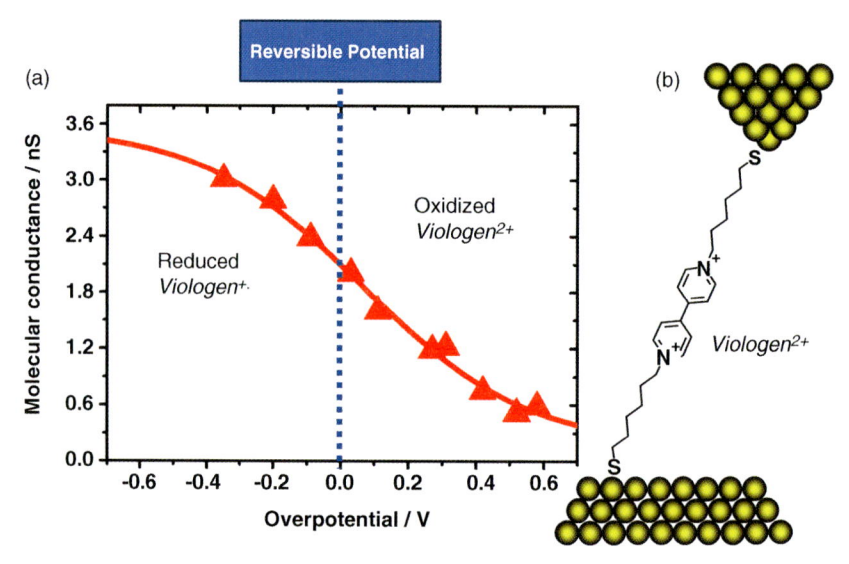

Figure 3.13 Single molecule conductance versus electrochemical overpotential data (a) for 6V6 molecular bridges (b). Data recorded in phosphate buffer electrolyte using the I(s)-based STM technique with $V_{bias} = 0.2$ V. The absence of a maximum in the conductance versus overpotential relation can be explained by the "soft gating" superexchange model of reference [61]. In this model the reduced viologen $V^{•+}$ state is not attained for the gold–viologen–gold junction even negative of the reversible $V^{2+}/V^{•+}$ potential [1,61,64,65]. Adapted with permission from Leary, E. et al., J. Am. Chem. Soc. **2008**, 130, 12204. Copyright 2008 American Chemical Society [65].

conductance of 6V6 versus overpotential. Positive overpotentials correspond to the oxidized form of the molecule (dication form, V^{2+}), while negative overpotentials correspond to the electrochemically reduced form (cation radical form, $V^{+\bullet}$). As the potential is moved negative from +0.6 V, through the reversible potential to −0.4 V, there is a marked but gradual increase in the conductance from 0.6 to 3 nS, that is, a factor of 5.

This gradual sigmoidal increase in the molecular conductance does not immediately accord with either the resonant tunneling model or the two-step hopping models described in the earlier sections. If, on the other hand, mono-thiolated forms of the viologen are self-assembled into monolayers on the gold substrate and probed with scanning tunneling spectroscopy (STS), then a bell-shaped conductance versus potential form, with a clear maximum around the reversible potential, is observed [51]. This case fits well to the Kuznetsov Ulstrup two-step model with partial vibrational relaxation [51]. There are notable differences between these two STM experiments that are likely to underlie the two very different conductance versus overpotential forms which are observed. In the STS determination the target viologen is more rigidly contained within a self-assembled monolayer, while in the single molecule conductance determination the target molecule is not contained in a self-assembled monolayer matrix. This may give the viologen bridge molecule more conformational mobility in the single molecule conductance determination. Another difference is the proximity of the STM tip to the molecular target. In the single molecule determination the tip is directly attached to the target molecule to form a gold–molecule-gold bridge. On the other hand in the STS determination the STM tip is above the self-assembled monolayer, with an electrolyte gap between the viologen adlayer and the tip. The fraction of the electrochemical potential at the site of the viologen redox group is likely to be different in these two different nanogap configurations.

The essence of the two-step process is that at, or close to, the equilibrium potential, both the oxidized and reduced forms contribute comparably to the current flow across the junction, hence giving a maximum in the current versus overpotential relation. Likewise, resonant tunneling modes are also expected to exhibit a maximum. The absence of a maximum in the molecular conductance versus overpotential relation of Figure 3.13 would indicate that resonance is not attained for the single molecule conductance configuration. Instead, electron transfer is by superexchange through a significantly off-resonance (with respect to the Au Fermi levels) empty redox level of the viologen moiety [61]. This empty or oxidized state of the viologen group is electrochemically tuned towards the Fermi levels of the Au electrodes as the potential is moved negative, but not sufficiently to achieve alignment and reduction of the viologen group (V^{2+} to $V^{+\bullet}$) in this configuration. The disparity between the applied electrochemical potential and the effective electrochemical potential experienced at the redox group can be accounted for by the redox center only experiencing a fraction of the substrate electrode–electrolyte potential drop. These features can rationalize the absence of a maximum. The off-resonant oxidized state (V^{2+}) thus dominates the behavior across the potential range of Figure 3.13 (hence the "reduced viologen," $V^{+\bullet}$, state marked on this figure, is then not achieved in this configuration). The energy of the oxidized state is nevertheless gated sufficiently by the electrode potential to account for the conductance increase towards negative electrode potentials [61]. Large configuration fluctuations also have to be introduced into the superexchange mode of electron transport to account for the conductance rise as seen in Figure 3.13. These configurational fluctuations of the molecule and its environment would bring the molecular bridge into non-equilibrium configurations which are much more conducive to electron tunneling across the molecular bridge. These thermally accessible configuration fluctuations pre-organize the redox bridge molecule into states favorable for electron tunneling. This has been termed "soft gating of the superexchange-based electron transfer" [61]. Computations and modeling of the 6V6 data in ref. 15 favor the role of low frequency, for instance torsional modes (and hence the term "soft") of the molecules in the pre-organization [61]. The theory and modeling in ref. [61] shows that all important aspects of the 6V6 data in Figure 3.13 can be rationalized with the soft gating superexchange model for physically sound values of the important parameters. The nature of these low frequency modes has been discussed [61,65], with a study by Leary et al. suggesting that twisting of the V^{2+} group about its central C—C bond could be a low-energy process which could promote the soft gating [65].

3.4.3
Osmium and Cobalt Metal Complexes

Osmium and cobalt transition metal complexes attached to Au(111) or Pt(111) single-crystal electrodes have proven to be very rich systems for studies of electron transfer using the *in situ* electrochemical STM [45,47–50]. These have been linked to the metal substrate through either pyridyl or thiol surface anchoring groups, as shown in Scheme 3.3 [48–50]. The different metal centers provide notably differing

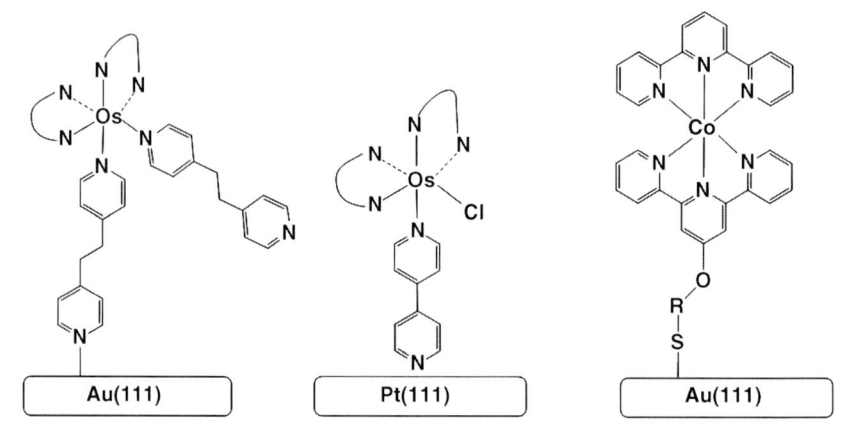

Scheme 3.3 Osmium and cobalt complexes discussed in Section 3.4.3 which have been the target of scanning tunneling spectroscopy measurements in electrochemical environments.

electron transfer rates as determined from electrochemical measurements. In particular, the electrochemical kinetics of the osmium complexes ($k_{et} > 10^6 \, s^{-1}$) are much faster than those of the cobalt complex ($k_{et} \sim 10^3 \, s^{-1}$) [50]. This difference has been attributed to larger intra-molecular reorganization energies and lower electronic coupling of the Co complex to the metal contacts [46].

As illustrated in Scheme 3.3 these complexes are just linked to the substrate electrode so the scanning tunneling spectroscopy configuration has been used. For these measurements the STM tip was held above the molecular layer at a constant height and with a relatively small bias voltage (values of V_{bias} between -0.2 and $+0.2$ V were employed) [48]. With the feedback loop disabled the electrochemical potential was rapidly swept and the tunneling current recorded. Figure 3.14 shows overlaid plots of the normalized tunneling current response as the electrode potential was swept negative through the equilibrium potential for reduction of the osmium complex (corresponding to the $[Os(bpy)_2(p2p)_2]^{2+/3+}$ couple) [48]. Curves for eight different bias voltage values are shown, with negative V_{bias} giving the four negative-going curves. All curves show a strong tunneling current enhancement corresponding to an "on–off" ratio of around 50 [48]. Notably, the lower bias voltages ($V_{bias} = \pm 50$ mV) gives the two curves which lie closest to the equilibrium potential ($\eta = 0$). This accords well with the Kuznetsov Ulstrup model for two-step electron transfer with partial vibrational relaxation. A plot of peak position of the STS feature against bias voltage gives a slope of about -0.5. From the relationship,

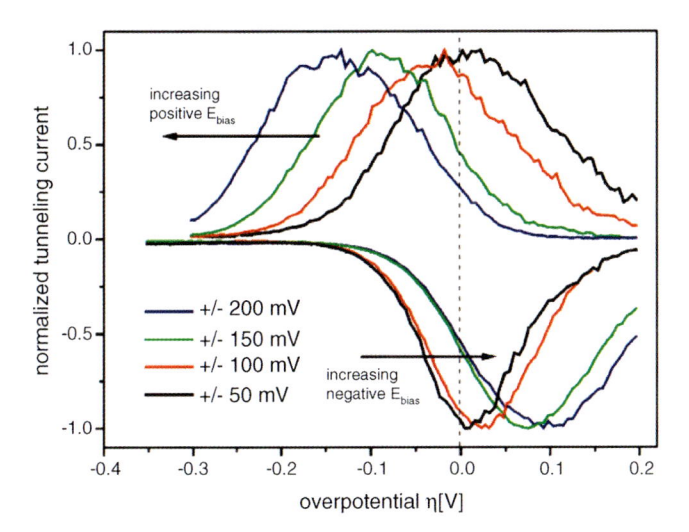

Figure 3.14 Scanning tunneling spectra of $[Os(bpy)_2(p2p)_2]^{2+/3+}$ on Au(111) recorded in 0.1 M HClO$_4$. Eight plots are shown for eight different bias voltage values; the four positive-going traces from left to right are $+200, +150, +100$ and $+50$ mV, respectively, while the four negative-going traces from left to right are $-50, -100, -150,$ and -200 mV, respectively. Reprinted with permission from Albrecht, T. et al. Nano Lett. **2005**, 5, 1451. Copyright 2005 American Chemical Society [48].

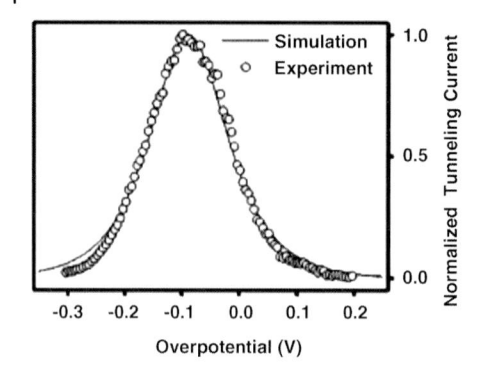

Figure 3.15 Circles: A scanning tunneling spectrum of [Os(bpy)$_2$(p2p)$_2$]$^{2+/3+}$ on Au(111) recorded in 0.1 M HClO$_4$ for $V_{bias} = +0.15$ V. Solid line: fit using the Kuznetsov Ulstrup two-step model with partial vibrational relaxation (Section 3.3.4). The fit parameters are $E_r = (0.3 \pm 0.05)$ eV, $\xi = (0.9 \pm 0.3)$ and $\gamma = (1 \pm 0.5)$. Reprinted with permission from Albrecht, T. et al. Nano Lett. **2005**, 5, 1451. Copyright 2005 American Chemical Society [48].

$\eta_{max} = \left(\frac{1}{2} - \gamma\right) V_{bias}$, where η_{max} is the overpotential where the scanning tunneling spectroscopy maximum occurs, a value of γ approaching unity is obtained. γ describes the fraction of the bias voltage experienced at the site of the redox center, implying that the bias voltage distribution is highly asymmetric in this nanojunction. The Kuznetsov Ulstrup equation has also been fitted to the complete STS curves. Such a fit is shown in Figure 3.15 for data with $V_{bias} = 0.15$ V [48]. The simulation to the experimental data was obtained with best fit parameters of $E_r = (0.3 \pm 0.05)$ eV, $\gamma = (1 \pm 0.5)$ and $\xi = (0.9 \pm 0.3)$ eV. The strong tunneling spectroscopic feature is a result of the current amplification in the two-step electron transfer mechanism. As discussed in Section 3.3.4 many reduction/oxidation cycles can occur per initial environmental pre-organization of the "hopping center," resulting in a large number of electron transfer events per initial thermal excitation. This provides an intensification of the tunneling current, with n_{el} characterizing the number of electrons transferred within a characteristic nuclear relaxation time. For complex 2 in Scheme 3.3, n_{el} was estimated to take on values in the range of 60–600 [49]. This attests to the electron transfer falling within the adiabatic regime and rationalizes the large current amplification occurring on resonance.

Scanning tunneling spectroscopy has also been used to study the cobalt analog (Complex 3 in Scheme 3.3) [50]. In a similar manner this complex also shows a peak in the tunneling current versus overpotential relation. However, the current amplification in this case is much weaker, with the enhancement being 100–400 times weaker for the Co-complex than the Os-complex [50]. This has been related to the much slower electrochemical electron transfer kinetics of the cobalt complex. It is proposed that the Co-complex may accord with the weak coupling, fully diabatic limit of two-step electron transfer. In this limit the redox center fully relaxes following the first electron transfer step. The electron is then transferred to the second contact, with only a single electron being transferred within the characteristic nuclear relaxation time. This mechanism would give much lower tunneling current enhancement factors than the adiabatic limit followed by the Os complexes [50].

3.4.4
PyrroloTTF (pTTF)

Although many scanning tunneling spectroscopy experiments of redox-active adsorbates have shown a bell-shaped tunneling current versus electrochemical potential form, with a clear maximum around the reversible potential, such behavior has been much more elusive in the "single molecule conductance" mode, in which a metal|molecule|metal junction is formed. For example, metal|viologen|metal STM junctions show a gradual sigmoidal increase in the molecular conductance as the electrochemical potential is scanned through the reversible potential. No maximum is observed in the molecular conductance around the reversible potential (as described in Section 3.4.2) and likewise metal|PTCDI|metal junctions show only evidence for a weak maximum (Section 3.4.6). This clearly does not accord straightforwardly with two-step mechanisms and reasons for this and the "soft gating" mechanism used to rationalize this behavior are discussed in Section 3.4.6. On the other hand Leary et al. [65] have more recently identified a metal|molecule|metal systems which fits well with the Kuznetsov Ulstrup (KU) model for two-step electron transfer across the junction. In contrast to the viologen (Section 3.4.2) and the PTCDI (Section 3.4.6) systems the conductance of 6pTTF6 (Scheme 3.4) single molecule junctions rises as it is electrochemically oxidized to 6pTTF6$^{\bullet+}$ with a clear and symmetrical peak at the 6pTTF6/6pTTF6$^+$ redox potential. As the electrochemical potential is then taken to more positive values the conductance falls. This behavior is shown in the experimental plot of single molecule conductance (G) versus overpotential in Figure 3.16 alongside schematic illustrations of the positions of the frontier orbitals ("$\varepsilon_{Reduced}$" and "$\varepsilon_{Oxidized}$") at different overpotentials [65]. The difference between the viologen and pTTF system has been tentatively rationalized in terms of their differing conformational dynamics [65]. V^{2+} molecules are twisted about the central ring which connects the two aromatic rings (Scheme 3.2). When reduced to V^+ the rings become coplanar as a result of the larger C=C character of the bond between the rings. There is thus a large structural change between V^{2+} and $V^{\bullet+}$, as well as accompanying changes in the conformational dynamics. The planarized ($V^{\bullet+}$ form) may be expected to offer a higher conductance than the twisted form (V^{2+}) and this may also in part contribute to the step-up in molecular conductance on electrochemically reducing V^{2+} to $V^{\bullet+}$ (Figure 3.13). By contrast, the pTTF group has coplanar rings in both its neutral and cation radical states, and thus would not be expected to experience the large structural and conformational dynamics changes of the viologen on redox switching. This may account for the more straightforward compliance of 6pTTF6 to the KU model for sequential two-step electron transfer across the junction. Recent quantitative fitting of the 6pTTF6 data (Figure 3.16) to the KU model for sequential two-step electron transfer (adiabatic limit) has resulted in an excellent fit to the data in Figure 3.16 with

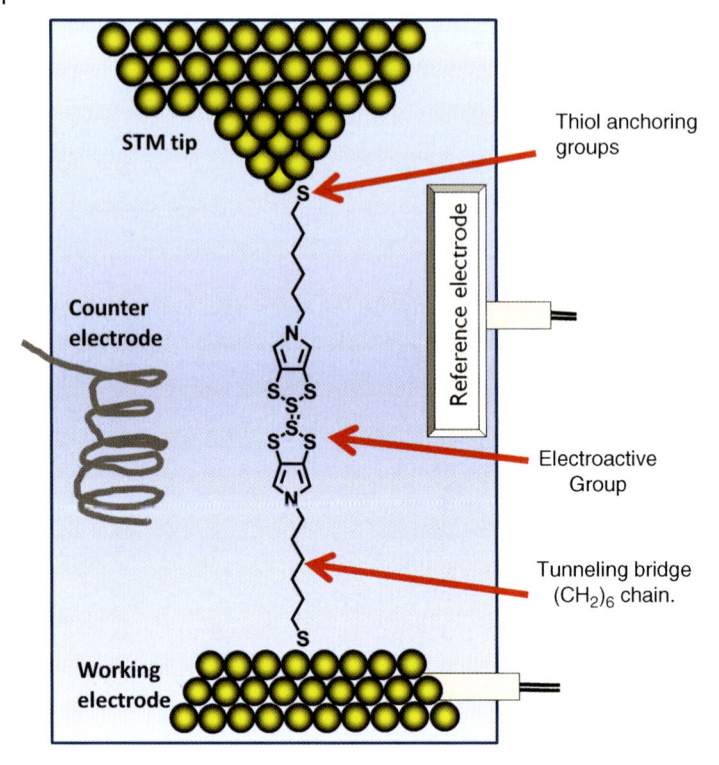

Scheme 3.4 The molecular structure of the pyrrolo-TTF molecular wire 6pTTF6, shown here in an electrochemically gated single molecule configuration. The "6"s in here in 6pTTF6 refers to the $-(CH_2)_6SH$ linkers on both sides on the molecule.

the values of $E_r = 0.4$ eV, $\gamma = 0.4$, and $\xi = 0.5$ [66]. Recently, such experiments have been performed in ionic liquid environments in the *in situ* STM single molecule junction configuration [66]. This has afforded the opportunity to study both the first and second electrochemical oxidations of pTTF; by contrast, only the first oxidation was observed in aqueous electrolytes. As well as observing single molecule conductance switching which followed sequential two-step electron transfer (KU model), reorganization energies could also be quantified for this ionic liquid environment and compared to those for aqueous environments. Reorganization energies of ~1.2 eV were obtained for the ionic liquid used, as compared to 0.4 eV for the aqueous environment. These differences were attributed to a large outer sphere reorganization energy for charge transfer across the single molecule junction in the ionic liquid [66].

Thiol-terminated oligoaniline molecular wires are another system which have shown a peak in the molecular conductance as a function of electrochemical potential [67]. The conductance of a number of aniline oligomers (trimers, pentamers, and heptamers; the molecular structure of the heptamer is shown in Scheme 3.5) has been measured as a function of the electrochemically controlled oxidation state of the molecules. These conductance measurements were performed

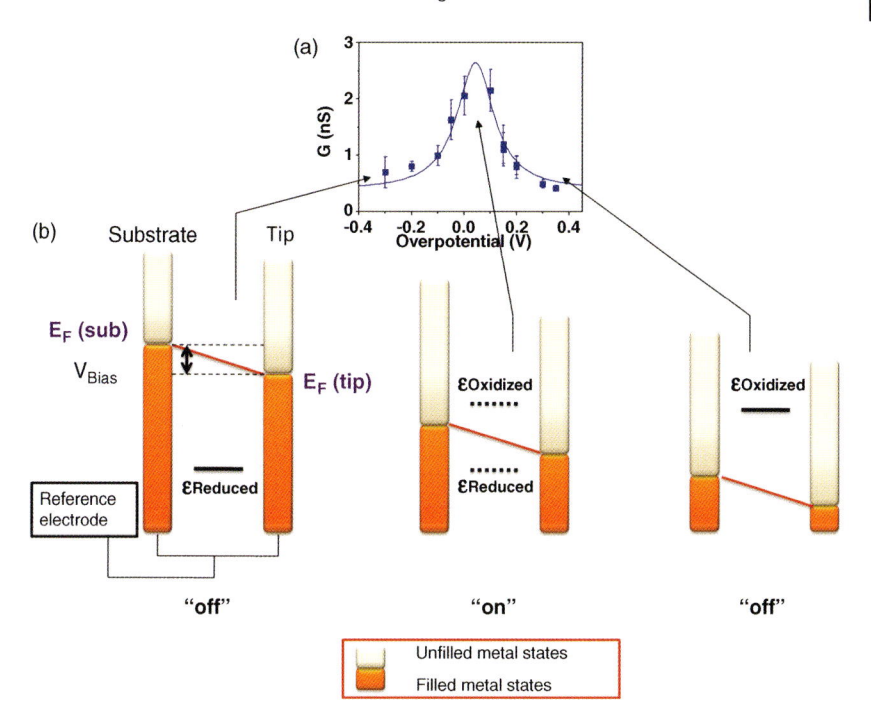

Figure 3.16 (a) Single molecule conductance versus overpotential for 6pTTF6, recorded by the $I(s)$ method [65]. These data follow the Kuznetsov Ulstrup (KU) model for sequential 2-step electron transfer (adiabatic limit) [66]. (b) Energy level diagrams with frontier molecular orbitals marked for the reduced and oxidized state of the molecules ("$\varepsilon_{Reduced}$" and "$\varepsilon_{Oxidized}$") with respect to the gold tip and gold substrate Fermi levels. The arrows schematically illustrate the coincidence of the redox potential for 6pTTF6/6pTTF6$^{\bullet+}$ with the maximum in the conductance versus overpotential relation (the "resonance condition" or "on" state). In the "off" state ("off resonance") the relevant orbitals are far away from the contact Fermi energies. Reprinted from Leary, E. et al., J. Am. Chem. Soc. **2008**, 130, 12204. Copyright 2008 American Chemical Society [65]. From Nichols et al., Phys. Chem. Chem. Phys., 2010, **12**, 2801–2815 [19] – reproduced by permission of the PCCP owner societies. http://pubs.rsc.org.ezproxy.liv.ac.uk/en/content/articlelanding/2010/cp/b922000c.

Scheme 3.5 Aniline heptamer in neutral state with thiol contacting groups at either end.

with the *in situ* BJ method, with Au(111) substrates and 50 mM H_2SO_4 electrolyte. A peak was seen in the vicinity of the first oxidation potential when the oligomer is in its half-oxidized and partially protonated ("emeraldine salt") state. The conductance measured at this peak was 5 nS for the heptamer (with an applied bias of 50 mV).

This compared to about 0.3 nS for the neutral form of the aniline heptamer measured in toluene [67], thus demonstrating a significant electrochemical gating of the molecular conductance.

3.4.5
Perylene Tetracarboxylic Diimides

A series of perylene tetracarboxylic diimides studied with the *in situ* BJ method with electrochemical potential control are shown in Scheme 3.6 [68]. Compounds 1 (top) and 3 (lower) have thiol anchor groups and their reversible electrochemical reduction can be seen when they are anchored onto gold electrodes. A relatively broad single voltammetric peak is seen for the surface immobilized species in aqueous electrolytes [69]. This corresponds to the electrochemical reduction of perylene tetracarboxylic diimide redox moiety by two electrons (two partially resolved one-electron redox waves can be seen in non-aqueous electrolytes) [69]. The dependence of the single molecule conductance on the electrode potential is shown in Figure 3.17 [68]. In the case of T-PTCDI a large increase of ~2 orders of magnitude is seen as the molecule is reduced. These conductance values were obtained from a conductance histogram analysis of a large number of conductance–distance traces recorded with the *in situ* BJ method. Each point in Figure 3.17 required that the electrode potential be held at the corresponding value for a time sufficiently long to record sufficient conductance–distance traces for the histogram analysis [68]. At the negative end of the electrochemical range studied, reductive desorption and the "hit rate" (number of tip retraction events showing molecular junction formation) were an issue. For this reason single junction forming events were also examined. In this method the STM tip was stopped during the tip retraction, with the molecular wire in place, and

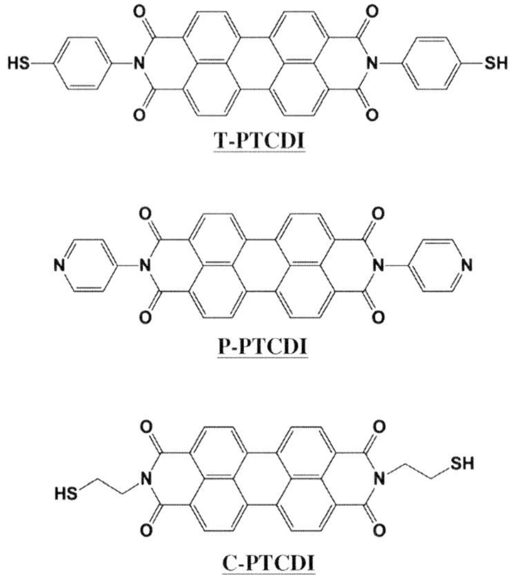

Scheme 3.6 Perylene tetracarboxylic diimide (PTCDI) derivatives discussed in Section 3.4.5.

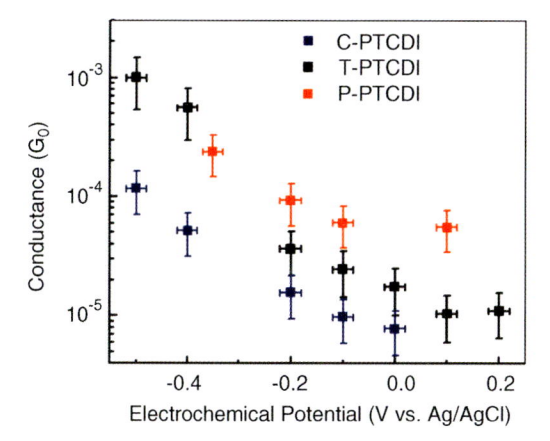

Figure 3.17 Single molecule conductance values for C-PTCDI, T-PTCDI and P-PTCDI as a function of electrochemical potential. Data obtained from conductance histograms obtained using the *in situ* BJ method in 0.1 M $NaClO_4$ and $V_{bias} = 0.1$ V. An increase in conductance is seen for all three compounds as the electrochemical potential is moved to negative potentials where the molecular target is electrochemically reduced. Adapted with permission from Li, X. L. *et al.*, *J. Am. Chem. Soc.* **2007**, *129*, 11535. Copyright 2007 American Chemical Society [68].

the electrochemical potential (electrochemical gate voltage, V_g) was rapidly swept so as to record tunneling current versus electrochemical potential for single junction formation events. An example of these retract-and-hold curves is shown in Figure 3.18 (solid black lines) [68]. These broadly accord with data obtained by the statistical conductance histogram evaluation. However, these curves are only for single and not statistically averaged junction formation events and, despite variability, evidence for peak formation in these current versus electrochemical potential curves was apparent. Data points from the statistical conductance histogram evaluation are overlaid alongside the retract-and-hold curve in Figure 3.18. Both plots show the ~2 orders of magnitude tunneling current enhancement on reduction of T-PTCDI. The appearance of a maximum towards the negative end of the retract-and-hold curves may be qualitatively in-line with a two-step sequential electron transport process. However, the absence of a full tunneling spectroscopic feature precludes a complete quantitative analysis with respect to two-step models and, therefore, precise details of the mechanism remain in question. A further interesting feature of the T-PTCDI system is the observation of temperature dependence of the molecular conductance for the oxidized form of the molecule [68]. Over a narrow temperature range (5–35 °C) a significant increase in single molecule conductance of the oxidized form was observed. This shows the importance of thermal activation in the electron transfer mechanism and, since electrochemical electron transfer processes are thermally activated, this would be in line with a two-step sequential process. However, at more negative electrochemical potentials, as T-PTCDI is reduced, the temperature dependence vanishes, which cannot be satisfactorily explained by the two-step model [68].

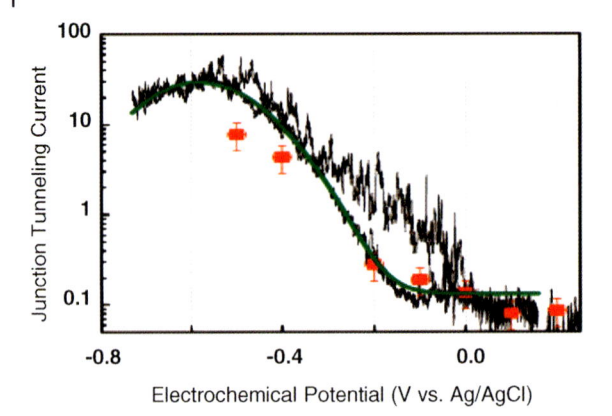

Figure 3.18 Tunneling current (in nano-amps) versus electrochemical potential for a T-PTCDI junction and a gold STM tip and substrate in 0.1 M NaClO$_4$ with $V_{bias} = 0.1$ V. Data recorded using the "retract-and-hold" method in which the STM tip was stopped during the tip retraction and the electrochemical potential was rapidly swept. Tunneling current versus electrochemical potential is consequently recorded for a single junction formation event (solid lines). The filled squares are, for comparison, data obtained from conductance histograms recorded by the conventional *in situ* BJ method. Adapted with permission from Li, X. L. et al., *J. Am. Chem. Soc.* **2007**, *129*, 11535. Copyright 2007 American Chemical Society [68].

3.4.6
Oligo(phenylene ethynylene) Derivates

Oligo(phenylene ethynylene) derivatives (OPEs) have been of particular interest in molecular electronics due to their effective π-conjugation and rod-like structure. OPEs have phenyl groups interlinked by alkyne groups, giving the linear conjugated structure (Scheme 3.7). Most molecular electronics studies of the electrical characteristics of this class of compounds have been in two terminal junctions, either at the single molecule limit or as molecular monolayers, and these studies have not involved electrochemical potential control. By contrast, Xiao *et al.* have examined the conductance of a nitro-substituted OPE derivative (OPE-NO$_2$) using the in-situ BJ method with electrochemical potential control [70]. This compound was of particular interest since a nitro-substituted OPE had featured in earlier molecular junction studies by Chen *et al.* in which a strong switching behavior (NDR, negative differential resistance) was observed [71].

Scheme 3.7 Oligo(phenylene ethynylene) derivatives, OPE-1 and OPE-NO$_2$.

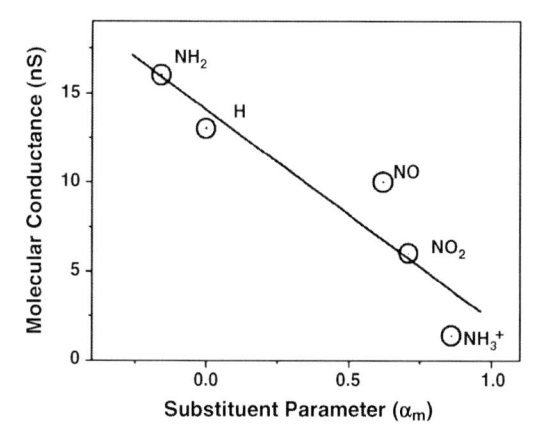

Figure 3.19 A plot of molecular conductance for 6 OPE derivatives versus the Hammett substituent parameter (α_m). Reprinted with permission from Xiao, X. Y. et al., J. Am. Chem. Soc. 2005, 127, 9235. Copyright 2005 American Chemical Society [70].

The study of OPE-NO$_2$ by Xiao et al. [70] significantly differs in concept to the examples highlighted in the preceding sections in which reversible electron switching between redox states (e.g., viologen^{2+} to viologen$^{\bullet+}$) was the focus. By contrast, the study by Xiao et al. exploited irreversible electrochemical reduction of the nitro group. By switching the electrochemical potential to successively more negative electrode potentials the —NO$_2$ group could be sequentially reduced to —NO and —NH$_2$, giving a series of OPR derivatives with the substituents R = —NO$_2$, —NO and —NH$_2$, respectively. Measurements in sufficiently acidic electrolytes also afforded the substituent R = —NH$_3^+$, while the reference compound OPE-1 provided R = H. In this way, electrochemistry aided the study of the single molecule conductance of a homologous series of OPE derivatives with different side chains with different electronegativity values [70]. Single molecule conductance was measured with the in situ BJ method, with the electrochemical reduction being achieved in situ. Figure 3.19 shows that the measured conductance decreases with Hammett substitution parameter σ. A large σ value corresponds to a highly electron-withdrawing substituent and a corresponding lower molecular conductance of the OPE wire.

3.5
Conclusions and Outlook

The measurement of single molecule electrical properties in electrochemical environments is providing new insights into charge transfer across electrical junctions. The in situ electrochemical STM has enabled such measurements to be made. There are two general modes of STM characterization of molecular electrical properties, both of which can be applied in electrochemical environments. Both can be referred to as "electrochemically gated molecular transistor configurations" since the electrochemical potential can be used to control ("gate") the molecular bridge state energetics, the redox state and the conductance of the molecular junction. In the first type, which is a STS configuration, the STM is held above a molecular monolayer. In this mode tunneling current is monitored either as a function of electrochemical potential or tip-to-sample bias voltage. In particular, the $I_{tunnelling}$ versus overpotential relation provides detailed insight into the mechanisms of charge transport across the junction, and the influence of the electrochemical potential and the redox state of the molecular bridge. Notable systems studied by STS methods, and discussed in this review, include iron protoporphyrin, viologen and ferrocene complexes and inorganic complexes containing osmium and cobalt centers. These systems have been theoretically considered with either resonant tunneling models, or sequential two-step electron transfer. As discussed in Section 3.4.3 the inorganic complexes containing osmium and cobalt centers provide a particularly interesting elucidation of the influence of the redox active metal center on the charge transfer mechanism. The Co system, which has much slower electron transfer kinetics, is found to accord with the weak coupling, fully diabatic limit of two-step electron transfer, while the osmium complexes follow the adiabatic limit of sequential two-step electron transfer [50].

The other mode of electrical characterization of single molecules using an electrochemical STM is the "single molecule conductance" mode, in which a metal|molecule|metal junction is formed, with the molecule tethered between the substrate surface and the STM tip. Such measurements are generally made with the in situ BJ technique or the $I(s)$ or $I(t)$ methods. As discussed in this chapter, systems studied by these methods with electrochemical potential control include viologens, pyrrolo-TTF, PTCDI, oligoanilines and redox-active OPE molecular wires. Most commonly, molecular conductance versus overpotential relations are recorded. Often, these do not straightforwardly accord with either resonant tunneling models, or sequential two-step electron transfer mechanisms, unless additional features such as soft gating of charge transfer are introduced [61], as discussed in Section 3.4.2 for the viologen systems. On the other hand, the system pyrrolo-TTF exhibits rather ideal behavior in its electrochemical dependence of molecular conductance, in that it can be fitted well to the Kuznetsov Ulstrup model for sequential two-step electron transfer (adiabatic limit) [65,66].

To date a rather small number of redox-active molecular bridges have been studied by STM in the "single molecule conductance" mode, in which metal|molecule|metal junctions are formed under electrochemical potential control. This area would benefit from the analysis of more redox-active molecular wire systems, to better develop the understanding of structure–property relationships in electrochemical gating. Great benefit could also be achieved from detailed computational models of such systems, which are able to include molecular details of both the junction and the immediate electrolytic environment and double layer. Even for the simpler two-terminal molecular electronic junctions, there are few examples of liquid environments being introduced into simulations, even though it is clear for some systems that this can have a large impact on the junction electrical properties. Cao et al. [72] have included water in a first principle calculation of charge transport in PTCDI junctions, while Leary et al. [73] have included water in detailed DFT computations of oligothiophene-containing molecular junctions. In both cases water was shown to markedly influence the junction conductance.

Another interesting area for further development is long-range charge transport in single molecular wires, and in particular how this can be controlled by the electrochemical potential. Most electrochemical single molecule conductance studies have involved relatively short molecular bridges. Oligo-porphyrins, peptides, DNA and organic oligomers, on the other hand, have all been targets for the study of longer range charge transport in single molecular bridges in two terminal junctions. For instance, charge transport in oligo-porphyrins has been studied in two terminal configurations using the *I(s)* and *in situ* BJ methods and also nano-fabricated junctions [28,74–80]. Interestingly, oligo-porphyrins single molecular wires have been shown to sustain phase coherent tunneling up to bridge lengths of at least 5 nm [28]. Oligo-porphyrins would be particularly attractive for future electrochemical single molecule studies, since their electrochemical behavior can be readily adapted through the choice of ligating metal ion in the porphyrin rings. The influence of multiple pendant redox-active centers on long-range charge transport could also be studied by the use of "molecular scaffolds." Such scaffolds could include DNA or peptides, which have already been the targets of interesting STM-based studies using the *in situ* BJ, *I(s)* or *I(t)* methods [5,81–93].

Ionic liquids also provide an attractive medium for the study of electrochemical gating of single molecule conductance. They offer a much wider electrochemical potential range than can be studied with aqueous electrolytes. Albrecht *et al.* have demonstrated that scanning tunneling spectroscopy can be conducted in an ionic liquid with electrochemical potential control and they have used this to study surface-immobilized osmium complexes [94]. Kay *et al.* have recently demonstrated that single molecule conductance can be determined in ionic liquid environments using the *I(s)* technique [95]. This opens up the attractive possibility of studying molecular conductance over very wide electrode potential windows, which span multiple redox states of the molecular bridge [66]. Another interesting development is the use of electrochemical gating to control quantum interference in molecular junctions at the single molecule level. Tao *et al.* have demonstrated that the conductance in a single anthraquinone-based norbornylogous bridge molecule can be controlled by electrochemical potential [96]. As the molecule is electrochemically reduced, the conductance is switched from the lower conductance anthraquinone state, which is cross-conjugated, to a higher conductance hydroanthraquinone state which is linearly conjugated. Although the conductance changes are just one order of magnitude in this case, there are theoretical predictions that molecular conductance can be very strongly suppressed as a result of quantum interference, with electrochemical gating potentially offering an attractive way of achieving large conductance modulation.

Acknowledgment

We gratefully acknowledge discussions and helpful feedback on this manuscript from Tim Albrecht.

References

1 Haiss, W., van Zalinge, H., Higgins, S.J., Bethell, D., Hobenreich, H., Schiffrin, D.J., and Nichols, R.J. (2003) *J. Am. Chem. Soc.*, **125**, 15294–15295.
2 Fan, F.R.F. and Bard, A.J. (1995) *Science*, **267**, 871–874.
3 Bard, A.J. and Fan, F.R.F. (1996) *Acc. Chem. Res.*, **29**, 572–578.
4 Fan, F.R.F., Kwak, J., and Bard, A.J. (1996) *J. Am. Chem. Soc.*, **118**, 9669–9675.
5 Chang, S., He, J., Kibel, A., Lee, M., Sankey, O., Zhang, P., and Lindsay, S. (2009) *Nat. Nanotechnol.*, **4**, 297–301.
6 Chang, S., Huang, S., He, J., Liang, F., Zhang, P., Li, S., Chen, X., Sankey, O., and Lindsay, S. (2010) *Nano Lett.*, **10**, 1070–1075.
7 Zevenbergen, M.A.G., Singh, P.S., Goluch, E.D., Wolfrum, B.L., and Lemay, S.G. (2011) *Nano Lett.*, **11**, 2881–2886.
8 Xu, B.Q. and Tao, N.J.J. (2003) *Science*, **301**, 1221–1223.
9 Haiss, W., Nichols, R.J., van Zalinge, H., Higgins, S.J., Bethell, D., and Schiffrin, D.J. (2004) *Phys. Chem. Chem. Phys.*, **6**, 4330–4337.
10 Cui, X.D., Primak, A., Zarate, X., Tomfohr, J., Sankey, O.F., Moore, A.L., Moore, T.A., Gust, D., Harris, G., and Lindsay, S.M. (2001) *Science*, **294**, 571–574.
11 Reed, M.A., Zhou, C., Muller, C.J., Burgin, T.P., and Tour, J.M. (1997) *Science*, **278**, 252–254.
12 Kergueris, C., Bourgoin, J.P., Palacin, S., Esteve, D., Urbina, C., Magoga, M., and Joachim, C. (1999) *Phys. Rev. B*, **59**, 12505–12513.
13 Weber, H.B., Reichert, J., Weigend, F., Ochs, R., Beckmann, D., Mayor, M., Ahlrichs, R., and von Lohneysen, H. (2002) *Chem. Phys.*, **281**, 113–125.
14 Smit, R.H.M., Noat, Y., Untiedt, C., Lang, N.D., van Hemert, M.C., and van Ruitenbeek, J.M. (2002) *Nature*, **419**, 906–909.
15 Gonzalez, M.T., Wu, S.M., Huber, R., van der Molen, S.J., Schonenberger, C., and Calame, M. (2006) *Nano Lett.*, **6**, 2238–2242.
16 Huber, R., Gonzalez, M.T., Wu, S., Langer, M., Grunder, S., Horhoiu, V., Mayor, M., Bryce, M.R., Wang, C.S., Jitchati, R., Schonenberger, C., and Calame, M. (2008) *J. Am. Chem. Soc.*, **130**, 1080–1084.
17 Shu, C., Li, C.Z., He, H.X., Bogozi, A., Bunch, J.S., and Tao, N.J. (2000) *Phys. Rev. Lett.*, **84**, 5196–5199.
18 Gruter, L., Gonzalez, M.T., Huber, R., Calame, M., and Schonenberger, C. (2005) *Small*, **1**, 1067–1070.
19 Nichols, R.J., Haiss, W., Higgins, S.J., Leary, E., Martin, S., and Bethell, D. (2010) *Phys. Chem. Chem. Phys.*, **12**, 2801–2815.
20 Wagenknecht, H.-A. (2006) *Charge Transfer in DNA*, Wiley-VCH Verlag GmbH, pp. 1–26.
21 Zhou, X.-S., Liu, L., Fortgang, P., Lefevre, A.-S., Serra Muns, A., Raouafi, N., Amatore, C., Mao, B.-W., Maisonhaute, E., and Schoellhorn, B. (2011) *J. Am. Chem. Soc.*, **133**, 7509–7516.
22 Tomfohr, J.K. and Sankey, O.F. (2002) *Phys. Rev. B*, **65**, 245105.
23 Wang, G., Kim, T.W., Jang, Y.H., and Lee, T. (2008) *J. Phys. Chem. C*, **112**, 13010–13016.
24 Engelkes, V.B., Beebe, J.M., and Frisbie, C.D. (2004) *J. Am. Chem. Soc.*, **126**, 14287–14296.
25 Akkerman, H.B., Naber, R.C.G., Jongbloed, B., van Hal, P.A., Blom, P.W.M., de Leeuw, D.M., and de Boer, B. (2007) *Proc. Nat. Acad. Sci. USA*, **104**, 11161–11166.
26 Choi, S.H., Kim, B., and Frisbie, C.D. (2008) *Science*, **320**, 1482–1486.
27 Hines, T., Diez-Perez, I., Hihath, J., Liu, H., Wang, Z.-S., Zhao, J., Zhou, G., Muellen, K., and Tao, N. (2010) *J. Am. Chem. Soc.*, **132**, 11658–11664.
28 Sedghi, G., Garcia-Suarez, V.M., Esdaile, L.J., Anderson, H.L., Lambert, C.J., Martin, S., Bethell, D., Higgins, S.J., Elliott, M., Bennett, N., Macdonald, J.E., and Nichols, R.J. (2011) *Nat. Nanotechnol.*, **6**, 517–523.
29 Simmons, J.G. (1963) *J. Appl. Phys.*, **34**, 1793–1803.
30 Holmlin, R.E., Ismagilov, R.F., Haag, R., Mujica, V., Ratner, M.A., Rampi, M.A., and Whitesides, G.M. (2001) *Angew. Chem. Int. Ed.*, **40**, 2316.
31 Davis, J.J., Peters, B., and Xi, W. (2008) *J. Phys.: Condens. Matter*, **20**, 374123.

32 Haiss, W., Martin, S., Scullion, L.E., Bouffier, L., Higgins, S.J., and Nichols, R.J. (2009) *Phys. Chem. Chem. Phys.*, **11**, 10831–10838.

33 Scullion, L.E., Leary, E., Higgins, S.J., and Nichols, R.J. (2012) *J. Phys.: Condens. Matter*, 24.

34 Quek, S.Y., Venkataraman, L., Choi, H.J., Loule, S.G., Hybertsen, M.S., and Neaton, J.B. (2007) *Nano Lett.*, **7**, 3477–3482.

35 Neaton, J.B., Hybertsen, M.S., and Louie, S.G. (2006) *Phys. Rev. Lett.*, **97**, 216405.

36 Koentopp, M., Burke, K., and Evers, F. (2006) *Phys. Rev. B*, **73**, 121403.

37 Ke, S.H., Baranger, H.U., and Yang, W.T. (2007) *J. Chem. Phys.*, **126**, 201102.

38 Wang, C., Batsanov, A.S., Bryce, M.R., Martin, S., Nichols, R.J., Higgins, S.J., Garcia-Suarez, V.M., and Lambert, C.J. (2009) *J. Am. Chem. Soc.*, **131**, 15647–15654.

39 Schmickler, W. (1996) *Interfacial Electrochemistry*, Oxford University Press.

40 Schmickler, W. and Tao, N.J. (1997) *Electrochim. Acta*, **42**, 2809–2815.

41 Schmickler, W. and Widrig, C. (1992) *J. Electroanal. Chem.*, **336**, 213–221.

42 Kuznetsov, A.M. and Ulstrup, J. (2000) *J. Phys. Chem. A*, **104**, 11531–11540.

43 Zhang, J., Chi, Q., Kuznetsov, A.M., Hansen, A.G., Wackerbarth, H., Christensen, H.E.M., Andersen, J.E.T., and Ulstrup, J. (2002) *J. Phys. Chem. B*, **106**, 1131–1152.

44 Zhang, J.D., Kuznetsov, A.M., and Ulstrup, J. (2003) *J. Electroanal. Chem.*, **541**, 133–146.

45 Zhang, J.D., Chi, Q.J., Albrecht, T., Kuznetsov, A.M., Grubb, M., Hansen, A.G., Wackerbarth, H., Welinder, A.C., and Ulstrup, J. (2005) *Electrochim. Acta*, **50**, 3143–3159.

46 Zhang, J.D., Kuznetsov, A.M., Medvedev, I.G., Chi, Q.J., Albrecht, T., Jensen, P.S., and Ulstrup, J. (2008) *Chem. Rev.*, **108**, 2737–2791.

47 Albrecht, T., Guckian, A., Ulstrup, J., and Vos, J.G. (2005) *IEEE T. Nanotechnology*, **4**, 430–434.

48 Albrecht, T., Guckian, A., Ulstrup, J., and Vos, J.G. (2005) *Nano Lett.*, **5**, 1451–1455.

49 Albrecht, T., Guckian, A., Kuznetsov, A.M., Vos, J.G., and Ulstrup, J. (2006) *J. Am. Chem. Soc.*, **128**, 17132–17138.

50 Albrecht, T., Moth-Poulsen, K., Christensen, J.B., Guckian, A., Bjornholm, T., Vos, J.G., and Ulstrup, J. (2006) *Faraday Discuss.*, **131**, 265–279.

51 Pobelov, I.V., Li, Z., and Wandlowski, T. (2008) *J. Am. Chem. Soc.*, **130**, 16045–16054.

52 Kuznetsov, A.M. (1995) *Charge Transfer in Physics, Chemistry and Biology*, Gordon & Breach, Reading.

53 Kuznetsov, A.M. and Ulstrup, J. (1998) *Electron Transfer in Chemistry and Biology, An Introduction to the Theory*, John Wiley & Sons Ltd, Chichester.

54 Zhang, J., Albrecht, T., Chi, Q., Kuznetsov, A.M., and Ulstrup, J. (2008) in *Bioinorganic Electrochemistry* (ed. O. Hammerich and J. Ulstrup), Springer, Netherlands, pp. 249–302.

55 Nichols, R., Haiss, W., Fernig, D., Zalinge, H., Schiffrin, D., and Ulstrup, J. (2008) (eds O. Hammerich and J. Ulstrup), in *Bioinorganic Electrochemistry*, Springer, Netherlands, pp. 207–247.

56 Tao, N.J. (1996) *Phys. Rev. Lett.*, **76**, 4066–4069.

57 Davis, J.J., Peters, B., Xi, W., Appel, J., Kros, A., Aartsma, T.J., Stan, R., and Canters, G.W. (2010) *J. Phys. Chem. Lett.*, **1**, 1541–1546.

58 Mishchenko, A., Abdualla, M., Rudnev, A., Fu, Y., Pike, A.R., and Wandlowski, T. (2011) *Chem. Commun.*, **47**, 9807–9809.

59 Rudnev, A.V., Pobelov, I.V., and Wandlowski, T. (2011) *J. Electroanal. Chem.*, **660**, 302–308.

60 Li, Z.H., Pobelov, I., Han, B., Wandlowski, T., Blaszczyk, A., and Mayor, M. (2007) *Nanotechnology*, **18**, 044018.

61 Haiss, W., Albrecht, T., van Zalinge, H., Higgins, S.J., Bethell, D., Hobenreich, H., Schiffrin, D.J., Nichols, R.J., Kuznetsov, A.M., Zhang, J., Chi, Q., and Ulstrup, J. (2007) *J. Phys. Chem. B*, **111**, 6703–6712.

62 Li, Z.H., Pobelov, I., Han, B., Wandlowski, T., Blaszczyk, A., and Mayor, M. (2006) International Conference on Nanoscience and Technology, Basel, Switzerland.

63 Li, Z., Han, B., Meszaros, G., Pobelov, I., Wandlowski, T., Blaszczyk, A., and Mayor, M. (2006) *Faraday Discuss.*, **131**, 121–143.

64 Haiss, W., van Zalinge, H., Hobenreich, H., Bethell, D., Schiffrin, D.J., Higgins, S.J., and Nichols, R.J. (2004) *Langmuir*, **20**, 7694–7702.

65 Leary, E., Higgins, S.J., van Zalinge, H., Haiss, W., Nichols, R.J., Nygaard, S., Jeppesen, J.O., and Ulstrup, J. (2008) *J. Am. Chem. Soc.*, **130**, 12204–12205.

66 Kay, N.J., Higgins, S.J., Jeppesen, J.O., Leary, E., Lycoops, J., Ulstrup, J., and Nichols, R.J. (2012) *J. Am. Chem. Soc.*, **134**, 16817–16826.

67 He, J., Chen, F., Lindsay, S., and Nuckolls, C. (2007) *Appl. Phys. Lett.*, **90**, 072112.

68 Li, X.L., Hihath, J., Chen, F., Masuda, T., Zang, L., and Tao, N.J. (2007) *J. Am. Chem. Soc.*, **129**, 11535–11542.

69 Li, C., Mishchenko, A., Li, Z., Pobelov, I., Wandlowski, T., Li, X.Q., Wurthner, F., Bagrets, A., and Evers, F. (2008) *J. Phys.: Condens. Matter*, **20**, 374122.

70 Xiao, X.Y., Nagahara, L.A., Rawlett, A.M., and Tao, N.J. (2005) *J. Am. Chem. Soc.*, **127**, 9235–9240.

71 Chen, J., Reed, M.A., Rawlett, A.M., and Tour, J.M. (1999) *Science*, **286**, 1550–1552.

72 Cao, H., Jiang, J., Ma, J., and Luo, Y. (2008) *J. Am. Chem. Soc.*, **130**, 6674–6675.

73 Leary, E., Hobenreich, H., Higgins, S.J., van Zalinge, H., Haiss, W., Nichols, R.J., Finch, C.M., Grace, I., Lambert, C.J., McGrath, R., and Smerdon, J. (2009) *Phys. Rev. Lett.*, **102**, 086801.

74 Sedghi, G., Esdaile, L.J., Anderson, H.L., Martin, S., Bethell, D., Higgins, S.J., and Nichols, R.J. (2012) *Adv. Mater.*, **24**, 653.

75 Li, Z. and Borguet, E. (2012) *J. Am. Chem. Soc.*, **134**, 63–66.

76 Qian, G., Saha, S., and Lewis, K.M. (2010) *Appl. Phys. Lett.*, 96.

77 Kiguchi, M., Takahashi, T., Kanehara, M., Teranishi, T., and Murakoshi, K. (2009) *J. Phys. Chem. C*, **113**, 9014–9017.

78 Sedghi, G., Sawada, K., Esdaile, L.J., Hoffmann, M., Anderson, H.L., Bethell, D., Haiss, W., Higgins, S.J., and Nichols, R.J. (2008) *J. Am. Chem. Soc.*, **130**, 8582.

79 Kang, B.K., Aratani, N., Lim, J.K., Kim, D., Osuka, A., and Yoo, K.H. (2006) *Mater. Sci. Eng. C-Biomimetic Supramol. Syst.*, **26**, 1023–1027.

80 Kang, B.K., Aratani, N., Lim, J.K., Kim, D., Osuka, A., and Yoo, K.H. (2005) *Chem. Phys. Lett.*, **412**, 303–306.

81 Hihath, J., Guo, S., Zhang, P., and Tao, N. (2012) *J. Phys.: Condens. Matter*, 24.

82 Hihath, J., Chen, F., Zhang, P.M., and Tao, N.J. (2007) *J. Phys.: Condens. Matter*, **19**, 215202.

83 Hihath, J., Xu, B.Q., Zhang, P.M., and Tao, N.J. (2005) *Proc. Nat. Acad. Sci. USA*, **102**, 16979–16983.

84 Xu, B.Q., Zhang, P.M., Li, X.L., and Tao, N.J. (2004) *Nano Lett.*, **4**, 1105–1108.

85 Huang, S., He, J., Chang, S., Zhang, P., Liang, F., Li, S., Tuchband, M., Fuhrmann, A., Ros, R., and Lindsay, S. (2010) *Nat. Nanotechnol.*, **5**, 868–873.

86 He, J., Lin, L.S., Liu, H., Zhang, P.M., Lee, M., Sankey, O.F., and Lindsay, S.M. (2009) *Nanotechnology*, **20**, 075102.

87 Chang, S.A., He, J., Lin, L.S., Zhang, P.M., Liang, F., Young, M., Huang, S., and Lindsay, S. (2009) *Nanotechnology*, **20**, 185102.

88 He, J., Lin, L., Zhang, P., and Lindsay, S. (2007) *Nano Lett.*, **7**, 3854–3858.

89 Sek, S. (2007) *J. Phys. Chem. C*, **111**, 12860–12865.

90 Sek, S., Misicka, A., Swiatek, K., and Maicka, E. (2006) *J. Phys. Chem. B*, **110**, 19671–19677.

91 Sek, S., Swiatek, K., and Misicka, A. (2005) *J. Phys. Chem. B*, **109**, 23121–23124.

92 Sek, S., Sepiol, A., Tolak, A., Misicka, A., and Bilewicz, R. (2004) *J. Phys. Chem. B*, **108**, 8102–8105.

93 Scullion, L., Doneux, T., Bouffier, L., Fernig, D.G., Higgins, S.J., Bethell, D., and Nichols, R.J. (2011) *J. Phys. Chem. C*, **115**, 8361–8368.

94 Albrecht, T., Moth-Poulsen, K., Christensen, J.B., Hjelm, J., Bjornholm, T., and Ulstrup, J. (2006) *J. Am. Chem. Soc.*, **128**, 6574–6575.

95 Kay, N.J., Nichols, R.J., Higgins, S.J., Haiss, W., Sedghi, G., Schwarzacher, W., and Mao, B.-W. (2011) *J. Phys. Chem. C*, **115**, 21402–21408.

96 Darwish, N., Diez-Perez, I., Da Silva, P., Tao, N., Gooding, J.J., and Paddon-Row, M.N. (2012) *Angew. Chem. Int. Ed.*, **51**, 3203–3206.

4
From Microbial Bioelectrocatalysis to Microbial Bioelectrochemical Systems
Uwe Schröder and Falk Harnisch

4.1
Prelude: From Fundamentals to Biotechnology

The foundation of microbial bioelectrochemistry dates back more than 100 years. Michael C. Potter [1] at Durham University (now Newcastle) was the first who reported the occurrence of an electromotive force between electrodes immersed in a sterile solution and a non-sterile solution containing bacteria. Yet, although Potter's results were 20 years later confirmed and elaborated by Cohen [2] and were revived in the 1960s [3,4] the induced research activity was rather limited. It was only at the turn of the millennium that an ever increasing interest and strongly enforced research and development activity in the field started (see e.g., Figure 4.1).

Thereby, this growing interest is not primarily motivated by the fascination of the still only partly understood fundamentals of microbial electrochemistry. In fact, the main driving force is the development of a platform technology for different sustainable applications, based on microbial bioelectrocatalysis, often referred to as "microbial bioelectrochemical systems" (BES) [5].

4.2
Microbial Bioelectrochemical Systems (BESs)

4.2.1
The Archetype: Microbial Fuel Cells (MFCs)

Probably more than 90% of current BES research is dedicated to microbial fuel cells (MFCs), the archetype microbial bioelectrochemical system. Although microbial fuel cells were long seen as a scientific peculiarity the first systematic MFC studies were performed in the framework of the NASA space program in the 1960s [3,4,6], later followed by Wilkinson and coworkers aiming at their application in robotics [7–9]. Figure 4.2 depicts the working principle of a microbial fuel cell. At the anode the oxidation of substrates, which may range from acetate to complex wastewater

Electrocatalysis: Theoretical Foundations and Model Experiments, First Edition.
Edited by Richard C. Alkire, Ludwig A. Kibler, Dieter M. Kolb, and Jacek Lipkowski.
© 2013 Wiley-VCH Verlag GmbH & Co. KGaA. Published 2013 by Wiley-VCH Verlag GmbH & Co. KGaA.

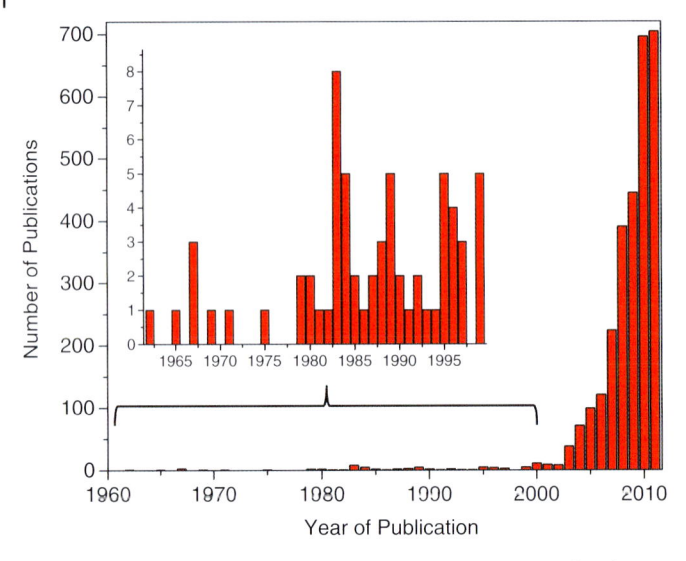

Figure 4.1 Number of publications from 1960–2011 containing the phrase "microbial fuel cell" (SciFinderScholar, April 2012).

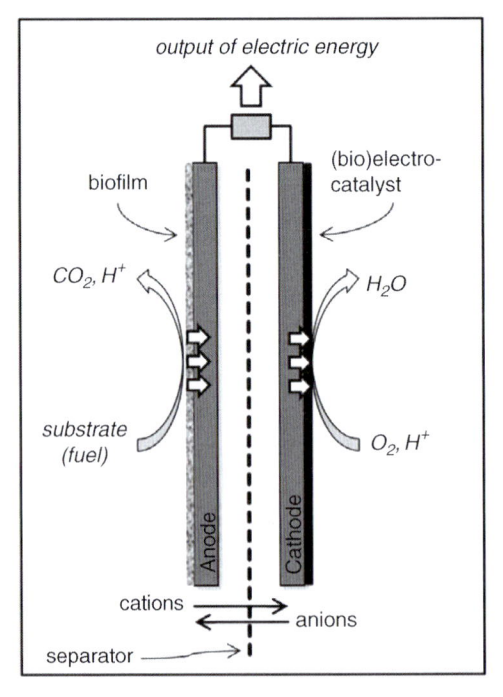

Figure 4.2 Principle sketch of a microbial fuel cell and its major reactions.

constituents (see [10] for an overview), takes place, whereas the oxygen reduction reaction (ORR) represents the common cathode reaction [11]. In principle both anode and cathode reactions may be microbially catalyzed; yet only for the anode does the necessity of microbial electrocatalysis exists, whereas for the oxygen reduction a number of abiotic electrocatalyst options are available [12].

Depending on the envisaged fields of application different concepts and types of microbial fuel cells have evolved. These include, for instance, benthic fuel cells (for the powering of marine sensors) [13,14], microbial fuel cells for autonomous robots [15,16], photomicrobial fuel cells [17,18], and the use of microbial fuel cells as biosensors [19,20]. The major emphasis, however, is certainly the development of MFC technology for wastewater treatment [5].

4.2.2
Strength Through Diversity: Microbial Bioelectrochemical Systems

For a long time BES research focused exclusively on microbial fuel cells. Recently, several further bioelectrochemical systems and new applications have been proposed (see also Figure 4.3). A major step was the introduction of microbial electrolysis cells (MEC) that in contrast to electricity-producing MFC use electric current to realize endergonic cell reactions, such as the synthesis and upgrading of chemical products, prospectively including fine and platform chemicals and biofuels [12,21]. One of the initial examples is the electrolytic hydrogen generation supported by the microbially catalyzed anodic oxidation of organic substrates (e.g., waste components) [22].

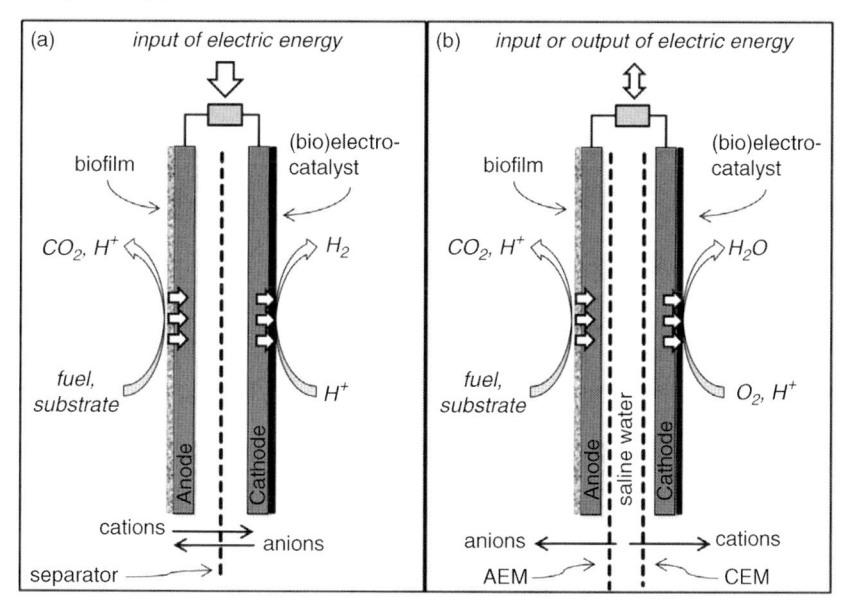

Figure 4.3 Sketch of a microbial bioelectrochemical system (a) for the production of hydrogen and (b) for desalination (AEM: anion exchange membrane, CEM: cation exchange membrane).

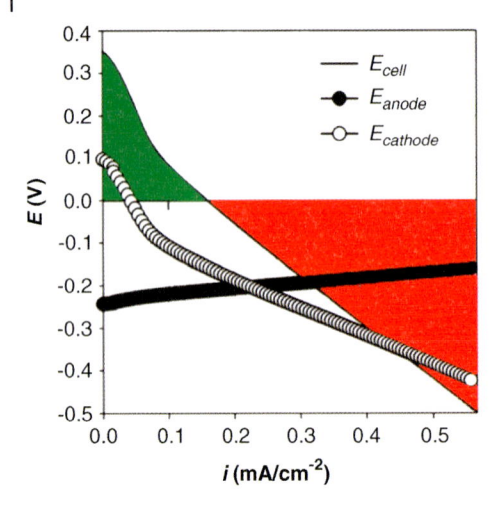

Figure 4.4 Cell voltage behavior of a hydrogen peroxide producing BES upon polarization (adapted from Rozendal et al. [23]. The green area denotes operation as galvanic cell (MFC) and thus electric energy gain, the red area denotes operation as electrolysis cell (MEC) and thus electric energy input (permission provided by Elsevier).

Considering the biological standard potentials $E^{0'}_{acetate}$ (anode) $= -0.290$ V and $E^{0'}_{H2/H+}$ (cathode) $= -0.414$ V the standard cell voltage is -0.124 V. In comparison to conventional water electrolysis (with a standard cell voltage of -1.23 V) only one tenth of electric energy is necessary to drive the hydrogen evolution reaction. In many cases the boundaries between MFC and MEC are blurred, as in the case of the cathodic production of hydrogen peroxide in a BES [23]: The cell reaction (anodic, microbially catalyzed oxidation of, e.g., acetate; and cathodic abiotic reduction of oxygen to H_2O_2) is exergonic. Thus, the production of hydrogen peroxide can be achieved in MFC mode, that is, under the generation of excess electric energy. Yet, in order to accelerate the reaction to an economically feasible rate the transition into MEC mode may become necessary (see Figure 4.4).

Similar situations may be observed for devices such as microbial desalination cells [24,25] – bioelectrochemical systems that can be understood as microbial fuel cells with an integrated desalination chamber (Figure 4.3b). Here, depending on the external conditions (e.g., rate and degree of desalination) a switching from MFC to MEC mode may become necessary.

4.3
Bioelectrocatalysis: Microorganisms Catalyze Electrochemical Reactions

In line with classical electrocatalysis, bioelectrocatalysts can be defined as moieties of biological origin leading to an increase in the rate constant (i.e., exchange current

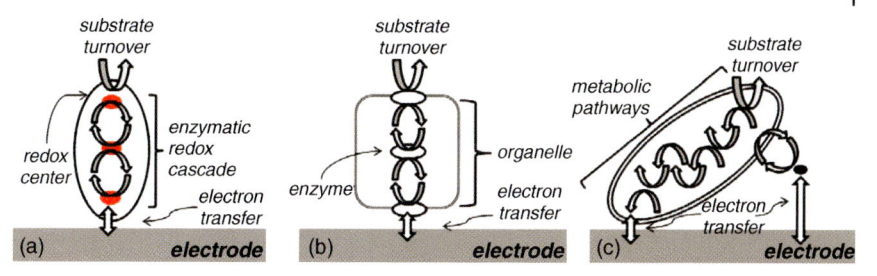

Figure 4.5 Sketches of the three different classes of bioelectrocatalysts: (a) enzymes, (b) organelles, (c) microbial cells (redrawn from [27]).

density) of a given electrode reaction, respectively, the decrease of its overpotential [26]. Thereby three different classes of bioelectrocatalysts can be distinguished: (i) enzymes, (ii) cell organelles and (iii) microbial cells (Figure 4.5).

Among these, microbial bioelectrocatalysts have recently attracted increasing attention due to a number of properties that clearly set them apart from enzyme-based bioelectrocatalysts. These comprise most importantly: self-reproduction and longevity, ability to catalyze multi-electron reaction steps (often with high specificity) and the ability to transform complex substrates and substrate mixtures. However, the mechanisms of the electron transfer between electrodes and microorganisms are only started to be understood (see Section 4.3.2).

4.3.1
Energetic Considerations of Microbial Bioelectrocatalysis

Microbial bioelectrocatalysis relies on living cells, which need a certain amount of energy for their maintenance, growth and proliferation. In order to generate this energy the cells use a certain share of the chemical energy of a substrate to build up internal potential gradients. From the electrochemical perspective this biological energy need leads to a loss of useful electromotive force of a corresponding bioelectrochemical cell (see Figure 4.6). This loss can be denoted as biological energy dissipation (ΔG_{biol}). Since the electron transfer from or to an electroactive microorganism occurs at distinct redox potentials, E_{ET}, (which are determined by the actual electron transfer pathway and the involved redox proteins or mediator compounds – see Section 4.3.2) the biological energy dissipation can be quantified (i) for the anode by the difference in the redox potential of the electron donor (e.g., the fuel of a biofuel cell) and $E^{0'}_{ET,anode}$, and for the cathode by the difference in the redox potential of the oxidant (e.g., oxygen) and $E^{0'}_{ET,cathode}$ (Figure 4.6).

The biological energy dissipation can be minimized, but it can never be totally avoided, since it is the actual driving force for the bacteria's voluntary participation in the electrochemical energy conversion process and assures their long term survival.

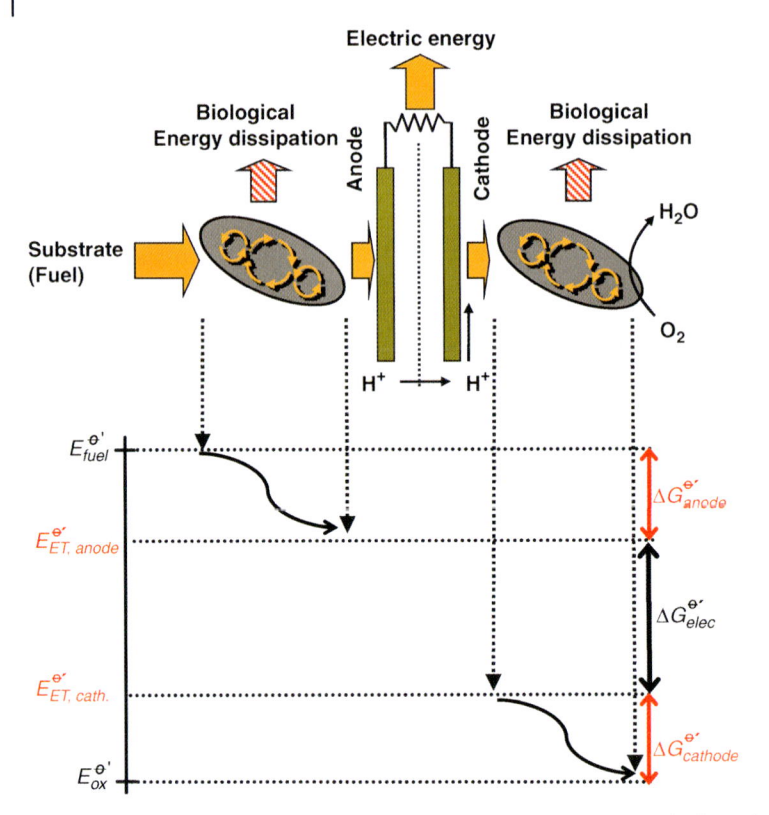

Figure 4.6 Schematic illustration of thermodynamic losses in a (fully microbially catalyzed) microbial fuel cell.

The interference of the organism in the net energy and mass balance leads to the conclusion that the term "bioelectrocatalysis" might, strictly speaking, not be entirely correct, and one should rather use the term "bio(electro)transformation". However, the term bioelectrocatalysis has been established in the literature and is, therefore, also used in this chapter.

The energetic consequences of the biological energy dissipation depend on the specific anode and cathode reaction and the involved microbial electron transfer redox species. In the following sections this will be illustrated by means of two selected case studies (i) the bioelectrocatalytic acetate oxidation by an anodic biofilm of the model organism family *Geobacteraceae* and (ii) the cathodic hydrogen evolution reaction (HER) catalyzed by a mixed culture microbial biofilm.

4.3.1.1 Case Study: The Anodic Acetate Oxidation by *Geobacteraceae*

Numerous studies have shown that by tailoring the growth conditions of anodic microbial cultures – that is, by a tailored electrochemically driven selection – a biofilm highly dominated by a specific species or family can be gained from highly diverse natural incoli [28,29]. Thus, a *Geobacter* dominated biofilm is formed at an electrode surface when acetate is the sole carbon and energy source and the electrode is poised at a positive potential (usually ranging between −0.2 and 0.6 V vs. SHE) to function as an electron acceptor.

For the following discussion let us consider the bioelectrocatalytic substrate oxidation at an acetate fed, *Geobacter*-based biofilm electrode, set at a potential of 0.400 V vs. SHE. The oxidation of acetate (shown in Reaction (4.1)) takes place at a biological standard potential of $E^{0'} = -0.290$ V.

$$C_2H_4O_2 + 2H_2O \rightarrow 2CO_2 + 8H^+ + 8e^- \tag{4.1}$$

(Note that Reaction (4.1) and derived reactions are, for simplicity reasons, based on the protonated acetate species.)

Taking into account the electrode potential of 0.4 V the potential difference between the electron donor (substrate) and the electron acceptor (electrode) is 0.690 V. This potential difference can easily be converted to a total Gibbs free energy (ΔG, with $\Delta G = -nF\Delta E$) of 532.6 kJ per mol of acetate or 66.6 kJ per mole of electrons. Yet, as illustrated in Section 4.1.1, the electron transfer of *Geobacter* to the electrode via its terminal electron transfer proteins takes place at a potential of −0.150 V, a value 140 mV more positive than that of the primary substrate. This decreases the maximum cell voltage of the fuel cell from 0.690 to 0.550 V and thus the energy gain to 53.1 kJ per mole of electrons. It leaves a share of about 13.5 kJ per mole of electrons as the biological energy dissipation to the *Geobacter* cells, for their growth, proliferation and heat production.

Similar treatises can be performed for further microbial species, electrode potentials and substrates. However, this is (so far) often not straightforward, as the thermodynamics of the microbial electron transfer is partially unknown and, even more importantly, the metabolic networks within single bacteria species as well as multi-species food-webs need further consideration (see also Section 4.3.3).

4.3.1.2 Case Study: The Cathodic Hydrogen Evolution Reaction (HER)

The hydrogen evolution reaction is an archetypal cathode reaction of microbial electrolysis cells – microbial bioelectrochemical systems that aim to generate chemicals rather than electricity (see Section 4.2.2). This reaction (Reaction (4.2)), possesses a biological standard potential of −0.414 V.

$$2H^+ + 2e^- \rightarrow H_2 \tag{4.2}$$

As discussed in Section 4.3.2 the electron transfer mechanisms of microbial biocathodes are so far mostly unknown and the same applies to the redox potential of the bacteria's electron transfer moieties. Therefore, we consider the following scenario: Following the study of Rozendal and coworkers [30] the bioelectrocatalytic HER commences at −0.650 V, which we now assume to represent the potential of the microbial electron transfer site. Analogous to Section 4.3.1.1, the potential difference between the electron donor (the cathode) and the electron acceptor (the hydronium ion) is 0.236 V, corresponding to 45.5 kJ per mole hydrogen, 22.8 kJ per mole of electrons. This cathodic biological energy dissipation is of the same order of magnitude as in the anodic acetate oxidation process (13.5 kJ per mole of electrons). However, there is a remarkable difference between the microbially catalyzed hydrogen evolution reaction and the anodic acetate oxidation. The anodic biofilm profits not only energetically from the substrate conversion process but also extracts organic carbon for heterotrophic growth and maintenance. The cathode biofilm on the other hand faces autotrophic growth conditions. Here, carbon dioxide fixation, a highly energy intensive growth process [31], has to be performed for biomass build-up and maintenance. Thus, one may be sceptical if a completely autotrophic growth is possible and favorable, or if a mixotrophic growth based on small amounts of organic acids (entering the cathode as the result of crossover processes from the anode side [32]) plays the decisive role for cathodic biofilm formation.

4.3.2
Microbial Electron Transfer Mechanisms

In the great majority of biological respiratory processes (aerobic and anaerobic) the terminal electron acceptor is molecular in nature. The reduction process takes place within the cells and requires the electron acceptor to be absorbed by the cell. Thus it is not surprising that – starting with the early studies by Potter [1,33] and Cohen [2] – it was for a long time believed that for the transfer of electrons from a microorganism to an electrode the addition of exogenous redox mediators was inevitable. This assumption, together with the fact that such mediators, like Neutral Red, were often toxic dyes, [34], surely contributed to the slow research progress in the field of microbial fuel cells and microbial bioelectrocatalysis.

As discovered by Kim and colleagues [35] at the turn of the last century, the use of artificial redox mediators is not necessary and is nowadays almost completely banned in the research field. It was discovered and started to be understood that nature has already created means for extracellular (often also referred to as exocellular) electron transfer. Two types of electron transfer pathways can be distinguished: direct electron transfer (DET) and mediated electron transfer (MET). Depending on the electron transfer abilities of the bacterial species involved the distance of electron transfer can reach up to hundreds of micrometers.

It is noteworthy that, so far, the electron transfer mechanisms have been almost exclusively studied on anodic biofilms; the present knowledge on the electron transfer of microbial cathodes is summarized in Section 4.3.2.3. Furthermore, it may be stated that all electron transfer mechanisms can, in principle, take place not only between a microorganism and an electrode but also between the microbial cells [36,37]. For the latter phenomenon the termini inter-cellular electron transfer and inter-species electron transfer, have been established.

4.3.2.1 Direct Electron Transfer (DET)
An obvious necessity for direct electron transfer (DET) is the physical contact of the electron transfer site and the electrode. In the great majority of cases this means a permanent cell–electrode connection by means of the formation of a bacterial

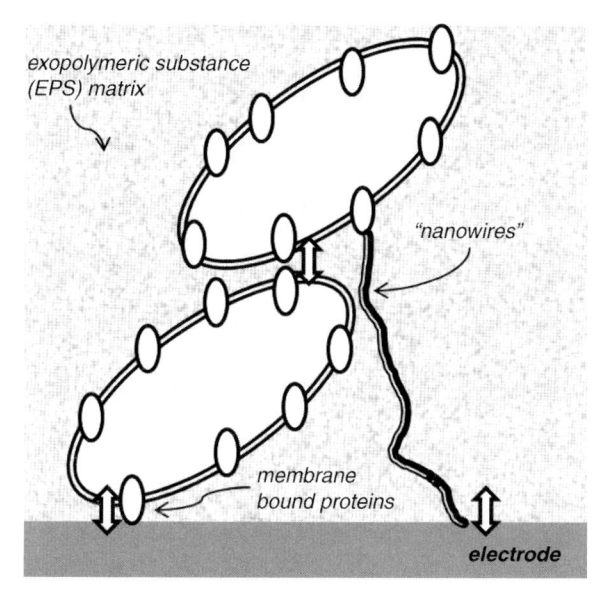

Figure 4.7 Sketch of DET via (a) membrane-bound DET proteins, (b) nanowires and (c) the EPS matrix to the electrode and between the bacteria (redrawn from [27]).

biofilm at the electrode (see also inset in Figure 4.10). However, DET via temporary cell–electrode contact (e.g., of *Shewanella* species) has been reported [38]. Figure 4.7 depicts the principle pathways of the direct electron transfer; here, three different possibilities of DET may be distinguished.

4.3.2.1.1 **DET Based on Trans-Membrane Electron Transfer Proteins** This is probably the most straightforward electron transfer path: the bacterium is adjacent to the electrode (or to another cell) and transfers the electrons via trans-membrane proteins, such as cytochromes. This electron transfer mechanism is ascribed to a number of microorganisms, with *Geobacter*, for example, [39,40], and *Shewanella*, for example, [41,42], species being its most studied representatives. Commonly the microorganisms possess a trans-membrane multi-protein redox cascade for the electron shuttling – which is shown in Figure 4.8 on the example of *Shewanella* [43].

4.3.2.1.2 **DET Based on Microbial Nanowires** The first indications for this pathway were discovered in 2005, where pilus-like structures (specifically formed for growth conditions requiring DET) were shown to possess electronic conductivity [44,45]. Later this was believed to be demonstrated not only for the transverse, but also the longitudinal direction [39,46].

The latest research on *Geobacter* indicates that there is strong interaction of this electron transfer pathway with DET based on trans-membrane proteins [47]. Simply speaking the nanowires provide the "motorways" for the electrons to reach the

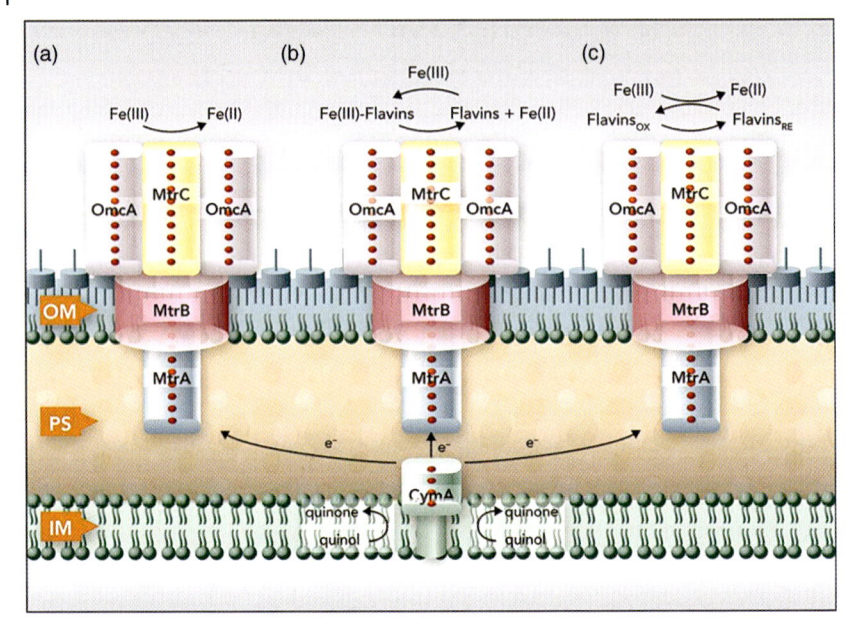

Figure 4.8 Example of the transmembrane complex for DET and MET of Shewanella (from Shi et al. [43]) with iron(II) as terminal acceptor: Proteins that are known to be directly involved in the reduction include (i) the inner membrane (IM) tetrahaem c-Cyt CymA that is a homolog of NapC/NirT family of quinol dehydrogenases, (ii) the periplasmic decahaem c-Cyt MtrA, (iii) the outer membrane (OM) protein MtrB and (iv) the OM decahaem c-Cyts MtrC and OmcA. Together, they form a pathway for transferring electrons from a quinone/quinol pool in the IM to the periplasm (PS) and then to the OM where MtrC and OmcA can transfer electrons directly to the surface of solid Fe(III) oxides (A). MtrC and OmcA might also reduce Fe(III) oxides indirectly by transferring electrons to either flavin-chelated Fe(III) (B) or oxidized flavins (C) Permission provided by Wiley.

bottom of the biofilm, where a specific redox protein (here OmcZ) acts as a terminal electron gate to the electrode.

4.3.2.1.3 DET Based on the EPS Matrix There is growing evidence that the matrix of exopolymeric substances (EPS) may play an important role in the microbial electron transfer [48,49]. The actual mechanisms and involved species are so far unknown (hence it is also not clear if this path relies on DET or MET principles) and thus this mechanism is clearly the most speculative among all electron transfer pathways.

4.3.2.2 Mediated Electron Transfer (MET)
Mediated electron transfer can be observed via (i) primary metabolites, that is, substances directly deriving from fermentation or anaerobic respiration of a substrate and (ii) secondary metabolites, that is, substances synthesized to act as reversible electron shuttles (Figure 4.9).

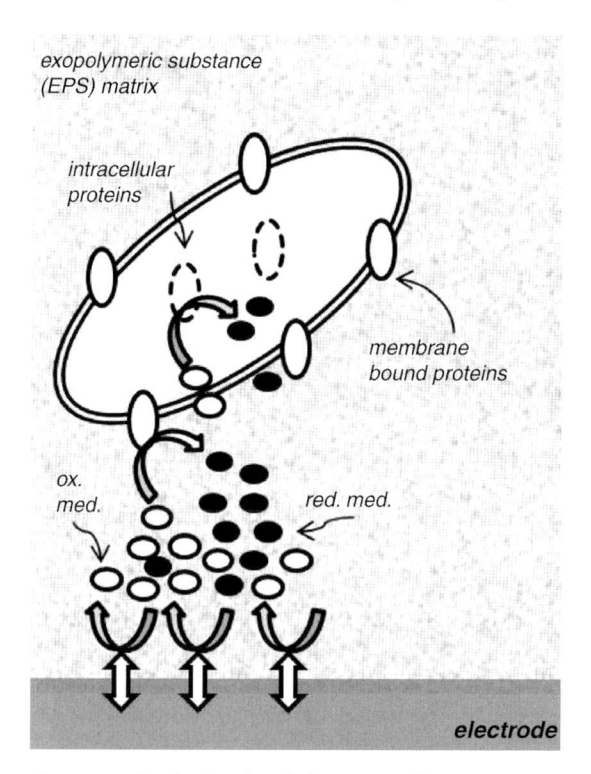

Figure 4.9 Sketch of mediated electron transfer mechanisms based on secondary metabolites for details see text (redrawn from [27]).

4.3.2.2.1 **MET Based on Secondary Metabolites** Whereas the use of exogenous, artificial mediators represents a generally abandoned concept, endogenous mediators, that is, redox-active substances synthesized by the microorganisms (see Figure 4.9), are intensively studied. Endogenous mediators are not formed during the catabolic substrate degradation but have to be synthesized by the organisms with a considerable energetic investment. In contrast to primary metabolites these mediators are produced only in small concentrations, the necessary electron transfer rate being achieved via a permanent recycling (reoxidation) and re-use of the mediator. Examples for endogenous redox mediators are pycocyanine [50] or 2-amino-3-carboxy-1,4-naphthoquinone [51]. They allow the electron transfer through biofilms, but also within suspensions of planktonic cells. Often, mediators produced by one species may be exploited also by other species in ecological networks (see also Section 4.3.3).

4.3.2.2.2 **MET Based on Primary Metabolites** The production of reduced primary metabolites is directly linked to the substrate oxidation process and serves the organism to dispose the electrons from substrate oxidation. The production of the primary metabolites is in stoichiometric relation to the rate of substrate oxidation. The metabolites may either derive from anaerobic respiration or from fermentation.

Hydrogen, ethanol and short chain acids, like formic acid and lactic acid, are important electron carriers produced by fermenting bacteria such as *Escherichia coli* or *Clostridia* [52–54]. On the other hand, sulfate reduction is a common respiratory path among anerobic bacteria [55], and especially in waste water-based microbial fuel cells [56] and in benthic fuel cells [57] sulfide oxidation can represent an important electron transfer mechanism.

$$SO_4^{2-} + 8H^+ + 8e^- \underset{\text{Anode}}{\overset{\text{Bacteria}}{\rightleftharpoons}} S^{2-} + 4H_2O \qquad (4.3)$$

To exploit primary metabolites in microbial bioelectrochemical systems they have to be oxidized at suitable electrode materials. This usually requires the use of appropriate electrocatalysts as has been demonstrated for hydrogen, formic acid and lactic acid [52,58] as well as sulfide [56,59].

4.3.2.3 Cathodic Electron Transfer Mechanisms

In contrast to anodic biofilms (discussed in Sections 4.3.2.1 and 4.3.2.2) very little is known about the electron transfer mechanisms at cathodes. Here, often exogenous mediators like methyl-viologen (e.g., [60,61]) are used, or a phenomenological description of the cathode process is given, see for example, [22,62,63]. A recent work by Rosenbaum and colleagues gives an interesting, speculative, insight into possible electron transfer mechanisms [64]. The authors discuss different pathways and their combinations, involving direct electron transfer moieties like cytochromes and yet unknown endogenous mediators and their coupling.

4.3.3
Microbial Interactions: Ecological Networks

The interactions between microorganisms of the same species as well as of different species are diverse and are only starting to be understood. Deriving from general biofilm research (e.g., in water ecology or medicine [65]) it seems logical and evident that the genetic and enzymatic activity of cells embedded in biofilms differs significantly from planktonic cells. Here, it has been shown that signal molecules – leading to quorum sensing – likely play an important role [66] in triggering the different activities.

When considering the biofilm at the anode as a complex microbial habitat two boundary cases have to be considered: (i) the cells in the upper layer(s) of the biofilm face high concentrations of the, often complex, microbial substrate but are apart from their terminal electron acceptor, and (ii) the cells in the vicinity of the electrode have good access to the electron sink but only limited access to the substrate.

Generally, the interactions of the microorganisms might be divided into interactions based on redox chemistry and interactions based on metabolic networks.

4.3.3.1 Interspecies Electron Transfer and "Scavenging" of Redox-Shuttles

As already depicted in Figures 4.7 and 4.9, the microbial electron transfer may not only take place to or from the electrode, but also between different microorganisms. This has been exemplarily shown by Summers *et al.* [36] on a co-culture of two

Geobacter species. Briefly, the interspecies electron transfer allowed the complete degradation of the substrate ethanol, which could not be achieved by a single species alone, and led to the formation of ordered aggregates. These microbial aggregates were assumed to possess an intrinsic conductivity; this was confirmed for mixed culture aggregates from wastewater digesters [37]. A further interesting finding is that the endogenous redox-shuttles can be used not only by the microorganisms that synthesized the respective compound, but also (and sometimes even more effectively) by other microbial species [67]. Whether this "scavenging" is a parasitic or even mutualistic effect is not unequivocal and clearly depends on further interactions among the microorganisms (see below).

4.3.3.1.1 **Establishment of Metabolic Networks** Complex substrates and their mixtures, for instance carbohydrates or proteins, cannot be metabolized by the archetypes of electroactive microorganisms. Therefore these microorganisms rely on the activity of other microbial species to break down these complex molecules into short chain carboxylic acids (e.g., acetate for *Geobacter* or lactate for *Shewanella*) that they can digest. Therefore, the establishment of food-webs within the biofilms seems necessary to make use of these substrates in microbial bioelectrochemical systems. This was exemplarily studied by Rosenbaum and coworkers [68] on a co-culture of *Shewanella oneidensis* and *Lactococcus lactis*. Only by the concerted action of both microbial species was the electricity generation from glucose at a coulombic efficiency of $\sim$17% possible. Interestingly, the authors found that the interaction between the two species was solely based on consecutive substrate degradation, whereas in a further study the same authors demonstrated that metabolite-based mutualism between microorganisms, here *Pseudomonas aeruginosa* PA14 and *Enterobacter aerogenes*, leading to enhanced bioelectrocatalysis, can occur as well [69].

4.4
Characterizing Anodic Biofilms by Electrochemical and Biological Means

Figure 4.10 shows a chronoamperometric curve of the growth and operation of a waste-water-derived mixed culture biofilm. The shape of the current curve is determined by the feed-batch mode, which is characterized by periodic substrate feeding/replenishment and substrate depletion. It is evident from Figure 4.10 that during the consecutive first three feed-cycles the maximum current density increases due to biofilm formation and growth and stabilizes thereafter at a certain level, here $250\,\mu\text{A}\,\text{cm}^{-2}$ – which is denoted as the maximum current density j_{max}.

The current density is often used as a characteristic phenomenological measure for a certain electrode/biofilm system and is used as a criterion for comparison within and between studies [12]. Thereby it has been shown that the current density is determined by several environmental variables, including, for example, pH [71] and temperature [72], and certainly the electrode material as the biofilm habitat.

Whereas the current density can be considered as a highly relevant phenomenological variable from the engineering perspective, it provides no information on the

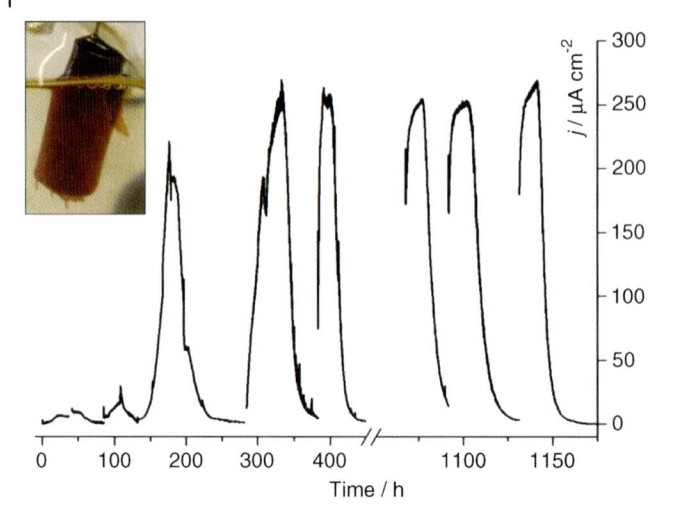

Figure 4.10 Chronoamperometric curve (at 0.2 V vs. Ag/AgCl) of a waste water derived biofilm in a fed-batch reactor and photograph of a biofilm covered anode (data derived from [70]).

composition, dynamics and processes at different hierarchical levels of the biofilm [73]. To gather this detailed information a plethora of electrochemical and hyphenated techniques has been employed. These techniques and their potential for biofilm studies are listed in Table 4.1.

Table 4.1 Selected techniques and methods (including selected references) used for the study of electroactive microbial biofilms (adapted from [73]).

Method	Brief description	Gained Information
Potentiostat/Galvanostatic methods		
Potentiostatic/galvanostatic operation [5]	The current at a constant potential or the potential at constant current of a biofilm is monitored over time.	The "macroscopic" performance of a biofilm system, for example, j_{max} and the coulombic efficiency at a given potential, can be assessed as function of time and external variables.
Linear sweep voltammetry (LSV)/linear galvanodynamic polarization [5]	The potential/current of the biofilm is changed linearly in time from an initial/start potential to an end/final potential, using a certain scan rate, that is, speed.	The dependence of applied potential and bioelectrocatalytic current can be assessed. This allows the generation of polarization curves for individual electrodes and/or whole BES.

Cyclic voltammetry (CV) [74,75]	The potential of the biofilm electrode is changed linearly in time from an initial/start potential to an end/final potential and back again to initial potential; multiple scans are advisable and different scan rates can be used. One distinguishes turn-over CV (presence of substrate) and non-turnover (substrate depletion).	The identification of the oxidation, reduction and formal potential of possible and *de facto* electron transfer sites is possible. By employing different scan rates the mass-transfer regime and nature of the electron transfer can be identified.
Further voltammetric techniques (e.g., square wave voltammetry (SWV), difference pulse voltammetry (DPV)) [5]	The potential of the WE is varied with particular time–potential regimes.	Information is in principle similar to CV, however the selectivity and/or sensitivity can be considerably different; allowing the identification of processes that cannot be analyzed using CV.
Electrochemical impedance spectroscopy (EIS) [76]	An AC-potential with varying frequency is applied to the electrode (hold at open circuit or a specific potential); the response of the electrode system is studied.	Allows the assessment of the resistance regime (mass transfer, ohmic and Warburg resistance) at the electrode surface and mass transfer regime.
Infrared (IR) spectroscopy [49]	Biofilms and cells can be characterized using their specific IR absorbance.	The sample is characterized on the chemical composition using IR-bands, for example, COOH-groups using a data base.
Fluorescence spectroscopy [77]	Using a (laser) light source the fluorescence spectra of the sample is detected.	The fluorescence spectra of biofilms and cells and their components are gained (especially important as only limited molecules are fluorescent).

Spectroelectrochemistry: Based on a potentiostat and a spectroscopy set-up

UV/vis spectroelectrochemistry [78]	Using transparent electrode UV spectra are measured.	UV spectra of the reduced and oxidized form can be obtained; characteristic bands like Soret-band (oxidized Cyt-c) and Q-band (reduced Cyt-c) can be measured.
UV/vis evanescent field wave spectroscopy [79]	Using an ITO-coated quartz waveguide as working electrode characteristic UV-spectra can be recorded in diffuse transmission mode employing evanescent field waves.	See above

(*continued*)

Table 4.1 (Continued)

Method	Brief description	Gained Information
Attenuated total reflectance surface enhanced infrared absorption spectroscopy (ATR-SEIRAS) [80]	Combining a tailored electrode leading to surface attenuation and infrared spectroscopy total reflectance spectra are obtained.	Evanescent field wave attenuated infrared spectra of the sample can be obtained and characteristic bands for oxidized and reduced forms can be assessed.
Surface enhanced Raman resonance scattering (SERRS) [81]	A laser light source hits the surface of a tailored electrode thus creating a surface enhancement effect; the Raman resonance is dependent on the combination of laser light wavelength and probe.	The direct vicinity of the electrode surface is probed; Raman spectra are obtained for oxidized and reduced conditions. Thus obtaining a "Raman-fingerprint" including the analysis of the coordination state of the catalytic center.
Confocal laser scanning microscopy (CLSM)		
Fluorescent dye-based imaging [82]	Biofilms can be visualized by labeling all cells, exopolymeric substances or adding indicator dyes.	Key aspects that can be learned are the biofilm structure and thickness, physico-chemistry (e.g., local pH) can be combined with (cryo)sectioning.
Fluorescent *in situ* hybridization (FISH) [83]	Biofilms are visualized by labeling specific cells based on identity or genetic content.	Key aspects that are learned are the biofilm structure, cell identity and stratification.
Microautoradiography-FISH (MAR-FISH) [84]	Biofilms are incubated with radiolabeled substrate and subsequently fluorescently labeled for FISH.	One can combine biofilm structure and cell identity information with activity of the cells.
Atomic force microscopy (AFM)		
In situ Surface analysis (using AFM) [85]	The AFM-tip is scanned over the surface and thus the topography is mapped.	Maps of the outer surface layer are gained.
Conductivity probing [44–46]	Conductivity between surface and AFM-tip or between two tips is measured.	Allows a conductivity mapping and determination of specific conductivities.
Protein probing [86]	AFM-tips functionalized with an antibody.	Allows the localization of a specific protein on a cell.
Confocal Raman microscopy (CRM) [87]		
	In focus Raman images on selected depths can be gained.	The Raman signal at a certain depth can be used to characterize biofilm components.

Electron microscopy [47]		
	The sample is imaged using an electron beam and a magnified image is gained; different types are distinguished based on the exact mode: scanning electron microscopy (SEM), transmission electron microscopy (TEM).	Magnified image of the sample and its underlying (electrode) material is gained.
Scanning electrochemical microscopy (SECM) [88]		
	Using an ultramicroelectrode and a (natural) redox-probe the redox-activity can be mapped.	Three-dimensional image of the oxidation/reduction ability of a sample can be gained.
Secondary Ion Mass Spectrometry (SIMS)		
Nano-SIMS [89]	High resolution SIMS particularly suited for biological research.	Allows determination of elemental composition at cell level for example, related to enzyme densities.
FISH-SIMS [90]	High resolution SIMS combined with FISH.	Allows determination of microbial identity in combination with elemental composition at cell level.
Community fingerprinting		
DNA based [91]	Main approaches are: • Pyrosequencing (including pyrotag) • Clone libraries • Phylogenetic microarrays • Denaturing gradient gel electrophoresis (DGGE) • Terminal Restriction Fragment Length Polymorphism (T-RFLP)	Allows identification at a genetic level of the microorganisms present in the biofilm
RNA-based [92]	Same as above however RNA-based.	Allows identification of the active microorganisms present in the biofilm at the genetic level.
Flow-cytometry [29]		
	Cells are stained with a dye and subsequently automatically characterized one-by-one counted on their light scattering and absorbing properties.	Allows the rapid determination of the abundance of microbial species within samples. Cells can be sorted for DNA/RNA analysis.

(continued)

Table 4.1 (Continued)

Method	Brief description	Gained Information
Genomics/transcriptomics/proteomics [93]	Via extraction of the DNA and/or RNA and/or proteins, and subsequent sequence analysis of these fractions, the genetic and proteomic content of multiple or even single cells can be evaluated.	Allows assessment of (i) the metabolic potential ("genomic"); (ii) which part of the metabolic potential is expressed by the cell ("transcriptomic") and (iii) which proteins are present and thus the true present ability of the cell to perform reactions ("proteomic").
Stable isotope probing (SIP)		
DNA-SIP [94]	DNA-based fractionation	Allows an assessment of which cells have taken up a labeled substrate over time.
RNA-SIP [95]	RNA-based fractionation	Allows an assessment of which cells are actively taking up substrate at the time of sampling.
Protein-SIP [96]	Protein-based fractionation	Allows an assessment of which cells are actively converting substrate.

Subsequently, examples of the use of cyclic voltammetry and *in situ* confocal Raman microscopy will be discussed in detail.

4.4.1.1 Case Study: On the use of Cyclic Voltammetry

One of the major research objectives in microbial electrochemistry is gaining deeper insights into the electron transfer, its mechanisms, thermodynamics and kinetics. For this, dynamic electrochemical techniques like cyclic voltammetry (CV) [74], square wave voltammetry [97] and electrochemical impedance spectroscopy [76] provide useful tools.

Figure 4.11a shows the cyclic voltammogram of a *Geobacter*-dominated anodic biofilm for non-turnover, that is, substrate depleted, conditions. One can clearly identify four redox couples, with the formal potentials E_1^f to E_4^f, all representing possible electron transfer sites. When adding substrate (here acetate) to the bacterial solution and recording the respective cyclic voltammetric curve under turn-over conditions, a typical S-shaped catalytic wave can be detected (Figure 4.11b). The inset of Figure 4.11b, showing the respective first derivative, clearly reveals that only the redox moieties represented by E_2^f and E_3^f are related to the bacterial electron transfer. Thus it can be derived from the measurements that the average potential level at which electrons are transferred to the electrode is about $-150\,mV$ (vs. SHE, equivalent to $-350\,mV$ vs. Ag/AgCl). This information is highly relevant for the discussion of the

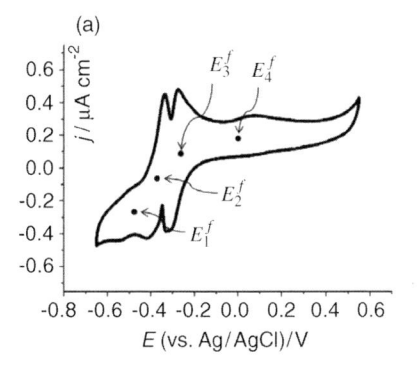

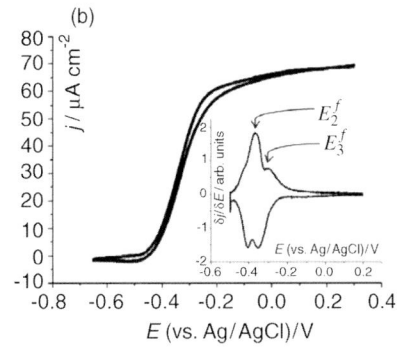

Figure 4.11 (a) Cyclic voltammogram of a *Geobacter* biofilm (grown at 0.2 V vs. Ag/AgCl on a graphite rod electrode) in substrate depleted (non turn-over) conditions E_1^f to E_4^f indicate formal potentials of the four detected redox couples of the biofilm; (b) CV of the same biofilm in the presence of acetate (turn-over conditions) the inset shows the first derivative of the CV curves and clearly reveals that only E_2^f and E_3^f are associated with the bacterial electron transfer. (scan rate in all cases 1 mV s^{-1}; data according to Fricke et al. [102]).

microbial electron transfer thermodynamics (see Section 4.3.1). Furthermore, CV can be used to draw mechanistic conclusions, for instance differentiating DET and MET processes. Note that the effect of a potentially rate-limiting counter ion/proton transfer within the biofilm has to be taken into consideration [74,98]. Cyclic voltammetry may also allow kinetic analysis. Yet, for this – as well as for mechanistic analysis – the development of new models beyond the "traditional" ones from enzyme and fundamental electrochemistry seems necessary, see, for example, [99–101].

4.4.1.2 Case Study: Raman Microscopy

Microscopic methods are widely employed for the imaging of biological aggregates and biofilms. However, most of these methods (except for optical microscopy) require a sample manipulation (e.g., introduction of dyes or marker genes), fixation or preparation, as, for example, scanning electron microscopy [103] or confocal laser scanning microscopy [82]. All these methods are invasive to the sample or can be performed only on cultures of known composition. Here confocal Raman microscopy, recently introduced by Virdis et al. [87], provides a promising tool for the rapid and non-invasive collection of morphological and compositional information. Figure 4.12a shows the individual spectra of a biofilm electrode in an aqueous environment: that of the biofilm itself (a *Geobacter* dominated biofilm), of the surrounding water and of carbon (the electrode material). Based on the selection of characteristic wavelengths (i.e., binning), cross-sectional images of the biofilm, as in Figure 4.12b, can be gained.

As the bacterial spectra can be specific for certain species and families, this method may not only allow to assess the morphological properties of the biofilms, but also to gain compositional information by distinguishing different microorganisms – based on their respective Raman spectra [104]. Furthermore these spectra can be recorded without noticeable biofilm destruction, which allows an almost *in situ* analysis. As the spectra of oxidized and reduced biofilms differ, most prominently in

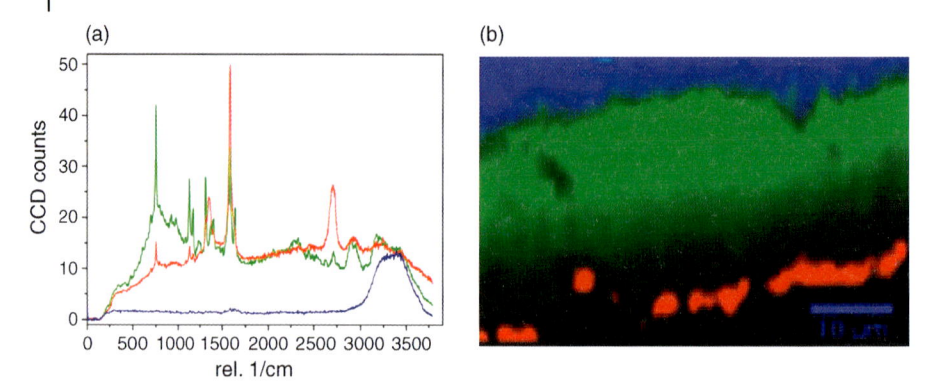

Figure 4.12 Raman spectra (a) and respective derived Raman-based image (b); color coding green: bacteria, red: carbon, blue: water, (retrieved from [87], where also details can be found).

the intensity of the cytochromes signal (due to the Raman resonance effect), the method potentially allows the detection of redox-gradients in the biofilm. However, the latter developments are still in their infancy and have to overcome a so far insufficient spectral and spatial resolution.

Acknowledgments

F.H. thanks the *Fonds der Chemischen Industrie* (FCI) for support. U. S. acknowledges the foundation of the professorship *Sustainable Chemistry and Energy Research* by the *Volkswagen AG* and the Verband der Deutschen Biokraftstoffindustrie e.V.

References

1 Potter, M.C. (1911) Electrical effects accompanying the decomposition of organic compounds. *Proc. Roy. Soc. London (B)*, **84**, 260–276.
2 Cohen, B. (1931) The bacterial culture as an electrical half-cell. *J. Bacteriol.*, **21**, 18–19.
3 Canfield, J.H., Goldner, B.H., and Lutwack, R. (1963) Utilization of human wastes as electrochemical fuels. NASA Technical report: Magna Corporation, Anaheim, CA, p. 63.
4 Bean, R.C., Inami, Y.H., Basford, P.R., Boyer, M.H., Shepherd, W.C., and Walwick, E.R. (1964) Study of the fundamental principles of bio-electrochemistry. NASA Final technical report, Research Laboratories, Philco Corporation, p. 107.
5 Rabaey, K., Angenent, L., Schröder, U., and Keller, J. (eds) (2010) *Bioelectrochemical Systems: From Extracellular Electron Transfer to Biotechnological Application*, IWA Publishing, London, New York.
6 Ellis, G.E. and Sweeny, E.E. (1963) Biochemical fuel cells. NASA Technical Report, The Marquardt Corporation (25,093).
7 Wilkinson, S. (2000) "Gastronome" – A Pioneering Food Powered Mobile Robot. Proceedings of the IASTED International Conference, Robotics and Applications, Honolulu.
8 Wilkinson, S. (2000) "Gastrobots"-benefits and challenges of microbial fuel cells in food powered robot. *Auton. Robot.*, **9**, 99–111.

9 Wilkinson, A.G. and Campbell, C. (1996) "Green" Bug Robots – Renewable Environmental Power for Miniature Robots. Proceedings of 4th IASTED International Conference, Robotics & Manufacturing, Honolulu.

10 Pant, D., Van Bogaert, G., Diels, L., and Vanbroekhoven, K. (2010) A review of the substrates used in microbial fuel cells (MFCs) for sustainable energy production. *Bioresour. Technol.*, **101** (6), 1533–1543.

11 Zhao, F., Harnisch, F., Schröder, U., Scholz, F., Bogdanoff, P., and Herrmann, I. (2006) Constraints and challenges of using oxygen cathodes in microbial fuel cells. *Environ. Sci. Technol.*, **40** (17), 5191–5199.

12 Harnisch, F. and Schröder, U. (2010) From MFC to MXC: chemical and biological cathodes and their potential for microbial bioelectrochemical systems. *Chem. Soc. Rev.*, **39**, 4433–4448.

13 Tender, L.M., Gray, S.A., Groveman, E., Lowy, D.A., Kauffman, P., Melhado, J., Tyce, R.C., Flynn, D., Petrecca, R., and Dobarro, J. (2008) The first demonstration of a microbial fuel cell as a viable power supply: Powering a meteorological buoy. *J. Power Sources*. doi: 10.1016/j.jpowsour.2007.12.123.

14 Tender, L.M., Reimers, C.E., Stecher, H.A., Holmes, D.E., Bond, D.R., Lowy, D.A., Pilobello, K., Fertig, S.J., and Lovley, D.R. (2002) Harnessing microbially generated power on the seafloor. *Nat. Biotechnol.*, **20**, 821–825.

15 Melhuish, C., Ieropoulus, I., Greenman, J., and Horsfield, I. (2006) Energetically autonomous robots: food for thought. *Auton. Robot.*, **21**, 187–198.

16 Ieropoulus, I., Melhuish, C., Greenman, J., and Horsfield, I. (2005) *Artificial symbiosis: Towards a robot–microbe partnership*. Proceedings of the Towards Autonomous Robot Systems Conference.

17 Rosenbaum, M., He, Z., and Angenent, L. (2010) Light energy to bioelectricity: Photosynthetic microbial fuel cells. *Curr. Opin. Biotechnol.*, **21**, 259–264.

18 Rosenbaum, M., Schröder, U., and Scholz, F. (2005) Utilizing the green alga *chlamydomonas reinhardtii* for microbial electricity generation: a living solar cell. *Appl. Microbiol. Biot.*, **68**, 753–756.

19 Di Lorenzo, M., Curtis, T.P., Head, I.M., and Scott, K. (2009) A single-chamber microbial fuel cell as a biosensor for wastewaters. *Water Res.*, **43**, 3145–3154.

20 Tront, J.M., Fortner, J.D., Ploetze, M., Hughes, J.B., and Puzrin, A.M. (2008) Microbial fuel cell biosensor for *in situ* assessment of microbial activity. *Biosens. Bioelectron.*, **24**, 586–590.

21 Rabaey, K. and Rozendal, R.A. (2010) Microbial electrosynthesis - revisiting the electrical route for microbial production. *Nat. Rev. Microbiol.*, **8**, 706–716.

22 Rozendal, R.A., Hamelers, H.V.M., Euverink, G.J.W., Metz, S.J., and Buisman, C.J.N. (2006) Principle and perspectives of hydrogen production through biocatalyzed electrolysis. *Int. J. Hydrogen Energ.*, **31** (12), 1632–1640.

23 Rozendal, R., Leone, E., Keller, J., and Rabaey, K. (2009) Efficient hydrogen peroxide generation from organic matter in a bioelectrochemical system. *Electrochem. Commun.*, **11** (9), 1752–1755.

24 Jacobson, K.S., Drew, D.M., and He, Z. (2011) Efficient salt removal in a continuously operated upflow microbial desalination cell with an air cathode. *Bioresour. Technol.*, **102** (1), 376–380.

25 Cao, X., Huang, X., Liang, P., Xiao, K., Zhou, Y., Zhang, X., and Logan, B.E. (2009) A new method for water desalination using microbial desalination cells. *Environ. Sci. Technol.*, **43** (18), 7148–7152.

26 Bard, A.J., Inzelt, G., and Scholz, F. (eds) (2008) *Electrochemical Dictionary*, Springer, Berlin-Heidelberg.

27 Harnisch, F. and Rabaey, K. (2014) Bioelectrochemical Systems, in *Key Materials in Fuel Cells: Volume II Low Temperature Fuel Cells* (ed. B. Ladewig), John Wiley & Sons.

28 Torres, C.I., Krajmalnik-Brown, R., Parameswaran, P., Marcus, A.K., Wagner, G., Gorby, Y.A., and Rittmann, B. (2010) Selecting anode-respiring bacteria based on anode potential: phylogenetic, electrochemcial and microscopic characterization. *Environ. Sci. Technol.*, **2009** (43), 9519–9524.

29 Harnisch, F., Koch, C., Patil, S.A., Hübschmann, T., Müller, S., and Schröder, U. (2011) Revealing the electrochemically driven selection in natural community derived microbial biofilms using flow-cytometry. *Energy Environ. Sci.*, **4** (4), 1265–1267.

30 Rozendal, R., Jeremiasse, A.W., Hamelers, B., and Buisman, C.J.N. (2008) Hydrogen production with a microbial biocathode. *Environ. Sci. Technol.*, **42**, 629–634.

31 Thauer, R.K., Jungermann, K., and Decker, K. (1977) Energy conversion in chemotrophic anaerobic bacteria. *Bacteriol. Rev.*, **41** (1), 100–180.

32 Harnisch, F. and Schröder, U. (2009) Selectivity versus mobility: separation of anode and cathode in microbial bioelectrochemical systems. *ChemSusChem*, **2** (10), 921–926.

33 Potter, M.C. (1931) Measurement of the electricity liberated during the downgrade reactions of organic compounds. *Nature*, **127** (3206), 554–555.

34 Schröder, U. (2007) Anodic electron transfer mechanisms in microbial fuel cells and their energy efficiency. *Phys. Chem. Chem. Phys.*, **9**, 2619–2629.

35 Kim, B.H., Park, D.H., Shin, P.K., Chang, I.S., and Kim, H.J. (1999) Mediator-less biofuel cell. US Patent 5,976,719.

36 Summers, Z.M., Fogarty, H.C., Leang, E., Franks, A.E., Malvankar, N.S., and Lovley, D.R. (2010) Direct exchange of electrons within aggregates of an evolved syntrophic co-culture of anaerobic bacteria. *Science*, **330**, 1413–1415.

37 Morita, M., Malvankar, N.S., Franks, A.E., Summers, Z.M., Giloteaux, L., Rotaru, A.E., Rotaru, C., and Lovley, D.R. (2011) Potential for direct interspecies electron transfer in methanogenic wastewater digester aggregates. *mBio*, **2** (4), e00159–00111.

38 Harris, H.W., El-Naggar, M.Y., Bretschger, O., Ward, M.J., Romine, M.F., Obraztsova, A.Y., and Nealson, K.H. (2010) Electrokinesis is a microbial behavior that requires extracellular electron transport. *PNAS*, **107** (1), 326–331.

39 Malvankar, N.S., Vargas, M., Nevin, K.P., Franks, A.E., Leang, C., Kim, B.-C., Inoue, K., Mester, M., Covalla, S.F., Johnson, J.P., Rotello, V.M., Tuominen, T.M., and Lovley, D.R. (2011) Tunable metallic-like conductivity in microbial nanowire networks. *Nat. Nanotech*. doi: 10.1038/NNANO.2011.119.

40 Lovley, D.R. (2008) The microbe electric: conversion of organic matter to electricity. *Curr. Opin. Biotechnol.*, **19**, 1–8.

41 Carmona-Martinez, A.A., Harnisch, F., Fitzgerald, L.A., Biffinger, J.C., Ringeisen, B.R., and Schröder, U. (2011) Cyclic voltammetric analysis of the electron transfer of Shewanella oneidensis MR-1 and nanofilament and cytochrome knock-out mutants. *Bioelectrochemistry*, **81**, 74–80.

42 Nealson, K.H. and Scott, J. (2006) Ecophysiology of the genus *Shewanella*, in *The Prokaryotes* (eds M. Dworkin, S. Falkow, E. Rosenberg, K.-H. Schleifer, and E. Stackebrandt), Springer, pp. 1133–1151.

43 Shi, L., Richardson, D.J., Wang, Z., Kerisit, S.N., Rosso, K.M., Zachara, J.M., and Fredrickson, J.K. (2009) The roles of outer membrane cytochromes of Shewanella and Geobacter in extracellular electron transfer. *Environ. Microbiol. Rep.*, **1**, 220–227.

44 Gorby, Y.A., Yanina, S., McLean, J.S., Rosso, K.M., Moyles, D., Dohnalkova, A., Beveridge, T.J., Chang, I.S., Kim, B.H., Kim, K.S., Culley, D.E., Reed, S.B., Romine, M.F., Saffarini, D.A., Hill, E.A., Shi, L., Elias, D.A., Kennedy, D.W., Pinchuk, G., Watanabe, K., Ishii, S.i., Logan, B., Nealson, K.H., and Fredrickson, J.K. (2006) Electrically conductive bacterial nanowires produced by Shewanella oneidensis strain MR-1 and other microorganisms. *PNAS*, **103**, 11358–11363.

45 Reguera, G., McCarthy, K.D., Mehta, T., Nicoll, J.S., Tuominen, M.T., and Lovley, D.R. (2005) Extracellular electron transfer via microbial nanowires. *Nature*, **435**, 1098–1101.

46 El-Naggar, M.Y., Wanger, G., Leung, K.M., Yuzvinskya, T.D., Southame, G., Yang, J., Lau, W.M., Nealson, K.H., and Gorby, Y.A.

(2010) Electrical transport along bacterial nanowires from Shewanella oneidensis MR-1. *PNAS*, **107** (42), 18127–18131.

47 Inoue, K., Leang, C., Franks, A.E., Woodard, T.L., Nevin, K.P., and Lovley, D.R. (2011) Specific localization of the c-type cytochrome OmcZ at the anode surface in current-producing biofilms of Geobacter sulfurreducens. *Environ. Microbiol. Rep.*, **3**, 211–217.

48 Nielsen, L.P., Risgaard-Petersen, N., Fossing, H., Christensen, P.B., and Sayama, M. (2010) Electric currents couple spatially separated biogeochemical processes in marine sediment. *Nature*, **463**, 1071–1074.

49 Cao, B., Shi, L., Brown, R.N., Xiong, Y., Fredrickson, J.K., Romine, M.F., Marshall, M.J., Lipton, M.S., and Beyenal, H. (2011) Extracellular polymeric substances from Shewanella sp. HRCR-1 biofilms: characterization by infrared spectroscopy and proteomics. *Environ. Microbiol.*, **13** (4), 1018–1031.

50 Rabaey, K., Boon, N., Verstraete, W., and Höfte, M. (2005) Microbial phenazine production enhances electron transfer in biofuel cells. *Environ. Sci. Technol.*, **39** (9), 3401–3408.

51 Hernandez, M.E., Kappler, A., and Newman, D.K. (2004) Phenazines and other redox-active antibiotics promote microbial mineral reduction. *Appl. Environ. Microbiol.*, **70** (2), 921–928.

52 Rosenbaum, M., Zhao, F., Schröder, U., and Scholz, F. (2006) Interfacing electrocatalysis and biocatalysis using tungsten carbide: a high performance noble-metal-free microbial fuel cell. *Angew. Chem. Int. Ed.*, **45**, 6658–6661.

53 Nießen, J., Harnisch, F., Rosenbaum, M., Schröder, U., and Scholz, F. (2006) Heat treated soil as convenient and versatile source of bacterial communities for microbial electricity generation. *Electrochem. Commun.*, **8**, 869–873.

54 Nießen, J., Schröder, U., and Scholz, F. (2004) Exploiting complex carbohydrates for microbial electricity generation - a bacterial fuel cell operating on starch. *Electrochem. Commun.*, **6**, 955–958.

55 Madigan, M.T., Martink, J.M., and Parker, J. (1999) *Brock Biology of Microorganisms*, Prentice Hall International, Inc.

56 Rabaey, K., van den Sompel, K., Maignien, L., Boon, N., Aelterman, P., Clauwaert, P., de Schamphelaire, L., Pham, H.T., Vermuelen, J., Verhaege, M., Lens, P., and Verstraete, W. (2006) Microbial fuel cells for sulfide removal. *Environ. Sci. Technol.*, **40**, 5218–5224.

57 Reimers, C.E., Tender, L.M., Fertig, S., and Wang, W. (2001) Harvesting energy from the marine sediment - Water interface. *Environ. Sci. Technol.*, **35** (1), 192–195.

58 Harnisch, F., Schröder, U., Quaas, M., and Scholz, F. (2008) Electrocatalytic and corrosion behaviour of tungsten carbide in ph neutral electrolytes. *Appl. Catal. B: Environ.*, **87**, 63–69.

59 Zhao, F., Rahunen, N., Varcoe, J.R., Chandra, A., Avignone-Rossa, C., Thumser, A.E., and Slade, R.C.T. (2008) Activated carbon cloth as anode for sulfate removal in a microbial fuel cell. *Environ. Sci. Technol.*, **42** (13), 4971–4976.

60 Aulenta, F., Canosa, A., De Roma, L., Reale, P., Panero, S., Rossetti, S., and Majone, M. (2009) Influence of mediator immobilization on the electrochemically assisted microbial dechlorination of trichloroethene (TCE) and cis-dichloroethene (cis-DCE). *J. Chem. Technol. Biot.*, **84**, 864–870.

61 Aulenta, F., Catervi, A., Majone, M., Panero, S., Reale, P., and Rossetti, S. (2007) Electron transfer from a solid-state electrode assisted by Methyl viologen sustains efficient microbial reductive dechlorination of TCE. *Environ. Sci. Technol.*, **41** (7), 2554–2559.

62 Logan, B.E., Call, D., Cheng, S., Hamelers, H.V.M., Sleutels, T.H.J.A., Jeremiasse, A.W., and Rozendal, R.A. (2008) Microbial electrolysis cells for high yield hydrogen gas production from organic matter. *Environ. Sci. Technol.*, 428630–8640.

63 Wagner, R.C., Regan, J.M., Oh, S.E., Zuo, Y., and Logan, B.E. (2009) Hydrogen and methane production from swine

wastewater using microbial electrolysis cells. *Water Res.*, **43**, 1480–1488.

64 Rosenbaum, M., Aulenta, F., Villano, M., and Angenent, L.T. (2011) Cathodes as electron donors for microbial metabolism: which extracellular electron transfer mechanisms are involved? *Bioresour. Technol.*, **102** (1), 324–333.

65 Flemming, H.-C. and Wingender, J. (2010) The biofilm matrix. *Nat. Rev. Microbiol.*, **8**, 623–633.

66 Venkataraman, A., Rosenbaum, M., Arends, J.B.A., Halitschke, R., and Angenent, L.T. (2010) Quorum sensing regulates electric current generation of Pseudomonas aeruginosa PA14 in bioelectrochemical systems. *Electrochem. Commun.*, **12** (3), 459–462.

67 Pham, T.H., Boon, N., Aelterman, P., Clauwaert, P., De Schamphelaire, L., Vanhaecke, L., De Maeyer, K., Höfte, M., Verstraete, W., and Rabaey, K. (2008) Metabolites produced by *Pseudomonas* sp. enable gram positive bacterium to achieve extracellular electron transfer. *Appl. Microbiol. Biotechnol.*, **77**, 1119–1129.

68 Rosenbaum, M., Bar, H.Y., Beg, Q.K., Booth, J., Segre, D., Cotta, M.A., and Angenent, L.T. (2011) Synergistic conversion of glucose into electricity with a co-culture of Shewanella oneidensis and Lactococcus lactis. *Bioresour. Technol.*, **102** (3), 2623–2628.

69 Venkataraman, A., Rosenbaum, M.A., Perkins, S.D., Werner, J.J., and Angenent, L.T. (2011) Metabolite-based mutualism between Pseudomonas aeruginosa PA14 and Enterobacter aerogenes enhances current generation in bioelectrochemical systems. *Energy Environ. Sci.*, **4** (11), 4550–4559.

70 Liu, Y., Harnisch, F., Fricke, K., Sietmann, R., and Schröder, U. (2008) Improvement of the anodic bioelectrocatalytic activity of mixed culture biofilms by a simple consecutive electrochemical selection procedure. *Biosens. Bioelectron.*, **24**, 1012–1017.

71 Patil, S.A., Harnisch, F., Koch, C., Hübschmann, T., Fetzer, I., Carmona-Martinez, A.-A., Müller, S., and Schröder, U. (2011) Electroactive mixed culture derived biofilms in microbial bioelectrochemical systems: the role of pH on biofilm formation, performance and composition. *Bioresour. Technol.*, published online.

72 Patil, S.A., Harnisch, F., Kapadnis, B., and Schröder, U. (2010) The role of temperature on the formation and performance of bioelectrocatalytic active mixed culture biofilms for microbial bioelectrochemical systems. *Biosens. Bioelectron.*, online.

73 Harnisch, F. and Rabaey, K. (2012) The diversity of techniques to study electrochemically active biofilms highlights the need for standardization. *ChemSusChem*, **5**, 1027–1038.

74 Harnisch, F. and Freguia, S. (2012) A basic tutorial on the use of cyclic voltammetry for the study of electroactive microbial biofilms. *Chem. Asian. J.*, **7**, 466–475.

75 Strycharz-Glaven, S.M. and Tender, L.M. (2012) Study of the mechanism of catalytic activity of G. sulfurreducens biofilm anodes during biofilm growth. *ChemSusChem*, **5**, 1106–1118.

76 He, Z. and Mansfeld, F. (2009) Exploring the use of electrochemical impedance spectroscopy in microbial fuel cell studies. *Energy Environ. Sci.*, **2**, 215–219.

77 Esteve-Nunez, A., Sosnik, J., Visconti, P., and Lovley, D.R. (2008) Fluorescent properties of *c*-type cytochromes reveal their potential role as an extracytoplasmic electron sink in *Geobacter sulfurreducens*. *Environ. Microbiol.*, **10** (2), 497–505.

78 Liu, Y., Kim, H., Franklin, R.R., and Bond, D.R. (2011) Linking spectral and electrochemical analysis to monitor c-type cytochrome redox status in living geobacter sulfurreducens biofilm. *ChemPhysChem*, **12** (12), 2235–2241.

79 Shibanuma, T., Nakamura, R., Hirakawa, Y., Hashimoto, K., and Ishii, K. (2011) Observation of *in vivo* cytochrome-based electron-transport dynamics using time-resolved evanescent wave electroabsorption spectroscopy. *Angew. Chem.*, **123**, 9303–9306.

80 Busalmen, J.P., Esteve-Nunez, A., Berná, A., and Feliu, J.M. (2008) C-type cytochromes wire electricity-producing

bacteria to electrodes. *Angew. Chem.*, **47**, 4874–4877.

81 Millo, D., Harnisch, F., Patil, S.A., Ly, K. H., Schröder, U., and Hildebrandt, P. (2011) In situ Spectroelectrochemcial Investigation of electrocatalytic microbial biofilms by surface-enhanced resonance raman spectroscopy. *Angew. Chem. Int. Ed.*, **50**, 2625–2627.

82 Franks, A.E., Nevin, K.P., Jia, H., Izallalen, M., Woodward, T.L., and Lovley, D.R. (2009) Novel strategy for three-dimensional real-time imaging of microbial fuel cell communities: monitoring the inhibitory effects of proton accumulation within the anode biofilm. *Energy Environ. Sci.*, **2**, 113–119.

83 Read, S.T., Dutta, P., Bond, P.L., Keller, J., and Rabaey, K. (2010) Initial development and structure of biofilms on microbial fuel cell anodes. *BMC Microbiol.*, **10**, 98.

84 Burow, L.C., Mabbett, A.N., Borras, L., and Blackall, L.L. (2009) Anaerobic central metabolic pathways active during polyhydroxyalkanoate production in uncultured cluster 1 Defluviicoccus enriched inactivated sludge communities. *FEMS Microbiol. Lett.*, **298**, 79–84.

85 Dufrene, Y.F. (2008) Towards nanomicrobiology using atomic force microscopy. *Nat. Rev. Microbiol.*, **6** (9), 674–680.

86 Lower, B.H. et al. (2009) Antiobody recognition force microscopy shows the outer membrane cytochromes omcA and MtrC are expressed on the exterior surface of Shewanella oneidensis MR-1. *Appl. Environ. Microbiol.*, **75** (9), 2931–2935.

87 Virdis, B., Harnisch, F., Batstone, D., Rabaey, K., and Donose, B. (2012) Non-invasive characterization of electrochemically active microbial biofilms using confocal Raman microscopy. *Energy Environ. Sci.*, **5**, 5231–5235.

88 Koley, D., Ramsey, M.M., Bard, A.J., and Whiteley, M. (2012) Discovery of a biofilm electrocline using real-time 3D metabolite analysis. *Proc. Natl. Acad. Sci.*, **108**, 19996–20001.

89 Finzi-Hart, J.A., Pett-Ridge, J., Weber, P. K., Popa, R., Fallon, S.J., Gunderson, T., Hutcheon, I.D., Nealson, K.H., and Capone, D.G. (2009) Fixation and fate of C and N in the cyanobacterium Trichodesmium using nanometer-scale secondary ion mass spectrometry. *Proc. Natl. Acad. Sci. USA*, **106**, 6345–6350.

90 Behrens, S., Losekann, T., Pett-Ridge, J., Weber, P.K., Ng, W.O., Stevenson, B.S., Hutcheon, I.D., Relman, D.A., and Spormann, A.M. (2008) Linking microbial phylogeny to metabolic activity at the single-cell level by using enhanced element labeling-catalyzed reporter deposition fluorescence *in situ* hybridization (EL-FISH) and NanoSIMS. *Appl. Environ. Microbiol.*, **74**, 3143–3150.

91 Rabaey, K., Boon, N., Siciliano, S.D., Verhaege, M., and Verstraete, W. (2004) Biofuel cells select for microbial consortia that self-mediate electron transfer. *Appl. Environ. Microbiol.*, **70** (9), 5373–5382.

92 Wrighton, K.C., Virdis, B., Clauwaert, P., Read, S.T., Daly, R.A., Boon, N., Piceno, Y., Andersen, G.L., Coates, J.D., and Rabaey, K. (2010) Bacterial community structure corresponds to performance during cathodic nitrate reduction. *ISME J.*, **4**, 1443–1455.

93 Holmes, D.E., Chaudhuri, S.K., Nevin, K. P., Mehta, T., Methe, B.A., Liu, A., Ward, J.E., Woodard, T.L., Webster, J., and Lovley, D.R. (2006) Microarray and genetic analysis of electron transfer electrodes in *Geobacter sulfurreducens*. *Environ. Microb.*, **8** (10), 1805–1815.

94 Chen, Y. and Murrell, J.C. (2010) When metagenomics meets stable-isotope probing: progress and perspectives. *Trends Microbiol.*, **18** 157–163.

95 Whiteley, A.S., Thomson, B., Lueders, T., and Manefield, M. (2007) RNA stable-isotope probing. *Nat. Protoc.*, **2**, 838–844.

96 Jehmlich, N., Schmidt, F., Taubert, M., Seifert, J., Bastida, F., von Bergen, M., Richnow, H.H., and Vogt, C. (2010) Protein-based stable isotope probing. *Nat. Protoc.*, **5**, 1957–1966.

97 Srikanth, S., Marsili, E., Flickinger, M.C., and Bond, D.R. (2008) Electrochemical characterization of *Geobacter sulfurreducens* cells immobilized on graphite paper electrodes. *Biotechnol. Bioeng.*, **99** (5), 1065–1073.

98 Torres, C.I., Marcus, A.K., and Rittmann, B.E. (2008) Proton transport inside the biofilm limits electrical current generation by anode-respiring bacteria. *Biotechnol. Bioeng.*, **100** (5), 872–881.

99 Torres, C.I., Marcus, A.K., Lee, H.-S., Parameswaran, P., Krajmalnik-Brown, R., and Rittmann, B.E. (2010) A kinetic perspective on extracellular electron transfer by anode-respiring bacteria. *FEMS Microbiol. Rev.*, **34**, 3–17.

100 Strycharz, S.M., Malanoski, A.P., Snider, R.M., Yi, H., Lovley, D.R., and Tender, L.M. (2011) Application of cyclic voltammetry to investigate enhanced catalytic current generation by biofilm-modified anodes of Geobacter sulfurreducens strain DL1 vs. variant strain KN400. *Energy Environ. Sci.*, **4**, 896–913.

101 Richter, H., Nevin, K.P., Jia, H., Lowy, D.A., Lovley, D.R., and Tender, L.M. (2009) Cyclic voltammetry of biofilms of wild type and mutant Geobacter sulfurreducens on fuel cell anodes indicates possible roles of OmcB, OmcZ, type IV pili, and protons in extracellular electron transfer. *Energy Environ. Sci.*, **2**, 506–516.

102 Fricke, K., Harnisch, F., and Schröder, U. (2008) On the use of cyclic voltammetry for the study of anodic electron transfer in microbial fuel cells. *Energy Environ. Sci.*, **1** (1), 144–147.

103 Franks, A.E., Nevin, K.P., Glaven, R.H., and Lovley, D.L. (2010) Microtoming coupled to microarray analysis to evaluate the spatial metabolic status of Geobacter sulfurreducens biofilms. *ISME J.*, **4**, 509–519.

104 Kniggendorf, A.K., Gaul, T.W., and Meinhardt-Wollweber, M. (2011) Hierarchical cluster analysis (HCA) of microorganisms: an assessment of algorithms for resonance raman spectra. *Appl. Spectrosc.*, **65**, 165–173.

5
Electrocapillarity of Solids and its Impact on Heterogeneous Catalysis[1]
Jörg Weissmüller

5.1
Introduction

This chapter begins by introducing selected aspects of the theory of solid mechanics into a phenomenological description of the electrode/electrolyte system. Solid mechanics analyzes the stress and the elastic deformation within a solid body at equilibrium, subject to the action of external forces. At first sight, this description offers no relation to chemical reaction kinetics. Why, then, introduce the subject of mechanics in a treatise on catalysis? The answer to that question resides in the link between the mechanical deformation of solid substrates – as parametrized by their "strain"– and the energetics of adsorption. The reactions of heterogeneous catalysis take place in an ensemble of adsorbed species. Adsorption enthalpies govern the population of these species and they are inherently linked to the energies of the transition states. Recent research has provided compelling evidence that even small elastic strain of the substrate may strongly impact the adsorption enthalpy. This opens the perspective of tuning the elastic deformation – in other words, tuning the lattice parameter – of catalysts to enhance their reactivity and/or to gain insight into the microscopic details of the reaction.

The phenomenological description of the coupling between strain and reactivity rests on the theory of the capillarity of solid surfaces at equilibrium. Its basis is a surface thermodynamic potential which embodies the combined effect of mechanical deformation and the chemical or electrochemical adsorption on the energetics of the surface. The phenomenological approach identifies the relevant state variables, yields equilibrium conditions, identifies the central materials parameters and interconnects them through Maxwell relations. In this way, the theory of capillarity of solid surfaces is the key element in the science of electro-chemo-mechanical

---

1) This chapter is dedicated to the memory of Dieter M. Kolb.

Electrocatalysis: Theoretical Foundations and Model Experiments, First Edition.
Edited by Richard C. Alkire, Ludwig A. Kibler, Dieter M. Kolb, and Jacek Lipkowski.
© 2013 Wiley-VCH Verlag GmbH & Co. KGaA. Published 2013 by Wiley-VCH Verlag GmbH & Co. KGaA.

coupling that underlies strain effects in catalysis. By identifying the state variables and materials parameters, it provides rigorous definitions and, thereby, a basis for quantifying effects and a language for communicating observations. These are prerequisites for a scientific discussion that affords comparability between experiment, theory and numerical modeling. Maxwell relations promote scientific insight by interlinking seemingly unrelated observations and, thereby, they suggest unconventional experimental and theoretical approaches to the characterization of the impact of mechanics on reactivity.

Once the state variables and materials parameters for strained (electrode-) surfaces at equilibrium have been identified, theory can be taken one step further into the realm of nonequilibrium processes, such as catalytic reactions. Models for the reaction kinetics will explore the impact of the electro-chemo-mechanical coupling terms on the catalytic reactivity, linking the reaction rate to the electrocapillary coupling coefficients that can be independently measured in studies of surface behavior near equilibrium.

This chapter aims to describe the fundamentals of the theory of electrocapillarity of solids with an eye on their application in the field of strain-dependent catalysis. Since readers may be more familiar with the thermodynamics of fluids than with that of solid surfaces, we begin with a detailed discussion of the distinguishing features of the latter.

5.2
Mechanics of Solid Electrodes

5.2.1
Outline – Surface Stress and Surface Tension

The distinction between capillary phenomena in solids and fluids was first pointed out by Gibbs, who advises his readers that a theory of solid surfaces requires the distinction between two fundamentally different work terms, the work spent in *forming* a new surface by adding atoms and that spent in *stretching* an existing surface by elastic strain (see Ref. [1] and discussion in Ref. [2]). Gibbs points out that the two work terms must not be taken as equivalent when "states of strain are to be distinguished" [1]. Studying that distinction means connecting thermodynamics to solid mechanics, an issue which Gibbs left for others to explore during the twentieth century. Among the results of this more recent work is the advent of a theory of the capillarity of solid surfaces. This theory, which originates in a seminal paper by Shuttleworth [3] (closely followed by a similar study by Herring [4]), inspects the difference between the two work terms in detail, confirming Gibbs' notion that they are fundamentally different. The consequence is a need for introducing suitable measures for elastic strain and stress as additional, energy-conjugate variables in the fundamental equation for the surface tension, γ, of a solid. In the most general case [5,6], the strain term is a projection, $\mathbb{E}$, of the bulk elastic strain tensor into the tangent plane of the surface, and the stress term is a tangential stress, $\mathbb{S}$, that is designated the surface stress.[2]) The Gibbs adsorption equation for the total differential in γ then needs to be supplemented by an elastic work term [13], in the present notation $d\gamma = \mathbb{S}d\mathbb{E}$, on top of the conventional terms.

Specifically, the relevant forms of the adsorption equation for polarizable electrode surfaces are

$$d\gamma = -\sigma dT - \Gamma_i d\mu_i - qdE \quad \text{(fluid)} \tag{5.1}$$

and

$$d\gamma = -\sigma dT - \Gamma_i d\mu_i - qdE + \mathbb{S}d\mathbb{E} \quad \text{(solid)}. \tag{5.2}$$

The variables have the following meaning: σ – superficial excess entropy, T – temperature, Γ_i – superficial excess in component i, μ_i – chemical potential, q – superficial charge density, E – electrode potential.

The introduction of the tangential elastic strain $\mathbb{E}$ in Eq. (5.2) allows the theory to represent the fundamental issue under discussion, namely the distinction between two modes of changing the physical area of a surface: (i) an elastic strain, $\delta\mathbb{E}$, of the surface at constant number of atoms ("stretching" in Gibbs terms), and (ii) a change, δA, in surface area A by adding atoms ("forming") at constant elastic strain $\mathbb{E}$. As will be discussed below in more detail, the relevant changes in the net free excess free energy associated with the surface of the solid are $\delta G^{excess} = A\mathbb{S}\delta\mathbb{E}$ for stretching the existing surface and $\delta G^{excess} = \gamma \delta A$ for forming a new surface.

The adsorption equation, Eq. (5.2), introduces surface stress as a phenomenological concept. Experiment and quantum-mechanical computation tell us that, in fact, the surface stress at a clean surface under standard conditions will take on a significant numerical value. Figure 5.1 gives a schematic illustration of the origin of that stress. Creating a surface, for instance by truncation of a bulk crystal as indicated, creates dangling interatomic bonds which relax to form strengthened bonds in the surface plane. As a consequence, the surface atoms will typically favor a

---

2) Restricting the surface stress to be tangential is equivalent to saying that the surface free energy functions, such as γ, of a solid cannot *explicitly* depend on the bulk strain component which is directed along the surface normal. This is indeed a forceful result of the mechanics of surfaces [5,6]. However, if the surface is in contact with a fluid at pressure P, then the surface excess free energy, ψ (see below), varies along with the surface excess volume, $\mathbb{V}$, according to $d\psi = -Pd\mathbb{V}$. The normal component, $\mathbf{Sn}$, of the bulk stress, $\mathbf{S}$, in the solid is coupled to the fluid pressure via $\mathbf{Sn} = -P\mathbf{n} + \text{div}_S\mathbb{S}$ where div_S denotes the surface divergence operator and $\mathbf{n}$ the surface normal vector [5]. Since Hookes law couples normal stress to normal strain, the surface tension variation is thus *implicitly* coupled to the normal strain in the crystal lattice. From the point of view of mechanics, changes in $\mathbb{V}$ relate to an excess (or jump) in the normal displacement at the surface. In an atomistic picture, variations in $\mathbb{V}$ at solid surfaces embody the ubiquitous phenomenon of surface relaxation [7], a local deviation of the distance between the outermost planes of the crystal lattice from the bulk interplanar spacing. The relaxation is coupled to the state variables of the surface, for instance, the superficial excess of adsorbate [8] or the superficial charge density [9,10]. The relevance of local deformation along the surface normal for the surface energetics was pointed out in an important analysis by Noziéres and Wolf [11] and later discussed in a more general context by Gurtin et al. [6] and by Weissmüller and Kramer [12]. In the interest of conciseness, we ignore surface relaxation throughout the present work.

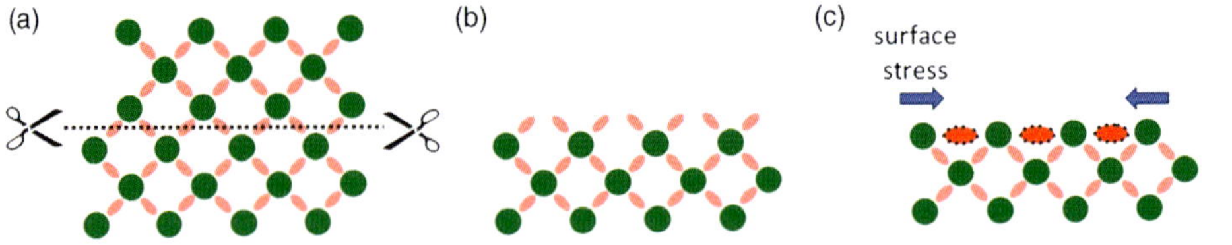

Figure 5.1 Microscopic origin of surface stress. (a) Illustrates cross-section through a bulk crystal with atoms (green circles) and interatomic bonds (pink ovals). Truncating the crystal as indicated creates a surface. The unrelaxed, as-truncated surface (b) exhibits dangling bonds that relax (c) to form new bonds in the surface plane, strengthening the in-plane bonding. As a consequence, the surface atoms favor a closer interatomic spacing than the bulk. This results in a stress on the bulk crystal, originating from the surface atoms. The surface stress quantifies that capillary force.

closer interatomic spacing than the atoms in the bulk of the underlying crystal. This reasoning is in line with the empirical trend of atomic bonds to be shortened at lesser coordination number (or bond order). Since the crystal lattice prevents the surface atoms from approaching each other, the strengthened in-plane bonding results in a stress on the bulk crystal. The stress originating from the surface atoms represents a capillary force, which is quantified by the surface stress. The tendency for *strengthened* bonding at clean metal surfaces is reflected by the general trend of *positive* surface stress of metals, which tends to *compress* the underlying lattice tangentially. Yet, negative surface stress values, expanding the underlying lattice, are thermodynamically allowed and are indeed observed at metal alloy surfaces [8,14].

Practically all modern review papers in the field embrace Shuttleworth's analysis and its consequences for the theory of capillarity of solid surfaces [15–20]. However, the agreement is not unanimous, and several authors continue to bring forward objections to the use of the concept of surface stress. The literature also testifies to the fact that it is not always obvious which of the two capillary parameters is relevant for a particular phenomenon under investigation.[3] Reference [20] presents a critical assessment of the points of debate, along with a summary of the experimental observations that request the separate introduction of surface tension – measuring the work spent in forming a new surface – and surface stress – measuring the work spent in stretching the surface.

This section discusses the different roles of surface tension and surface stress for equilibria at electrode surfaces.

---

3) Terminology can be an origin of confusion, and this applies here in the author's native tongue. Where, in English, one distinguishes the stress from the tension of a surface, the German language offers the same translation, "Spannung", for stress as well as tension. In fact, no translation of "surface stress" appears to have been established in the German literature. The author uses "elastische Flächenspannung" [21,22], emphasizing that the processes associated with the action of the surface stress involve small atomic rearrangements in the elastic regime, whereas no such restriction applies to processes which do work against the surface tension ("Oberflächenspannung").

5.2.2
Solid Versus Fluid

The state of a body of condensed matter depends on the mechanical forces acting on its surface. In *fluids*, the effect of an external load is a pressure in the bulk of the fluid. Thermodynamic potentials of fluids, such as their Gibbs free energy, therefore depend on the state variable, the pressure, or on the energy-conjugate variables volume or (atomic) density. In this chapter we shall inspect *solids*, with an eye on the impact of mechanics for the electrochemical and catalytic phenomena.

A solid is distinguished from a fluid by its ability to support shear stress. The immediate consequence is that the states of stress and strain require a more complex description than in fluids. Rather than working with the scalar variables pressure and volume (or density), we here consider the thermodynamic potentials to depend on stress and strain variables that take the form of second rank tensors, allowing for locally anisotropic deformation. The substitution of stress and strain for the more familiar state variables pressure and density has drastic consequences for the behavior of condensed matter, including the phenomena linked to the presence of a surface. Thus, the capillarity of solids differs qualitatively from that of fluids. This difference feeds specifically into the behavior of electrodes.

Body forces, such as gravity, can typically be ignored in the thermodynamics of heterogeneous equilibria. In fluids, the pressure is then a constant within each phase. By contrast, the stress in solids is generally nonuniform. This results from the presence of shear stress. A central task of mechanics is to analyze the spatial variation of stress and strain at equilibrium. In doing so, mechanics strives for local details that are excluded from the more familiar description of Gibbsian thermodynamics, which works with intrinsic state variables that are uniform in each phase and which focusses on the total values of the extrinsic state variables, ignoring their local details. The theory of capillarity of solid surfaces is, therefore, adequately derived from a description of the energetics of the solid in terms of free energy densities for bulk and surface, which depend on the position and which, thereby, can represent local variations of stress and strain.

5.2.3
Free Energy of Elastic Solid Surfaces

Here we briefly introduce a thermodynamic description of electrochemical surface mechanics. More details can be found in Refs. [12,20].

Consider a polarizable solid metallic electrode B in contact with a fluid electrolyte F. The solid acts as an elastic substrate that conducts charge via electronic transport and that does not exchange matter with the electrolyte. The energetics of the bulk phases are represented by free energy densities (per volume), Ψ^B for the electrode and Ψ^F for the electrolyte, and by the superficial free energy density, ψ (per area of the electrode surface, S). The free energy densities are defined so that the net free energy, $\mathfrak{F}$, obeys

$$\mathfrak{F} = \int_B \Psi^B dV + \int_F \Psi^F dV + \int_S \psi dA \tag{5.3}$$

where V and A denote volume and surface area, respectively.

"Lagrangian" coordinates are used throughout this work. As will be explained in more detail in Section 5.2.4, this convention simply means that V and A are measured not in the actual, strained state of the solid but in a strain-free "reference" configuration (see below). Therefore, A does not change during a purely elastic deformation. Furthermore, densities such as ψ or the superficial charge density, q, for any state of the surface are defined as the energy or charge in the actual state, per area of the surface in its undeformed state.

As the constitutive assumption for the (Helmholtz-type) free energy density function of the electrode we take $\Psi^B = \Psi^B(T, \mathbf{E})$ with $\mathbf{E}$ the strain tensor. With attention to processes at constant temperature we omit explicit display of the state variable T throughout the remaining text. The fundamental equation for Ψ is then

$$d\Psi^B = \mathbf{S} d\mathbf{E} \tag{5.4}$$

with $\mathbf{S}$ the stress tensor.

The constitutive assumption for the electrolyte is $\Psi^F = \Psi^F(T, \rho_i)$ with ρ_i the densities of the components i, again omitting the further display of T. The fundamental equation is here

$$d\Psi^F = \mu_i d\rho_i \tag{5.5}$$

where μ_i is the chemical potential of component i and the sum convention (summation over indices occurring twice) applies.

The central step in a thermodynamic analysis of the surface is to identify the appropriate state variables. The most general approach is to let the superficial free energy density ψ depend on the state of the abutting bulk phases through the limiting values (upon approach to the surface) of their state variables, the strain $\mathbf{E}$ in the solid and the densities ρ_i in the fluid. Furthermore, ψ is allowed to depend on additional composition parameters which are only relevant at the surface, the superficial excess, Γ_i, of the species in the electrolyte. The Γ_i characterize the adsorbate population and, thereby, the charge density q.

For surfaces with a sufficiently high symmetry, the acting capillary forces are isotropic in the plane [3]. The work done against the capillary forces by any arbitrary deformation will then depend on the strain only through the change in surface area, and will be independent of the shear. Therefore, a suitable strain variable, e, for use with isotropic surfaces is the relative change in physical area, $\tilde{A}$, during the deformation:

$$\delta e = \delta \tilde{A}/\tilde{A} \tag{5.6}$$

(see Section 5.2.4 for more details). Specifying absolute values of the strain requires that one particular state of the surface is chosen as the reference state, that is, the state for which $e = 0$. That choice is arbitrary here, as long as the reference state is well defined.

For fluids sufficiently far from a critical point it is typically a good assumption to ignore the dependence of the bulk free energy of the fluid on gradients in the densities ρ_i. One can then show [12] that $\partial \psi/\partial \rho_i = 0$, in other words, the surface free energy cannot explicitly depend on the composition of the fluid. As notable counterexamples, where gradient energy terms [23] do control the interfacial free energy, we mention spinodal decomposition [24] and critical point wetting [25]. These phenomena are of no interest in the present context.

The Γ_i are defined as excess (per area) of the actual amount of species i in the system over the amount in the same volume of fluid if there was no surface. In other words, the net amount of component i in the system is

$$N_i = \rho_i V^F + \int_S \Gamma_i dA \tag{5.7}$$

with V^F the volume of the fluid. By definition, the Γ_i include adsorbed species on the surface as well as the extra ions in the diffuse space-charge layer. The laws of electrostatics prohibit that surface regions carry a net charge. Restricting attention to net charge-neutral surfaces, one finds that q and the Γ_i are linked everywhere on the surface via

$$0 = q + z_i F \Gamma_i \tag{5.8}$$

with z_i the valency of the ions in solution and F the Faraday constant.

The considerations so far suggest a general constitutive equation for the surface in the form

$$\psi = \hat{\psi}(\Gamma_i, e) \tag{5.9}$$

A variational approach to energy minimization can be based on Eq. (5.3) with the superficial free-energy density of Eq. (5.9), postulating for the open system (exchange of charge, Q, and of components, amount N_i, with the environment) at equilibrium that

$$\delta \mathfrak{F} = E \delta Q + \mu_i \delta N_i \tag{5.10}$$

where E denotes the electrode potential. Among the results of the variational analysis [12] is the equilibrium condition for electrosorption (that is, adsorption on the surface of an electrode),

$$\frac{d\psi}{d\Gamma_i}\bigg|_{e,\Gamma_{j\neq i}} = \mu_i - z_i FE \tag{5.11}$$

If the μ_i are all held constant, as we shall assume throughout this work unless otherwise stated, then q as well as the Γ_i at equilibrium for any given state of strain are monotonic functions of E. It is then sufficient to take E as the state variable, rather than considering the Γ_i separately. Furthermore, the energy-conjugate variable to E is simply q. One can, therefore, in particular take $\psi = \psi(q, e)$ with the fundamental equation

$$d\psi = E dq + f de \tag{5.12}$$

where the energy-conjugate variable to e is the scalar surface stress, f.

The state function $\psi(q,e)$ is related to the surface tension, γ, by a Legendre transform:

$$\gamma(E,e) = \psi(q,e) - qE \tag{5.13}$$

and the fundamental equation for γ is therefore

$$d\gamma = -q\,dE + f\,de \tag{5.14}$$

Equation (5.14) is the variant of the Gibbs adsorption equation – see Eq. (5.2) for a more general form – that is relevant for the simplifying assumptions considered in this work: processes at constant temperature and constant chemical potentials of the adsorbing species, and surfaces which are sufficiently symmetric for the surface stress to be isotropic in the surface plane.

A consequence of the above considerations is that elastic deformation at constant area changes the net surface free energy of a uniform section of surface by $A\,d\psi = Af\,de$. By contrast, if the referential area of the section of surface is changed by varying its number of surface atoms at constant strain, then the change in free energy is ψdA or γdA, depending on whether q or E are held constant. The distinction between those variations and the associated work terms is the source of confusion in parts of the literature, and we now inspect and explain the issue.

5.2.4
Deforming a Solid Surface

As an illustration of surface deformation, the top part of Figure 5.2 shows a thin, planar slab of solid in cross-section. Let us call this particular state of the solid the "undeformed" state, and let us further assume that forces are applied to the edges of the slab, acting to extend the solid in the plane.

The two side-by-side images in the centre of Figure 5.2 illustrate two states of deformation which may result from the loading. In both states, the physical area, $\tilde{A}$, of each of the two opposing surfaces of the slab has increased by the same amount, $\delta\tilde{A} = \varepsilon\tilde{A}$ with ε a measure for the amount of area-change. Yet, the two modes of deformation result in two distinctly different states of the surface. In the left-hand part of Figure 5.2, atoms have diffused from the interior of the slab towards the surface and have attached there. In this way, the slab has deformed so that its surface area is increased through the attachment of extra atoms. The atomic structure of the surface, as characterized by the crystallography and the interatomic spacing, has remained identical. By contrast, the deformation in the right-hand part of Figure 5.2 leaves the number of atoms in the surface unchanged. Here, the increase in area is achieved by an elastic strain, which increases the interatomic spacing. The two modes of deformation exemplify the two distinct ways of changing the surface area that Gibbs refers to as "forming" and "stretching".

Since the two processes illustrated in Figure 5.2 – namely changing the number of surface atoms at constant structure versus changing the interatomic spacing at constant number of atoms – result in different states of the surface, it must be

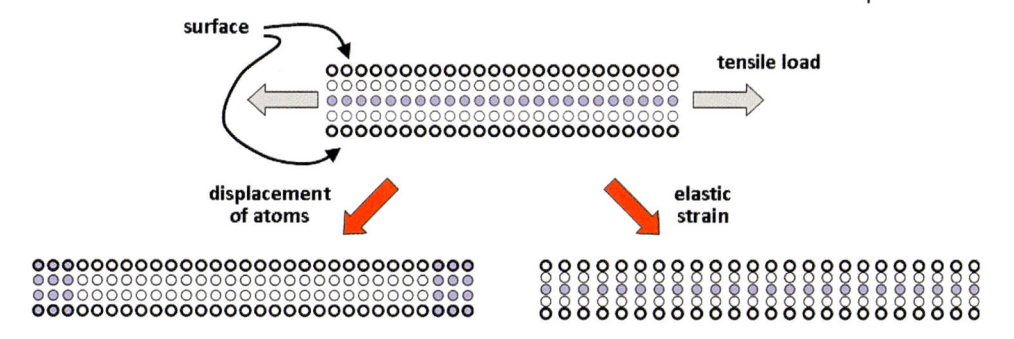

5.2 Mechanics of Solid Electrodes | 171

Figure 5.2 Two non-equivalent ways of increasing the physical area, $\tilde{A}$, of a crystal surface by the identical relative amount, ϵ. The graphs illustrate the processes, The table shows values of selected parameters. Circles represent atoms, bold open circles denoting surface atoms. Left side of graphics shows process where atoms (tagged blue) from the bulk have been rearranged by diffusion so as to enhance the number of surface atoms. This process enhances both the physical ($\tilde{A}$) and the Lagrangian (A) surface area. The number of electrode surface atoms thus increases from N_0 to $N_0(1+\epsilon)$. The process requires no elastic strain and leaves the interatomic spacing in the solid constant. The variation in excess Gibbs free energy associated with the electrode surface scales with γ, the surface tension, $\delta G = \gamma A_0 \epsilon$. By contrast, elastic strain of the crystal (right side of graphics) enhances $\tilde{A}$ by increasing the in-plane interatomic spacing from r_0^{NN} to $r_0^{NN}(1+\epsilon/2)$, while leaving A and the number of surface atoms constant. The variation in excess Gibbs free energy here scales with f, the surface stress, as $\delta G = f A_0 \epsilon$.

	SURFACE:	
$\tilde{A} = (1+\epsilon) A_0$	physical area	$\tilde{A} = (1+\epsilon) A_0$
$N = (1+\epsilon) N_0$	number of surface atoms	$N = N_0$
$A = (1+\epsilon) A_0$	Lagrangian area	$A = A_0$
$e = 0$	elastic strain	$e = \epsilon$
$r^{NN} = r_0^{NN}$	interatomic distance (in-plane)	$r^{NN} = (1+\tfrac{1}{2}\epsilon) r_0^{NN}$
$\gamma = \gamma_0$	surface tension	$\gamma = \gamma_0 + f\epsilon$
$\delta G = \gamma \, \delta A = \gamma A_0 \epsilon$	variation in excess free energy	$\delta G = A_0 \delta\gamma = f A_0 \epsilon$

admitted that the respective changes, δE^{excess}, in total surface excess energy may be different. As already mentioned in Section 5.2.1, embodying this difference in the thermodynamic potentials of the surface requires the introduction of an appropriate set of state variables. In the notation of Shuttleworth, these variables are the physical area $\tilde{A}$ of the surface – as measured in the laboratory and not Lagrangian coordinates – and the elastic strain, e. A drawback of that notation is that a purely elastic strain changes both parameters $\tilde{A}$ and e. Typically it is more convenient to choose truly independent state variables. This is achieved by using the above-mentioned Lagrangian coordinates, where the area is measured in the undeformed (reference) state. We here adhere to that notation, which was introduced into the field by Cahn [26]. Here, the surface states of Figure 5.2 are characterized by the pair of state variables $\{A = A_0, e = 0\}$ in the initial state,

$\{A = (1+\varepsilon)A_0, e = 0\}$ in the 'plastically' deformed state (left-hand side of Figure 5.2), and $\{A = A_0, e = \varepsilon\}$ in the elastically deformed state (right-hand side of Figure 5.2). In other words, creating extra surface area by adding atoms at constant structure changes the Lagrangian area A while leaving the elastic strain e a constant, whereas purely elastic straining leaves A invariant while changing e.

The two conventions for specifying the area of the surface – in terms of $\tilde{A}$ or of A – result in different variations of γ with e, since the identical amount of total surface excess energy is referred to two different values of the area. As a consequence, different definitions of the surface stress emerge. Working with laboratory coordinates ($\tilde{A}$) leads to the Shuttleworth equation in its original form [3],

$$f = \gamma + \frac{d\gamma}{de}$$

whereas Lagrange coordinates (A) lead to [26]

$$f = \frac{d\gamma}{de} \qquad (5.15)$$

Both expressions embody the same physics and they lead to identical numerical values of f for the unstrained surface. Their difference resides exclusively in the convention for specifying the surface area. The use of Lagrangian coordinates in the present work is motivated by the superior conceptual clarity and the drastic simplification of the notation that is achieved with that notation.

5.2.5
Case Study: Thought Experiment in Electrowetting

Consider a patch of electrode surface that is formed where a droplet of liquid electrolyte wets the planar surface of a solid metal, Figure 5.3. A counter-electrode in the droplet, connected to a voltage source, affords control over the charge density, q on the electrode. Assume that q has just been changed and that the droplet now rearranges to find a new equilibrium configuration. That configuration will be governed by work which, during the rearrangement, is done against the relevant capillary forces. It is, therefore, of interest to inspect the work terms.

The net free energy of the system (electrolyte plus metal) may be decomposed into contributions from the bulk solid and liquid and from the various interfaces. We focus on the excess free energy, $G^{sl} = \gamma^{sl} A$, that is associated with the (solid–liquid–) electrode surface of Lagrangian area A and surface tension γ^{sl}. In the present context, A is simply the product of the number of atoms in the electrode surface and of the physical surface area per atom in the reference state (i.e., at no elastic strain).

Specifically, we inspect the variation, δG^{sl}, of G^{sl} when the physical surface area, $\tilde{A}$, of the electrode is increased by the relative amount ε while the electrode potential, E, is held constant. The two distinct processes under consideration achieve the variation of $\tilde{A}$ in two nonequivalent ways. These processes are chosen because of their relevance to two important and quite different experimental situations.

5.2 Mechanics of Solid Electrodes

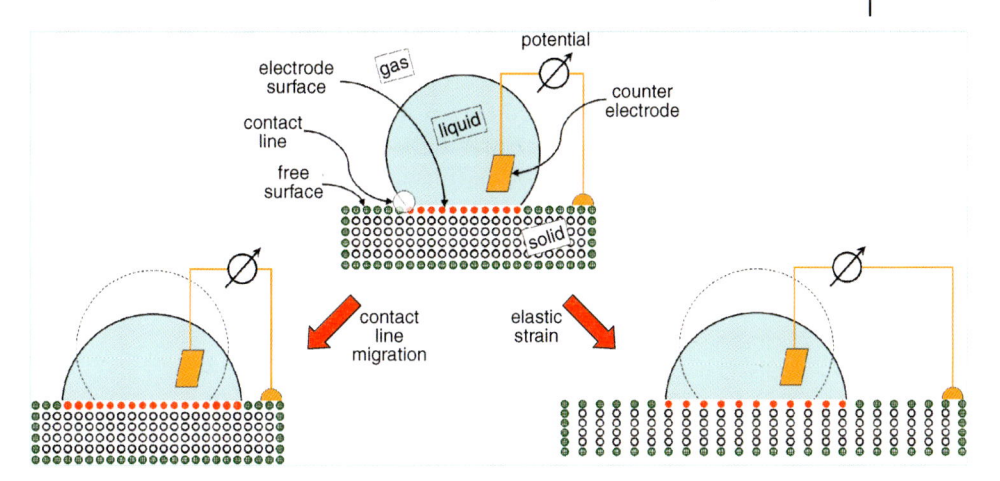

Figure 5.3 Two non-equivalent ways of varying the surface area of a solid electrode even when there is no diffusion in the solid. The electrode surface is formed where a droplet of liquid electrolyte wets a metal surface. Circles represent atoms, red solid circles denoting electrode surface atoms. Migration of the solid–liquid–gas triple line (left side of graphics) enhances both the physical (A) and the Lagrangian ($\bar{A}$) surface area by exposing more metal atoms to the electrolyte. Work $\delta G = \gamma A_0 \epsilon$ is done against the surface tension. By contrast, elastic strain of the metal (right side of graphics) enhances $\bar{A}$ by increasing the in-plane interatomic spacing from r_0^{NN} to $r_0^{NN}(1+\epsilon/2)$, while leaving A and the number of surface atoms constant. Here, the work $\delta G = f A_0 \epsilon$ is done against the surface stress.

As the first process (left-hand side of Figure 5.1) we consider the migration of the solid–liquid–gas contact line along the surface of a rigid solid, exposing extra metal atoms to the electrolyte and thus increasing the physical and Langrangian areas of the electrode surface. This phenomenon is observed in electrowetting experiments when a droplet changes its wetting angle subsequent to a change in the electrode potential. Since no elastic strain is involved, the contact line displacement leaves γ^{sl} invariant. The variation in G^{sl} is therefore entirely due to a change in Lagrangian area,

$$\delta G^{sl} = \gamma^{sl} \delta A = \gamma^{sl} A \varepsilon \quad \text{(contact line migration)}. \tag{5.16}$$

The dependence of the free-energy change on the surface tension and not on the surface stress implies that γ^{sl} is the relevant capillary parameter for the process under consideration. This is reflected in the statement of equilibrium, the Young equation, which relates the wetting angle, θ, at equilibrium to the surface tensions via [27]

$$\cos\theta = \frac{\gamma^{sv} - \gamma^{sl}}{\gamma^{lv}} \tag{5.17}$$

where γ^{sv} and γ^{lv} denote the solid–vapor and liquid–vapor surface tensions.

The second process under consideration (right-hand side of Figure 5.1) is an elastic strain, e, of the electrode in the tangent plane, with the strain magnitude

$e = \varepsilon$. This might be brought about by mechanical forces applied to the metal, as in the experiments of Ref. [28]. The contact line is taken to remain stationary in the Lagrangian frame, in other words, the line is displaced along with the surface atoms. Here, the physical area is varied by the relative amount ε while the number of surface atoms and, consequently, the Lagrangian area of the electrode remain constant. The variation in G^{sl} is then entirely due to a change in the surface tension, $\delta G^{sl} = A_0 \delta \gamma^{sl}$. The variation in γ^{sl} will not generally vanish, since the elastic strain changes the interatomic spacing between the metal atoms at the electrode surface, thereby modifying the state of the surface. In fact, experiment as well as *ab initio* computer simulation put the derivative, $f = \partial \gamma / \partial e$, of the surface tension of solids with respect to the elastic strain, at a similar numerical magnitude as the surface tension itself. The quantity f is termed the surface stress. In terms of f, we have here

$$\delta G^{sl} = A \delta \gamma^{sl} = A \frac{\partial \gamma^{sl}}{\partial e} e = f^{sl} A e \quad \text{(elastic strain)} \tag{5.18}$$

In contrast to the first process considered above, the relevant capillary parameter is here not the surface tension but rather its strain-derivative, the surface stress. In other words, the impact of the surface on the mechanical equilibrium state of a solid electrode is governed by f^{sl} and not γ^{sl}. An exemplary statement of mechanical equilibrium is Stoneys equation [13,29] for the equilibrium curvature, κ, of a compliant substrate when a stress acts tangentially at one of its surfaces. This equation is readily adapted to surface stress measurement and routinely used in experiment (see references in Section 5.3.10) in its form

$$\Delta \kappa = -\frac{6 \Delta f}{M h^2} \tag{5.19}$$

with M the biaxial modulus and h the cantilever thickness. Equation (5.19) here embodies the trend of cantilevers to bend when f is varied, for instance due to capacitive charging of the electrode surface or to adsorption. As will be discussed in Section 5.3.10, cantilever bending provides the most widespread experimental approach towards surface stress measurement.

Our thought experiment involved two separate variations, contact line migration on a rigid substrate and elastic strain with a contact line attached to the substrate. We concluded that two different capillary parameters, surface tension and surface stress, govern the relevant equilibrium. In a real-world experiment, both forces will act simultaneously in order to establish the respective equilibria. This serves to emphasize that both surface tension and surface stress can be relevant to the equilibrium at solid surfaces. Conversely, one may ask about the role of surface stress for fluids. In that context it is sometimes stated that surface stress and surface tension become identical for the fluid. Yet, that statement is not compatible with the defining equation of f as the derivative of γ with respect to tangential strain. Since the strain is not a meaningful state variable for the fluid, surface stress can simply not be defined there in a meaningful way. This was already recognized by Gibbs, who noted that fluid surfaces can be formed but not stretched [1].[4)]

The impact of the capillary terms for the experimentally accessible phenomena of surface-induced pressure and deformation in fluids and solids is described by capillary equations, which will now be inspected.

5.2.6
Capillary Equations for Fluids and Solids

At mechanical equilibrium, the forces which act at the surface of a solid need to be balanced by compensating stress in the bulk. It is widely known that the equilibrium condition for the case of a fluid droplet is the Young–Laplace equation [27,30], which links the jump, $\Delta P = P - P^{\text{extern}}$, in pressure across a surface between two fluids to the surface tension and to the surface mean curvature, κ, via

$$P - P^{\text{extern}} = 2\gamma\kappa \quad (5.20)$$

The origin of the Young–Laplace pressure is immediately accessible to intuition: the bulk fluid can diminish its volume and, thereby, its surface area by increasing its bulk density. Since the decrease in surface area reduces the net surface excess free energy, there is a thermodynamic driving force for densification; this driving force is embodied in the Young–Laplace pressure. Note that the compression of the bulk fluid leaves the surface atomic structure invariant, since the surface atoms can rearrange to find their energetically most favorable arrangement. Since the fluid does not transmit shear stress, the rearrangement recovers the essentially identical surface structure and energy, independent of the density of the underlying bulk fluid. This implies that the variation in surface area by bulk compression in a fluid diminishes the number of surface atoms while leaving γ invariant.

As illustrated in Figure 5.2, the situation in a solid is quite different from that in a fluid. Even when atoms can be rearranged by diffusion (at high temperature or in very slow processes), the atomic density at the surface is always coupled to that in the underlying bulk because the solid does support shear. In other words, the surface structure is coupled to that of the bulk by crystallographic relations or – in the case of a glassy solid – by a network constraint. Densification or expansion of the solid in the direction tangential to the surface will, therefore, always change the surface atomic structure, while leaving the number of surface atoms constant. The relevant balance

---

4) One might object that the free energy of the fluid depends on the atomic density, ρ, and that changes in the density upon compression or expansion of the fluid may provide a scalar equivalent to the tensor strain of a solid. Yet, the density is not a parameter in the Gibbs adsorption equation for fluids, Eq. (5.1). This may be traced back to the fact that – for lack of the lattice constraints or epitaxy conditions that act in a solid – atomic rearrangements at the surface of a fluid are not constrained by the structure of the underlying bulk. The surface atomic structure and density at equilibrium can, therefore, take on that configuration which minimizes the surface free energy, irrespective of the density of the underlying bulk fluid. This is equivalent to saying that γ does not depend on ρ.

176 | 5 Electrocapillarity of Solids and its Impact on Heterogeneous Catalysis

equation for P has been derived by Weissmüller and Cahn in Ref. [2]. In its scalar form, this generalized capillary equation for solids states that

$$3V\langle P - P^{\text{extern}}\rangle_B = 2A\langle f\rangle_S \qquad (5.21)$$

The angular brackets designate averages over the volume of the bulk solid ("B") and over the entire surface area ("S"), while V and A denote the respective values of net volume and area.

Figure 5.4 shows the two balance equations and highlights their salient differences. Most importantly, the relevant capillary parameters for fluid and solid differ. For the fluid, the pressure-jump across the surface scales with the surface tension. By contrast, the surface-induced pressure in the solid scales with the surface stress. As is emphasized, the two capillary parameters obey different sign restrictions. While the surface tension cannot be negative in stable states of matter [31], the surface stress obeys no such restriction. Even though most experimental and computational surface stress values are positive, negative values have been confirmed in experiment [8,14]. The relevant geometry parameters in the capillary equations are also different. While the pressure in a fluid droplet scales with the

	fluid	solid
balance law	$\Delta P = 2\gamma\kappa$	$3V\langle P\rangle_B = 2A\langle f\rangle_S$
		$\langle\rangle$ - averages
capillary parameter	γ surface tension	$f = \partial\gamma/\partial e$ surface stress
	+	±
geometry parameter	κ curvature	A/V area per volume
	±	+
charge-dependence	$\gamma = \gamma_0 - q^2/c$	$f = f_0 + \varsigma q$
	−, quadratic	±, linear

Figure 5.4 In the bulk of small objects, large stresses act to balance the capillary forces. Depending on the nature – fluid or solid – of the object, the surface-induced stresses depend in distinctly different ways on the geometry, charge density and on the thermodynamics. The figure summarizes the most important differences. The relevant balance law for fluids is the Young–Laplace equation, whereas the balance in solids is described by the generalized capillary equation for solids. The relevant capillary parameter for fluids, the surface tension γ, is positive defined, whereas the surface stress, f, relevant for solids, can take either sign. Similarly, the relevant geometry parameter mean curvature (κ, fluids) can take either sign whereas area per volume (A/V, solids) must be positive by definition. Yet another qualitative difference is the charge dependence, quadratic for γ (fluids) versus linear for f (solids). The summary emphasizes in particular that the validity of the Young–Laplace equation is restricted to fluids, and that its erroneous application to solids would lead to qualitatively incorrect predictions for the surface-induced stress.

mean curvature, that in a solid scales with surface area per volume. Here again, the sign conventions differ. The mean curvature can be of either sign (convex or concave surface), whereas area per volume is positive by definition. As will become apparent in Section 5.3, the variation of the two capillary parameters – surface stress and surface tension – is also fundamentally different. Taken together, the considerable differences between capillary effects in fluids and solids emphasize that the respective capillary equations must not be confused. In particular, and in contrast to what is incorrectly attempted even in many recent studies, the Young–Laplace equation must not be applied to derive predictions for the mechanical equilibrium in solids. The results obtained with that equation have no relation to the actual stresses and can be grossly erroneous.

We now turn to a case study that exemplifies the role of the geometry for the surface-induced pressure in solids.

5.2.7
Case Study: Molecular Dynamics Study of Surface-Induced Pressure

A recent atomistic simulation study [32] explores the surface-induced pressure in microstructures of different geometry. Using molecular statics, the study considers (i) spherical particles with convex surfaces and a positive mean curvature, (ii) an extended solid body containing an array of spherical voids with concave surfaces and, hence, negative mean curvature, and (iii) a bicontinuous network structure of interpenetrating solid phase and voids that has, on average, zero mean surface curvature. In each case, the mean pressure within the solid is recorded after elastic equilibration.

Figure 5.5 displays the mean pressures within the different solid microstructures, plotted versus the specific surface area. Two observations are notable: First, the pressure is always positive, even though κ can be positive, negative or zero. Obviously, the mean curvature is not an appropriate geometry parameter for the capillary pressure in solids. Secondly, it is observed, that the mean pressure scales linearly with area per volume. This agreement confirms the generalized capillary equation, Eq. 5.21, and it rules out the Young–Laplace equation, which does not appropriately describe the stress in solids.

5.3
Electrocapillary Coupling at Equilibrium

5.3.1
Outline – Polarizable and Nonpolarizable Electrodes

Within a phenomenological thermodynamics description, the key parameter connecting electrode processes to the mechanics is the response, dE/de, of the electrode potential to elastic strain. In this section we inspect the relation of that response parameter to other quantities.

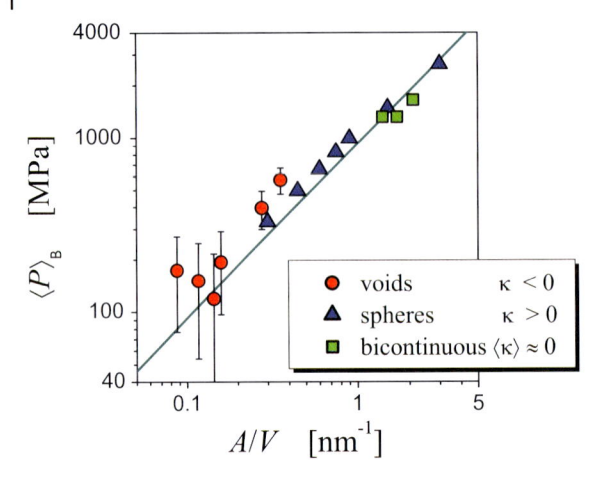

Figure 5.5 Results of an atomistic computation of the surface-induced mean pressure, $\langle P \rangle_B$, in solids with different geometry. Volumetric average, $\langle P \rangle_B$, of the mean pressure is plotted versus specific surface area, A/V. Symbols: results of simulation for a bulk solid with an array of nanovoids, for a set of isolated spheres, and for a bicontinuous nanoscale network structure. Line: fit with Eq. (5.21), using the value $f = 1.4\text{N m}^{-1}$. Note that pressure is always positive, even though the mean curvature, κ, of the solid surface takes on either sign, positive (convex) for the spheres and negative (concave) for the voids. The bicontinuous structures have regions of positive and negative mean curvature, with the average value near zero, independent of A/V. The finding illustrates that mean curvature is not an appropriate geometry parameter for the surface-induced pressure in solids. This emphasizes that the Young–Laplace equation does not apply for solids and that the generalized capillary equation, Eq. (5.21), must be used instead. After Ref. [32].

As a preliminary, it is emphasized that the considerations of this section require a *polarizable* electrode. In other words, it is presupposed that there exists a finite range of the electrode potential in which the electrode acts pseudocapacitively, in the sense that charge transport to or from the electrode serves exclusively for changing a recoverable charge density, q, which – at equilibrium – is a function of the state variables. In Section 5.5 we shall lift this restriction and allow in addition the presence of (Faraday-) reaction current, which does not contribute to changing q and which, in magnitude, depends on the potential and on the strain.

An extreme case that is ignored here is the *nonpolarizable* electrode, for instance, a metal in contact with an electrolyte containing the metal ion in solution. At a nonpolarizable electrode, the exchange current density of a reaction (e.g., metal dissolution/deposition) near equilibrium is large enough to fix the electrode potential at the value given by the Nernst equation. The derivative dE/de then represents the strain-dependence of the Nernst potential, E_{Nernst}. That latter quantity depends on the stored mechanical energy density in the electrode, which is a quadratic function of the strain relative to a stress-free reference state. Therefore, the derivative dE_{Nernst}/de is for all practical purposes zero when measured at a nonpolarizable electrode. Comparative experiments [33,34] with an Ag electrode in NaF solution (polarizable) and in AgNO$_3$ solution (nonpolarizable) provide a striking demonstration of the difference between the potential–strain response of polarizable and nonpolarizable electrodes.[5]

5.3.2
Lippmann Equation and Electrocapillary Coupling Coefficient

At a polarizable electrode, the capillary parameters surface tension and surface stress respond to changes in the electrode potential or in the superficial charge density. The response of the free energy function γ is embodied in the fundamental equations, Eqs. (5.1) and (5.2), as

$$\frac{d\gamma}{dE}\bigg|_{T,\mu_i} = -q \quad \text{(fluid)} \tag{5.22}$$

and

$$\frac{d\gamma}{dE}\bigg|_{T,\mu_i,\mathbb{E}} = -q \quad \text{(solid)} \tag{5.23}$$

Equation (5.22) is the Lippmann equation [13,36] as it applies to fluids, and Eq. (5.23) is the corresponding expression for a solid surface.[6]

---

5) As is emphasized in a footnote in Section 5.2.1, the discussion in this chapter ignores phenomena related to changes in the pressure, P, in the electrolyte. In the present context, however, it is important to note that the electrode potential, E, reacts in a fundamentally different way to changes in P as compared to changes in the tangential strain e. Note that P and e can be independently varied. When one metal atom of atomic volume Ω is removed from a *nonpolarizable* electrode and converted into an ion of volume $\Omega + \Delta\Omega$, the change in internal energy due to the mechanical work done by the pressure acting on the electrolyte is $-P\Delta\Omega$. Therefore E may be expected to change with pressure according to $\delta E = \delta P \Delta\Omega/(zF)$. This is a finite potential–pressure response coefficient, even though the Nernst potential does not respond to tangential strain. Measurements on solid electrolytes have documented this difference, and for oxygen exchange between ZrO_2 and gas the coefficient was found at 3×10^{-11} V Pa^{-1} [35]. The impact of electrolyte pressure on the electrode potential of *polarizable* electrodes has also been discussed. A Maxwell relation equates that coefficient with the change in surface excess volume with charge density. In other words, $dE/dP|_{e,q} = d\mathbb{V}/dq|_{e,P}$ [12]. Density functional theory computation puts that value at -8×10^{-12} V Pa^{-1} for the (111) surface of gold [10]. As noted previously, the above effects are ignored here. This simplification serves conciseness and it is permitted when considering – as we do – processes with the electrolyte at constant pressure.

6) Couchman and Davidson [37] have given a different expression for the Lippmann equation of a solid with in-plane isotropic surface stress. In our notation (Lagrange coordinates), their expression – which is sometimes referred to as the "generalized Lippmann equation" – would read $d\gamma/dE|_{T,\mu_i} = -q + fde/dE|_{T,\mu_i}$. The second term on the right-hand side accounts for mechanical work when the electrode undergoes spontaneous strain in response to the variation of f. This is appropriate in the special case where no external tractions act, so that the strain varies freely subject to the internal mechanical equilibrium conditions in the electrode. The strain can then be considered as an internal parameter instead of an independent state variable. That is not generally true, as is illustrated by experiments where the electrode is strained by external action, as in Sections 5.4.3 and 5.5. Here, e does take the role of an independent state variable. As compared to the fundamental equation, Eq. (5.14), the so-called "generalized Lippmann equation" is, therefore, actually the less general formulation for $d\gamma$. Furthermore, it has been shown that in experiments where the strain does vary freely, Couchman and Davidson's extra term that depends on the strain is typically small and can be neglected [13]. Within the context of the present work, the term has no special significance and can safely be ignored.

For the special case of a potential-independent value of the capacitance, c, integrating the Lippmann equation at constant T, μ_i and (for solids) e yields a quadratic variation of the surface tension with charge,

$$\gamma = \gamma_{zc} - \frac{q^2}{c} \tag{5.24}$$

This expression, which applies equally to fluids and solids, embodies the "electrocapillary maximum" of γ at the potential of zero charge (pzc).

While the free energy function γ is related to the state variables by the fundamental equations (5.1 and (5.2)) via the relevant *first* derivatives of γ, the dependence of the state variable f on the state of the electrode involves *second* derivatives of the free energy densities γ or ψ. In the formal structure of phenomenological thermodynamics, the second derivatives represent materials parameters. An example is the capacitance,

$$c = \frac{\partial^2 \gamma}{\partial E^2} = \frac{dq}{dE}\Big|_{T,\mu_i,e} \tag{5.25}$$

Similarly, and in view of Eq. (5.12), the variation of the surface stress with the charge density and the electrode potential is given by

$$\varsigma = \frac{\partial^2 \psi}{\partial e \partial q} = \frac{df}{dq}\Big|_{T,\mu_i,e} \tag{5.26}$$

The quantity ς is central to the present work. Gokhshtein [38,39] first systematically discussed this parameter and termed it the "estance", alternately motivating the term by an analogy of Gokhshteins measurement scheme to impedance measurements [39] and as a condensed form of "elastic surface tension alteration" [40]. The term has not stuck, and indeed the definition of ς in terms of Eq. (5.26) offers no obvious link to either, impedance or surface tension. Besides the fundamental definition in terms of Eq. (5.26), in this section we discuss various alternate expressions for ς. Their common trait is the connection between the capillarity and the state variables linked to electric charge or potential. We therefore refer to ς as the electrocapillary coupling coefficient.

In general, the materials parameters – and in particular c and ς – are functions of the state variables. Specifically, experiments measuring these parameters explore their variation with the electrode potential. In other words, experiment aims to empirically determine functions such as $c(E)$ and $\varsigma(E)$. As an important distinction, the phenomenological considerations – while predicting an electrocapillary maximum for γ – impose no restriction on sign or functional variation of f. In other words, the surface stress near the pzc may be approximated in a first order series expansion as

$$f = f_{zc} + \varsigma_{zc} q \tag{5.27}$$

with ς_{zc}, the value of the surface stress–charge coefficient at the pzc, in general not zero.

Since the Young–Laplace equation relates the pressure in a small fluid droplet to its surface tension, the potential-induced pressure variation in a liquid metal (e.g., mercury) droplet electrode affords an experimental observation of the electrocapillary maximum of Eq. (5.24). This has long been established as a reliable experimental signature of the pzc of the liquid metal. Solids with a high specific surface area (area per volume) will undergo measurable deformation in response to potential changes. As discussed in Ref. [20], such experiments have been explored in relation to an electrocapillary maximum in γ and to the pzc. This presupposes an equivalence between solids and fluids, where analogous experiments are well established. Yet, our considerations so far provide no basis for such an equivalence. Instead, by combining the generalized capillary equation for solids, Eq. (5.21), with Eq. (5.27) for the variation of f one finds that the pressure varies linearly with E or q at the pzc. Maxima in the surface-induced pressure in solids have been observed at various potentials, and can be traced to changes in the prevailing electrode process, for instance, a switch between capacitive charging and hydrogen underpotential deposition. These phenomena have no obvious link to the pzc. We shall now present an experiment that exemplifies the distinction between the electrocapillary maximum in γ and the linear variation of $f(q)$ near the pzc.

5.3.3
Case Study: Cantilever-Bending Experiment in Electrolyte

Here we evaluate data from a cantilever-bending experiment published in Ref. [41], where details can be found. The electrode is a thin gold film deposited on an electrically insulated silicon cantilever as the substrate. The entire cantilever is immersed in a 7 mM aqueous solution of NaF. The cantilever bends when the electrode potential is varied, testifying to a change in the capillary forces. The curvature change is measured from the deflection of an array of laser beams reflected at the electrode surface. Stoneys equation, Eq. (5.19), then provides a link between the changes in curvature and in surface stress, identifying surface stress and not surface tension as the capillary force inducing the mechanical deformation. No wetting angle is observed, since the electrode is completely immersed. Yet, the change in surface tension can be equated with the electrical work (per area) done when charging the electrode [13]. Therefore, the electric and mechanical signals provide two separate experimental signatures that allow one to independently derive the variation of the two separate quantities, surface tension and surface stress [18].

Figure 5.6a displays the two independent experimental data sets, the cyclic voltammogram of current density j versus potential E during a potential scan and the diagram of cantilever curvature κ versus E measured simultaneously with the voltammogram [41]. The variation of the electrode charge density, q, with E can be determined by first integrating the experimental $j(E)$ and then using Lippmans equation in its form $d\gamma = -qdE$ [13] to obtain data for $\Delta\gamma(E) = \gamma(E) - \gamma_0$ by a second integration. The integration constant for q is provided by the known value of the pzc [41]. The value of γ at the pzc is here denoted by γ_0. A small ohmic potential

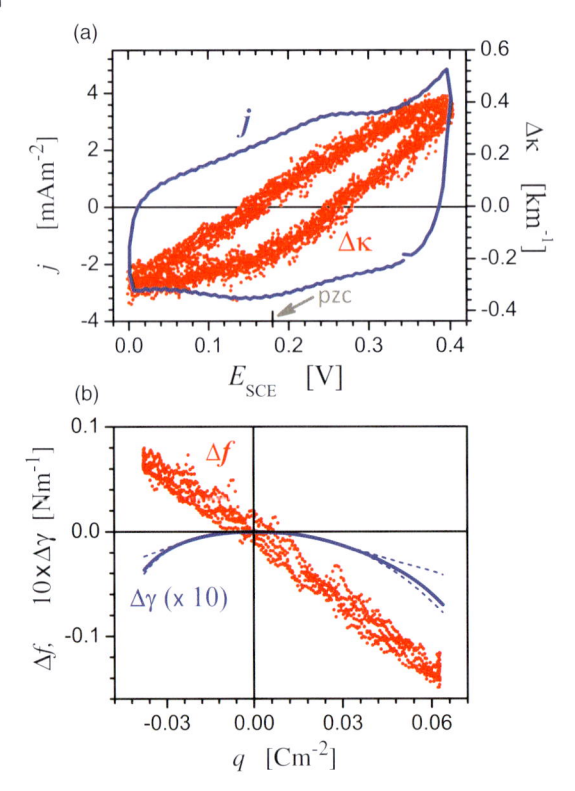

Figure 5.6 Experimental data from a cantilever-bending experiment with a gold electrode in aqueous NaF (7 mM) [41] and their implication for surface stress and surface tension. (a) Cyclic voltammogram (blue line) of current density j versus electrode potential E (specified versus the standard calomel electrode, SCE) during cyclic scans at 10 mV s^{-1}. Also shown is the variation in cantilever curvature, κ (red dots), with E measured simultaneously with the voltammogram, pzc is indicated. (b) Variation of the capillary parameters surface stress, f (red dots), and surface tension, γ (blue solid line, data multiplied by 10 for better readability), with superficial charge density q during the potential scans. The parameter γ is computed from the voltammogram data using Lippmanns equation, while f results from the curvature data and Stoneys equation. Blue dashed lines show estimated error in integration of Lippmanns equation. Note the fundamentally different variation of the two capillary parameters with q.

drop between reference and working electrode was corrected for, in order to remove the dissipative work done against the uncompensated solution resistance. The change, $\Delta f(E) = f(E) - f_0$, in surface stress can be evaluated from the cantilever curvature data using Stoneys equation [41].

Figure 5.6b shows the results for $\Delta\gamma$ and Δf, plotted versus q. It is seen that the two capillary parameters react to changes in the electrode potential in fundamentally different ways. While γ exhibits the well-known "electrocapillary maximum" [20], decreasing quadratically with q when the potential deviates from the pzc, the experiment finds f to vary linearly with q and by a much larger amount than γ. For use below we note that the slope of a straight line of best fit to $f(q)$ in Figure 5.6b gives the value of the electrocapillary coupling coefficient as $\varsigma = -1.95 \pm 0.1$ V [41]. Prompted by the debate about the necessity to distinguish between surface stress and surface tension, we emphasize that the presence of an electrocapillary maximum in γ at the pzc has been established since the nineteenth century, and that the lack of such a maximum even in the primary data for the curvature versus potential (Figure 5.2a) prohibits an identification of the responsible capillary force with surface tension.

The finding of a qualitatively different variation – linear versus quadratic – of surface stress and surface tension in response to changes in the electrode potential emphasizes that a clear distinction is mandatory and that grossly erroneous predictions may result if the two parameters are confused.

5.3.4
Important Maxwell Relations for Electrocapillarity

The fundamental equation (5.12) for the surface free energy density has the immediate consequence that, at equilibrium, a Maxwell relation relates the potential–strain response to the surface–stress charge response coefficient ς of Eq. (5.26): Since

$$\frac{\partial^2 \psi}{\partial e \partial q} = \frac{\partial^2 \psi}{\partial q \partial e} \tag{5.28}$$

it follows that

$$\frac{df}{dq}\bigg|_e = \frac{dE}{de}\bigg|_q \tag{5.29}$$

and with the definition (see Eq. (5.26))

$$\varsigma = \frac{df}{dq}\bigg|_e \tag{5.30}$$

also

$$\varsigma = \frac{dE}{de}\bigg|_q \tag{5.31}$$

In other words, the response of the electrode potential to elastic strain at constant charge density is numerically equal to the response of the surface stress to changes in the superficial charge density at constant strain.

An additional Maxwell relation is derived from the fundamental equation for the surface tension, Eq. (5.14). Here,

$$\frac{\partial^2 \gamma}{\partial e \partial E} = \frac{\partial^2 \gamma}{\partial E \partial e} \tag{5.32}$$

implies

$$\frac{df}{dE}\Big|_e = -\frac{dq}{de}\Big|_E \qquad (5.33)$$

If we further consider the definition of ς, Eq. (5.26), along with the identity $(df/dq)|_e = (df/dE)|_e \times (dE/dq)|_e$, where the last term on the right-hand side is the capacitance, we find that the expressions of Eq. (5.33) are related to ς via

$$\varsigma = \frac{1}{c}\frac{df}{dE}\Big|_e \qquad (5.34)$$

and

$$\varsigma = -\frac{1}{c}\frac{dq}{de}\Big|_E \qquad (5.35)$$

The Maxwell relation (5.29) was first pointed out by Gokhshtein [38]. The equation has a central role in the context of this book on catalysis, since it represents the aforementioned coupling between mechanics and electrochemistry in a phenomenological description and links it to the electrocapillarity of solid surfaces. The electrocapillary coupling coefficient ς, is open to quantitative and precise determination by experiment and by numerical simulation. The impact of ς goes beyond the immediately obvious finding of the variation of an electric potential with strain. At equilibrium, the electrode potential is coupled to the chemical potential of the electrons, thereby potentially affecting electron transfer processes. Furthermore, E is also coupled to the chemical potential of adsorbing species via the equilibrium condition, Eq. (5.11), while adsorption enthalpies can vary significantly with strain. The following subsections will inspect the relevant scenarios in more detail.

5.3.5
Electrocapillary Coupling During Electrosorption

Let us consider an electrosorption feature in a small potential interval, and assume that $q(E)$ in that interval is dominated by a single adsorption process, ignoring capacitive processes in the diffuse layer and the exchange of other adsorbates. Equation (5.8) for the charge neutrality then implies that the charge density and the superficial excess of adsorbate are linked via

$$dq = -zFd\Gamma \qquad (5.36)$$

Equation (5.36) implies that, in the idealized context of single-species electrosorption, one may exchange the state variable q with Γ. In view of Eq. (5.30), the electrocapillary coupling during electrosorption is then linked to the response of f to adsorption from gas via

$$\varsigma = -\frac{1}{zF}\frac{df}{d\Gamma}\Big|_e \qquad (5.37)$$

This expression is of interest since it connects the electrocapillary coupling coefficient to the adsorbate coverage. This is a prerequisite for comparing the mechanics of electrosorption processes to that of adsorption phenomena at surfaces of a solid in contact with gas.

5.3.6
Coupling Coefficient for Adsorption from Gas

Whereas some relevant experiments explore catalysis at electrode surfaces, theory – and specifically numerical studies using density-functional theory – typically considers adsorption from gas. Besides electrode surfaces, a discussion of adsorption–strain coupling must, therefore, include solid surfaces in contact with gas.

An appropriate constitutive assumption for the surface energy variation during the adsorption from gas can be based on the superficial free energy function ψ^{SG} for the solid–gas interface:

$$\psi^{SG} = \psi^{SG}(\Gamma, e) \tag{5.38}$$

with the fundamental equation

$$d\psi^{SG} = \mu^{SG} d\Gamma + f de \tag{5.39}$$

Here

$$\mu^{SG} = \frac{\partial \psi^{SG}}{\partial \Gamma} \tag{5.40}$$

is an adsorbate potential that, at equilibrium, satisfies

$$\mu^{SG} = \mu \tag{5.41}$$

with μ the chemical potential of the adsorbing species in the gas phase. For simplicity, we again consider a single adsorbing species. The analog to Eq. (5.29) is here

$$\frac{d\mu}{de}\bigg|_\Gamma = \frac{df}{d\Gamma}\bigg|_e \tag{5.42}$$

It is of interest to consider the case – which is at this point hypothetical – where the adsorbate–substrate bonding configuration for adsorption of A on the free surface of B is closely similar to that for electrosorption of A on a B electrode surface. The surface stress changes and, hence, the coefficients $df/d\Gamma$ are then also closely similar for the free surface and the electrode surface. One would here conclude an equivalence between ς, measured during electrosorption, and the coefficients defined for the solid–gas interface (superscript SG) according to

$$\varsigma \approx -\frac{1}{zF}\frac{d\mu}{de}\bigg|_\Gamma^{SG} \tag{5.43}$$

and

$$\varsigma \approx -\frac{1}{zF}\frac{df}{d\Gamma}\bigg|_e^{SG} \tag{5.44}$$

Equation (5.44) is the equivalent of Eq. (5.37) for surfaces in contact with gas.

5.3.7
Coupling Coefficient for the Langmuir Isotherm

The coupling between mechanics and adsorption can be illustrated by inspection of a microscopic model. To this end we consider the Langmuir model, with the isotherm

$$\mu^{SG} = \mu_0 + \Delta h^{ad} + RT \ln \frac{\theta}{1-\theta} \tag{5.45}$$

for the adsorbate potential μ^{SG} and with, consistently, the coverage at equilibrium ($\mu^{SG} = \mu$)

$$\theta = \left(1 + \exp \frac{\mu - \mu_0 - \Delta h^{ad}}{RT}\right)^{-1} \tag{5.46}$$

Here μ_0 denotes the chemical potential of the adsorbate in its standard state and Δh^{ad} the molar enthalpy of adsorption. The adsorbate occupies a two-dimensional lattice with a fixed number, Γ_0, of sites per area. The ad-species do not interact, and sites are randomly occupied with a probability θ (the fractional "coverage") that satisfies $\theta = \Gamma/\Gamma_0$.

A simple example of a free energy function which, when differentiated with respect to Γ, yields the isotherm of Eq. (5.45) while at the same time embodying surface stress is

$$\psi^{SG} = \Gamma_0 \Big(\theta\mu_0 + \theta\Delta h^{ad} + RT(\theta \ln \theta + (1-\theta)\ln(1-\theta))\Big) + f_0 e \tag{5.47}$$

where f_0 is the surface stress of the clean surface.

At any given value of the coverage (represented by Γ), the configurational entropy of the model adsorbate is a constant. One may then approximate the strain-dependence of μ at constant Γ as dominated by that of the (molar) enthalpy of adsorption, ignoring the impact of strain on the vibrational entropy. In other words, it is admitted that $\Delta h_{ads} = \Delta h_{ads}(e)$. For some scenarios, this assumption is corroborated by *ab initio* computation [42]. By inspection of Eq. (5.45) one then sees that

$$\frac{d\mu}{de}\Big|_\Gamma = \frac{d(\Delta h^{ad})}{de} \tag{5.48}$$

As a consistency check, taking derivatives of Eq. (5.47) first with respect to e and then to Γ yields $df/d\Gamma|_e$, which is found to agree with $d\mu/de|_\Gamma$ of Eq. (5.48), as required by the Maxwell relation, Eq. (5.42).

By virtue of Eq. (5.43), the electrocapillary coupling coefficient for the analogous electrosorption process is predicted to be

$$\varsigma = -\frac{1}{zF}\frac{d(\Delta h^{ad})}{de} \tag{5.49}$$

While this expression is here motivated by the specific adsorption isotherm, Eq. (5.45), it can be used quite generally – and irrespective of the details of the adsorption isotherm and of the charge transfer – as the formal definition for an apparent electrocapillary coupling coefficient, ς^{ad}, for adsorption. That coefficient has the same units as ς and it describes the shift in electrosorption potential with strain in important limiting cases [43]. Just to give a number, $+1$ V in electrocapillary coupling coefficient here translates to $-zF \times 1\text{V} = -96.5$ kJ mol^{-1} in enthalpy–strain response, assuming $z = +1$.

5.3.8
Case Study: Strain-Dependent Hydrogen Underpotential Deposition

In what may be the most meaningful experimental study of strain effects in electrosorption to date, Kibler and coworkers [44] have investigated the underpotential deposition (UPD) of hydrogen on atomic monolayers of palladium that were strained by epitaxy to single crystal substrates with different lattice parameters. The total interval in the area-strain e covered by this approach is as large as 0.14. As can be seen in Figure 5.7a the variation in the hydrogen electrosorption potential, E_H, in those studies is also large, covering a potential interval of 200 mV. The data from Ref. [44], which is reproduced in the figure, shows that the experimental change in electrosorption potential agrees well with predictions of the d-band shift from *ab initio* density functional theory (DFT) in Ref. [45].

The above agreement between experiment and DFT supports a link between the electronic band structure and electrosorption. However, the observation does not afford a discrimination between two possible origins of the d-band shift, namely the elastic strain and the so-called ligand effect, that is, the electron exchange between Pd adlayer and substrate. This issue was explored in a subsequent study [43] which, plotting E_H versus the strain (Figure 5.7b), found a good linear correlation with a slope of $\Delta E_\text{H}/\Delta e = 1.4 \pm 0.2$V. In view of Eq. (5.31), the potential variation with strain implies that the coupling coefficient has the same value, $\varsigma = 1.4 \pm 0.2$V. The correlation also suggests that the shift of the electrode potential can indeed be entirely explained as the result of electrocapillary coupling. That notion received support from an independent investigation into the coupling during electrosorption of H [43]. Here, dilatometry studies on nanoporous Pd and Pt electrodes (Figure 5.8) show a linear correlation of surface stress to charge density with $\varsigma = 1.5$V, consistent with the coefficient inferred from the graph of $E_\text{H}(e)$ in Figure 5.7b.

Considering the Maxwell relation, Eq. (5.29), each of the two above studies of H electrosorption on Pd is expected to measure the identical electrocapillary coupling coefficient ς. The numerical agreement confirms that expectation, suggesting that the observed variation of the electrosorption potential with strain is indeed a manifestation of the electrocapillary coupling. That conclusion receives further, independent support from a consideration of the chemo-mechanical interaction of H with thin layers of bulk Pd.

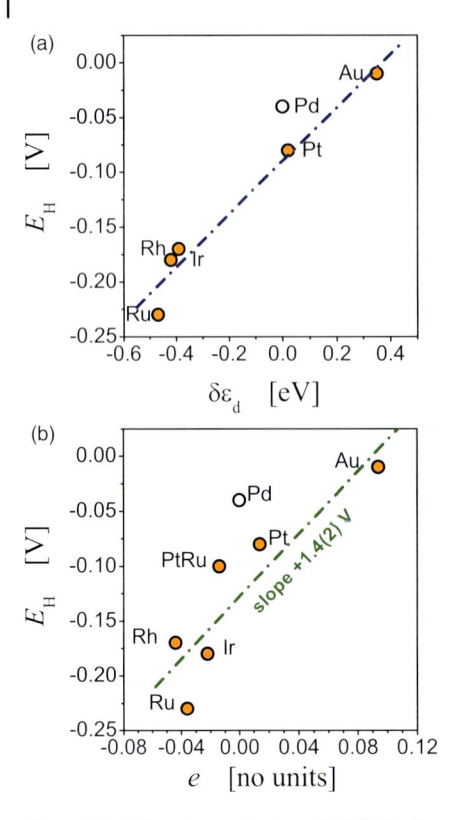

Figure 5.7 1 Experimental values (Ref. [44]) of the hydrogen desorption potential, E_H (quoted against the saturated calomel electrode), for Pd monolayers on various single crystal substrates as indicated by labels. In (a) the data are plotted versus *ab initio* density theory values (Ref. [45]) for the d-band shift, ϵ_d, in the Pd layer. In (b), the same data are plotted versus the area-strain, e, of the Pd layer. Note that both graphs show good correlation. Dash-dotted lines denote best straight-line fit, with slope indicated. Also shown are data for bulk Pd. Here, the strain value is uncertain due to possible hydrogen absorption in the bulk. This data was ignored when fitting. Figure reproduced from Ref. [43].

The properties of bulk alloys of Pd and H have been extensively studied by experiment [46], exploiting the fact that bulk Pd is readily equilibrated with H_2 gas at controlled partial pressure. Within a rigid-band picture of the alloy, the change in lattice parameter upon alloying may be understood as the result of H donating an electron into the electronic system of the Pd host metal. The extra electron contributes to filling the antibonding upper d-band states just above the Fermi level of Pd, thereby causing expansion. It is then not unreasonable to assume that similar processes act when H adsorbs on the surface of Pd. With that in mind, let us inspect – within a continuum approach – the change in apparent surface stress when H is alloyed into (the bulk of) a thin superficial layer of Pd, attached to a rigid bulk substrate [47]. The dissolved Pd atoms represent centers of dilatation, and the appropriate elastic boundary conditions are constant tangential strain and constant normal stress. Lemier

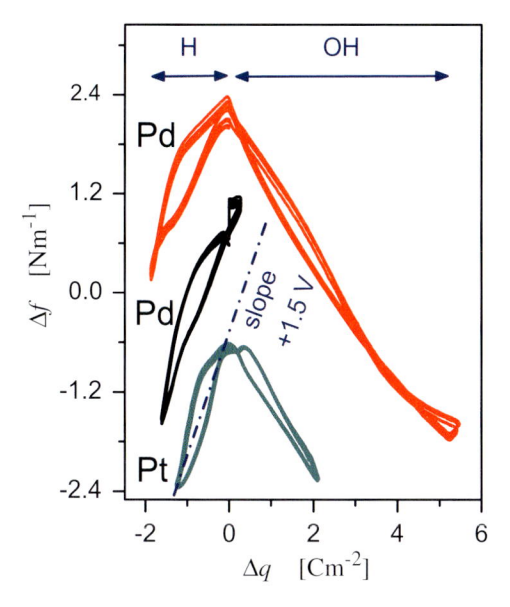

Figure 5.8 Results of *in situ* dilatometry experiments using nanoporous (np-) Pt and Pd in aqueous 0.7 M NaF for studying the surface stress–charge response. Red and black graphs show variation, Δf, of surface stress with surface charge density, Δq, for np-Pd in two different potential intervals, including oxygen species adsorption/desorption (red) and restricted to hydrogen underpotential (upd) deposition/stripping (black). Graph for np-Pt is shown by cyan line. Note that Δf and Δq are specified relative to arbitrary reference values. The slopes in the upd regime are closely compatible for all three graphs, between +1.4 and +2.0 V. Each graph shows at least five scans superimposed. Figure reproduced from Ref. [43].

and Weissmüller have analyzed that scenario in their discussion of the surface stress change during hydrogen segregation to grain boundaries [8], and found that the tangential stress within the layer is isotropic in the plane and of magnitude

$$S_\| = -\frac{Y^{Pd}}{1-\nu^{Pd}} \frac{1}{t} \frac{\eta}{\rho^{Pd}} \Gamma \tag{5.50}$$

Here, t is the layer thickness and Y^{Pd}, ν^{Pd}, and ρ^{Pd} denote, respectively, the Young's modulus, Poisson number and atomic density of Pd. The parameter η represents a concentration–strain coefficient for H in bulk Pd, which can be related to the partial atomic volume, Ω^H, of H in Pd via $\eta = 1/3\Omega^H \rho^{Pd}$. Since the surface stress can be defined in terms of force per line length at a cut surface in the interfacial plane, one finds that $\Delta f = S_\| t$ and, hence,

$$\Delta f = -\frac{Y^{Pd}}{1-\nu^{Pd}} \frac{\Omega^H}{3} \Gamma \tag{5.51}$$

for the apparent surface stress change due to H absorption in a thin layer of bulk near the surface. If we further compute an equivalent superficial charge density, exploiting

Eq. (5.36) in its form $q + q_0\Gamma = 0$ (with $q_0 = +1.6022 \times 10^{-19}$ C the elementary charge), we obtain the apparent electrocapillary coupling parameter from df/dq as

$$\varsigma = \frac{Y^{Pd}}{1 - \nu^{Pd}} \frac{\Omega^H}{3q_0} \tag{5.52}$$

By using the materials parameters for bulk Pd and for H in bulk Pd, as listed in Ref. [8], one finds that $\varsigma = +1.2V$ [48].

By comparison to the experimental data, one sees that the predictions of continuum mechanics for the apparent electrocapillary coupling parameter for H in a subsurface layer of Pd agrees remarkably well with the observations exploring H UPD in experiment. This is confirmed in a recent experimental study by Viswanath [48]. The finding is consistent with our initial assumption that the elastic interaction of H with the Pd lattice originates in charge transfer and bond weakening that acts similarly for H in the bulk or H at the surface. The agreement may also be taken as a further confirmation that the experimental ς value for H electrosorption on Pd from Ref. [43] is physically reasonable.

The partial atomic volumes of H absorbed in the bulk of various transition metals are quite similar, and always positive. Since the elastic parameters also have no dramatic variation between the transition metals, it appears not unrealistic to expect the electrocapillary coupling parameters for H on transition metal surfaces to be similar and with a value not too different from H on Pd.

5.3.9
Coupling Coefficient for Potential of Zero Charge and Work Function

Experiments so far (see Section 5.3.10) indicate that the magnitude of the electrocapillary coupling coefficient ς is largest when the electrode behaves capacitively, with negligible adsorption. The relevant studies typically focus on clean metal surfaces near their potential of zero charge, but capacitive processes can also be studied on surfaces covered by stable chemisorbed monolayers, for instance involving oxygen species.

If ς is measured at $q = 0$, then its magnitude measures the strain-dependence of the pzc, E_{zc}. We may therefore introduce the parameter ς_{zc}, which is defined by

$$\varsigma_{zc} = \frac{dE_{zc}}{de} = \frac{dE}{de}\bigg|_{q=0} \tag{5.53}$$

The pzc of a metal electrode may be related to the work function, W, of the otherwise identical metal surface in vacuum. The work function measures the work required to remove an electron from the metal. Empirically, W and E_{zc} are found to be linked according to [49]

$$E_{zc} = \frac{W}{q_0} + \text{constant} \tag{5.54}$$

Kolb [50] has shown that the scaling between electrode potential and work function continues to hold, analogously to Eq. (5.54), even when the electrode is emersed from electrolyte into vacuum while at potentials different from the pzc.

In view of the empirical relation, Eq. (5.54), one may be led to expect a relation between ς and the work-function strain response. Equation (5.54) here suggests the form

$$\frac{dE_{zc}}{de} = \frac{1}{q_0}\frac{dW}{de} \tag{5.55}$$

Equation (5.54) is of particular relevance for numerical *ab initio* computation of the electro-mechanical coupling at metal surfaces. The computation of W by DFT is routine and implemented in standard codes. The response dW/de is therefore readily accessible. By contrast, the computation of df/dq would require the introduction of excess charge in the DFT computation. This can be done [10,51–53], but it is neither routine nor computationally efficient.

Let us now consider dW/de from the point of view of surface physics and introduce the chemical potential, μ_e, of the electron as the derivative of the total free energy $\mathfrak{F}$ with respect to the number, N_e, of electrons. In other words,

$$\mu_e = \partial \mathfrak{F}/\partial N_e \tag{5.56}$$

In a metal, electric fields are screened by space-charge layers of atomic extension immediately at the surface. This implies that the bulk is charge neutral, that excess charge resides at the surface only, and that its ramifications are suitably embodied in a phenomenological description by means of a charge-dependent superficial free energy density, $\psi^{MV}(q,e)$, of the metal surface in vacuum. We here have $\delta N_e = -A\delta q/q_0$, so that – by Eq. (5.56) – the electron chemical potential is

$$\mu_e = -q_0 \frac{d\psi^{MV}}{dq}\Big|_e \tag{5.57}$$

The work function, W, of the charge-neutral metal surface is related to μ_e by [54]

$$\mu_e = -W = \phi_\infty - \varepsilon_F \tag{5.58}$$

where ϕ_∞ is the asymptotic vacuum value of the electrostatic potential and ε_F is the energy of the highest occupied single particle state (Fermi energy) of the metal. It is useful to measure ε_F relative to the average electrostatic potential in the bulk. The Fermi energy is then simply the sum of the kinetic energy and of the exchange correlation energy. Here, W can be expressed as the difference between the surface dipole, $\Delta\phi_D$, and ε_F. That is [55],

$$W = \Delta\phi_D - \varepsilon_F \tag{5.59}$$

That equation can be understood in the following way: removing an electron from the metal requires that energy is spent in crossing the surface dipole. That energy is partly compensated by the energy which is gained when the kinetic and exchange correlation energies of the bound state are disposed of. When the solid is strained, both $\Delta\phi_D$ and ε_F will generally vary. The work function, therefore, depends on the strain through a surface effect, the change in surface dipole, and through a bulk

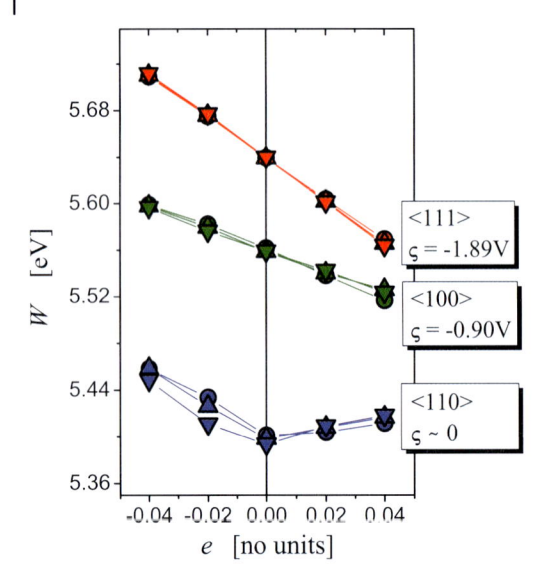

Figure 5.9 Results of numerical evaluation using ab initio DFT for the variation of the work-function, W, of Au surfaces as a function of the tangential strain, e. Surface crystallographic orientation is indicated by labels. The slopes determine the electrocapillary coupling coefficient ς of the charge-neutral surface (indicated in the figure) according to Eq. (5.61). Data from Ref. [56].

effect, the change in Fermi energy. One notes a subtle conflict with the phenomenological description of electrocapillarity, where the electrocapillary coupling coefficient is determined entirely by the surface free energy function $\psi(q, e)$. This function is defined as a local property of the surface, which does not explicitly depend on the strain in the bulk. Nonetheless, computations of the work-function strain response by DFT achieve excellent agreement with experimental data for ς_{zc}, confirming Eq. (5.55). The relevant data will be discussed in Section 5.3.10 below. As an example, Figure 5.9 presents work-function data from DFT for Au surfaces of varying orientation, plotted versus the tangential strain e [56].

The Maxwell relation for $\psi^{MV}(q, e)$ is

$$\frac{df}{dq}\Big|_e = -q_0^{-1} \frac{d\mu_e}{de}\Big|_q \tag{5.60}$$

In view of the relation between μ_e and W, Eq. (5.60), for the special case near $q = 0$ can be alternatively written by substituting $-W$ for μ_e. The surface-stress–charge coupling coefficient for the metal surface in vacuum near $q = 0$ therefore obeys

$$\varsigma_{zc} = q_0^{-1} \frac{dW}{de} \tag{5.61}$$

Equation (5.61) links the empirical relation, Eq. (5.55), to the definition of the work function in surface physics.

5.3.10
Empirical Data for the Electrocapillary Coupling Coefficient

Experimental studies of the potential–strain coupling of electrodes were first described by Gokshtein in the 1960s and 1970s [38,57,58]. Gokshtein's "estance" method (cf. Section 5.3.2) was based on a piezoelectric element – connected mechanically to an electrode that was partly submersed in electrolyte – which served either for creating cyclic, kHz frequency strain in the electrode or for detecting potential change. In different variants, this approach afforded the measurement of potential variation with strain and that of surface stress variation with potential. Gokshtein also presented the underlying thermodynamics that is outlined in Section 5.3.4 [38]. While Gokshtein's theory has been influential, his experimental approach appears not to have been picked up by others. In fact, the relevant publications – partly in Russian – do not provide easy reading, even when one is assisted by an expert translator. It is not always obvious exactly how the electrode processes were converted into electric signal, how signal strength was calibrated, what is the role of the meniscus created at the electrode–electrolyte–air triple line, and whether the reported data refer to the real or imaginary part of the signal. Nonetheless, Gokshtein's results, as far as they are presented as quantitative data, check out well with more modern findings.

At around the same time as Gokhshtein, Beck and coworkers studied electrocapillary effects by measuring the longitudinal extension of metal foil electrodes as the function of their electrode potential [59,60]. After the 1970s, the field of electrocapillary coupling lay dormant for many years. In the mid 1980s, Seo and coworkers explored a quasi-static piezoelectric detection scheme [61,62], but surface stress was not quantified and the approach appears to be no longer in use today. Wider spread and sustained interest started with the advent of cantilever-bending experiments of surface stress variation during potential cycles in the mid 1990s [63–69]. In fact, the cantilever-bending technique remains the standard tool in the field today [70–76]. The experimental approach originally used scanning-probe microscopy type set-ups. Instrumentation using the reflection of multiple laser beams off cm-size cantilevers, originally developed for the *in situ* measurement of stress during thin film growth [77], is now commercially available, while highly sensitive single-beam set-ups have also been developed [71]. The latter allow fast sampling, the frequency range being limited by the cantilever resonances [78], see Section 5.4 below.

Based on cantilever-bending experiments, Haiss and coworkers [18,67] pointed out the fundamental relevance of the surface stress–charge coupling coefficient and suggested that its value could provide information on the electron transfer during anion electrosorption and on the 'strength of adsorption'. Viswanath and coworkers [79] have confirmed Haiss' concept by observing the increase in coupling strength while the impact of anion adsorption was gradually diminished in dilution series. These authors used nanoporous metals as the electrode and dilatometry, rather than the standard technique of cantilever bending, as the detection scheme. The extremely large number of surfaces amplifies the strain created by the potential-induced variation of the capillary forces, suggesting the use of electrocapillary coupling in sensing [80] and actuation [81–84]. By means of demonstration, Kramer *et al.* [82] have coated a cantilever with nanoporous gold, achieving a tip displacement of several mm when the potential was cycled. Inasmuch as the action of the capillary forces at a solid surface became for the first time visible to the naked eye, the experiment may be seen as a milestone in the field of surface stress.

The amount of bending of a cantilever depends on the surface stress change via elastic parameters of the cantilever material and via the geometry. Some early experiments may have missed corrections to Stoneys equation (Eq. (5.19)) incorporating the impact of clamping constraints [74,85], and more recent work has highlighted the important role of surface roughness [76,86]. While the clamping issues are now routinely dealt with by suitable geometry, corrections for roughness – even though they are viable in principle [86,87] and successfully model phenomena such as the (line-) stress of step edges [88] – remain to be implemented into cantilever experiments. Some experiments using nanoporous solids face an unsolved issue in mechanics, namely a missing link between the mean surface-induced pressure (Eq. (5.21)) and the macroscopic volume change of the porous body [32,83,89]. Even for those studies where there is uncertainty in the scaling factors which link the experimental observable, namely the deformation, to the surface stress change, reported trends such as the variation of the experimental ς with strength of adsorption or with anion concentration in solution are expected to be valid.

As there are different measures for 'strength of adsorption', the above-mentioned link between the value of ς near the pzc and the type and concentration of anions is not forceful. With an eye on the Maxwell relation, Eq. (5.29), Sieradzki and co-workers suggested that the key experiment would directly determine the variation of the pzc with strain, assuming that this would exclude effects of adsorption on the value of ς [74]. Since pzc values typically come with a significant uncertainty, large strain values are required for shifting E_{zc} in a measurable way. In a study similar to that of Section 5.3.8, Kibler and coworkers have presented such an experiment, using thin Ag overlayers strained by epitaxy with different misfitting substrates to create strain of the order of several percent [90]. This does afford a measurable shift of E_{zc}, yet issues of anion adsorption are not completely ruled out, and new issues concern the question of strain relaxation by misfit dislocations as well as the so-called "ligand effect", in which the electronic structure of thin (few monolayers) metal films are affected not only by strain but also by electron exchange with the substrate crystal.

Potential–strain coupling coefficients can be explored in isolation, without the influence of adsorption or of ligand effects, by means of computer experiments. Using DFT, Weigend *et al.* [53] showed that charged Au clusters in vacuum deform in agreement with expectation based on empirical data for ς of gold in weakly adsorbing electrolytes. These authors proposed a model for electrocapillary coupling that focusses on a combination of out-of-plane relaxation in response to Hellmann–Feynman forces between the space-charge and the surface ion cores and transverse mechanical coupling that links relaxation to the surface stress in the plane. Umeno et al. [10] confirmed that the charge-induced relaxation of the outermost atomic planes of Au crystals in DFT agrees with experimental data by Nichols et al. [9] for relaxation of Au single crystals in electrolyte. In two separate studies [56,91], Umeno and coworkers also reported values of ς_{zc} for different single crystal surfaces based on DFT. The authors pointed out that the empirical finding of a scaling of the pzc of an electrode surface with the corresponding work function in vacuum links ς_{zc} to the work-function strain response (see Section 5.3.9). Since the computation of work functions by DFT is routine, electrocapillary coupling coefficients for surfaces of various transition metals could be computed.

Recent work in our laboratory has returned Gokshthein's quest of measuring the electrode-potential variation with strain directly during cyclic straining of the electrode. To this end, Dominik Kramer, Maxim Smetanin and Qibo Deng have developed an experimental set-up for dynamic electro-chemo-mechanical analysis (DECMA) [28,33,34], see Section 5.4.3. As explained there, DECMA uses a piezo-actuator to deform the electrode, which is a thin metal film – typically (111)-textured – supported on a polymer substrate. In this way, strain amplitudes $\hat{e}$ in the elastic regime (typically $\hat{e} = 10^{-4} - 10^{-3}$) at frequencies up to 100 Hz can be imposed. The approach measures $\varsigma(E)$ for all electrode processes that are accessible to cyclic voltammetry, including capacitive processes on clean or adsorbate-covered surfaces, electrosorption and underpotential deposition.

Electrocapillary coupling coefficients can be defined for various electrode processes, including, specifically, capacitive charging, adatom adsorption, and underpotential deposition. Tables 5.1 and 5.2 compile selected values for two specific types of processes, (nominally) capacitive charging and adsorption of protons and oxygen species. Data for surface stress variation during adsorption on metal surfaces in vacuum or during the underpotential deposition of metals can be found in the literature and may, in principle, be discussed in terms of apparent values of ς, quite analogously to what is done in the tables.

It is emphasized that Tables 5.1 and 5.2 cannot summarize all experiments on surface stress variation in electrolyte published so far; the focus is on those values for which the ς are reported or for which those parameters can be estimated based on the available data. Furthermore, we here ignore the important body of results on surface stress variation during the UPD of metals, as investigated by Stafford and Bertocci at NIST, by Seo at Hokkaido University and by others.

The data for nominally capacitive charging in Table 5.1 reveal the following trends:

- The Au(111) surface is best studied. Experimental values of ς are available from three independent approaches, namely the change in pzc during reconstruction, the variation in surface stress with superficial charge density as measured by cantilever bending, and the variation in electrode potential with strain as measured by DECMA. All experiments give the same value of ς, with a precision of $\pm 5\%$. Within that confidence limit, DFT data agree with experiment. By confirming the Maxwell relation, the experiment provides a strong confirmation of the phenomenological theory of electrocapillary coupling.

Table 5.1 Values of the electrocapillary coupling coefficient ς for nominally capacitive processes.

Surface	ς [V]	Method	Eq.	Ref.
Au (111) in HClO$_4$	-2	$\Delta E_{zc}/\Delta e$ during reconstruction	(5.53)	[74]
Au (111-tex.) in HClO$_4$	-2.0 ± 0.1	$f(q)$ cantilever bending	(5.30)	[41]
Au (111-tex.) in HClO$_4$	$-2.0 \cdots -2.4$	df/dq dynamic cantilever bending	(5.31)	[78]
Au (111-tex.) in HClO$_4$	-1.83	dE/de DECMA	(5.31)	[33]
	-1.9 ± 0.2		(5.31)	[28]
Au (111-tex.) in NaF	-1.95 ± 0.1	$f(q)$ cantilever bending	(5.30)	[41]
Au (111-tex.) in H$_2$SO$_4$	-1.8 ± 0.2	dE/de DECMA	(5.31)	[28]
Au (111) in vacuum	-1.86 ± 0.02	$W(e)$ DFT	(5.61)	[56]
Pt (pc) in NaF	-1.9 ± 0.2	$f(q)$ nanoporous metal dilatometry	(5.30)	[79]
Pd (111-tex.) in H$_2$SO$_4$	-1.3	dE/de DECMA	(5.31)	[92]
Ag (thin film) in H$_2$SO$_4$	~ -1.5	dE/de DECMA	(5.31)	[34]
Cu (111) in vacuum	-2.32	$W(e)$ DFT	(5.61)	[91]
Rh (111) in vacuum	0.94			
Pd (111) in vacuum	-0.98			
Ag (111) in vacuum	-2.31			
Ir (111) in vacuum	-0.61			
Pt (111) in vacuum	-1.00			
Au (100) in vacuum	-0.90 ± 0.02	$W(e)$ DFT	(5.61)	[56]
Au (110) in vacuum	~ 0			
Cu (100) in vacuum	< 0	$W(e)$ DFT	(5.61)	[93]
oxygen-covered Pt (pc) in NaF	$+1.6$	$f(q)$ nanoporous metal dilatometry	(5.30)	[79]
oxygen-covered Au (pc) in NaF and HClO$_4$	$+2.0 \cdots +4.7$			[89]
oxygen-covered Pd (pc) in H$_2$SO$_4$	$+0.5$	dE/de DECMA	(5.31)	[92]

Table lists (i) selected values from experiments (typically) in the vicinity of the pzc and using weakly adsorbing ions or (ii) values obtained by numerical computations via DFT for surfaces in vacuum. Focus is on metals with face-centered cubic (fcc) crystal lattice structure and on the more densely packed surfaces. Data for other surface orientations and for other crystal structures are available, see in particular Ref. [91]. Symbols and abbreviations: tex – textured, ΔE_{zc} – pzc, f – surface stress, E – electrode potential, e – area strain, q – superficial charge density, W – work function. Top section of table lists ς for Au(111) from different experimental approaches and theory. Note the excellent agreement.

- The coupling coefficients for clean transition metal surfaces are consistently negative. However, capacitive processes on oxygen-covered surfaces are characterized by a positive-valued ς.
- The DFT data for Au surfaces of different crystallographic orientation testify to a strong dependence of the coupling coefficient on the atomic structure of the surface.

As a background for the discussion of strain effects on catalysis it is also of interest to inspect the electrocapillary coupling associated with adsorption processes. To this

Table 5.2 Values of apparent electrocapillary coupling coefficients ς for adsorption of hydrogen and of oxygen species on noble metal surfaces.

Adsorbate	ς [V]	Method	Eq.	Ref.
H on Pd	$+1.4 \pm 0.2$	$E_{ad}(e)$ pseudomorphic layers	(5.31)	[43]
	$+1.7 \pm 0.3$	$f(q)$ nanoporous metal	(5.30)	[43]
	$+1.1$	$E(e)$ DECMA	(5.31)	Figure 5.14
	$+1.2$	theory H in surface layer	(5.52)	[48]
H on Pt	$+1.5$	$f(q)$ nanoporous metal	(5.30)	[79]
	$+1.9$	$f(\Gamma)$ DFT, 1 ML on (111)	(5.44)	[43,94]
O on Pt	-0.8 ± 0.1	$f(q)$ nanoporous metal	(5.30)	[79]
	-2.7	$f(\Gamma)$ DFT, 0.25 ML on (111)	(5.44)	[94]
O on Au	-1.6 ± 0.3	$f(\Gamma)$ DFT	(5.44)	[43,95]
	< 0	decoration of strain fields	—	[43,96]
	< 0	cantilever bending	(5.30)	[97]

end, Table 5.2 shows selected data for the adsorption of hydrogen and of oxygen species on noble metal surfaces. It is seen that the values which have been reported so far differ more than those for capacitive processes. Yet, a number of significant observations emerge:

- Experiments agree that ς for H adsorption on noble metal surfaces is positive-valued and of the order of 1–2 V. Here again, fundamentally different experimental approaches provide values of the coupling coefficient that agree within error bars, and DFT data are in reasonable agreement.
- The data for oxygen species adsorption agree that the coupling coefficient is negative. As compared to H adsorption, the data exhibit a much larger scatter. This may relate to the greater complexity of the adsorption process, where different species can adsorb (OH or O) and where oxygen and metal may undergo place exchange. In the absence of systematic experiments, it is not obvious which process dominates in any one of the different experimental studies.
- All data for oxygen as well as hydrogen are compatible with the concept that tensile strain makes the surface more binding for those adsorbates.
- One previously reported trend is not compatible with the data of Table 5.2: Ibach [68], working with data for adsorption from gas, considered how the surface stress variation connects to the expected charge transfer between adsorbate and substrate during the formation of chemical bonds. The data in that study suggest that "the sign of the change in the surface stress induced by adsorption depends on the direction of the charge donation involved in the adsorption process. The induced stress is compressive and tensile for a charge depletion and enrichment in the substrate surface, respectively." Inspecting the table, one would consider H as enriching the surface in electrons, yet $\varsigma > 0$ implies that the surface stress becomes more compressive ($f < 0$) during adsorption. Apparently, the interaction of adsorbates with the electronic structure of the surface defies simple concepts based on charge exchange.

In summary, even though correlations between the electronic structure of the elements and their electrocapillary coupling behavior have been pointed out, it seems fair to say that the state of the art allows predictions on the sign or magnitude of the electrocapillary coupling only – if at all – in exceptional cases.

5.4
Exploring the Dynamics

5.4.1
Outline

Cantilever-bending experiments, as discussed above, measure surface stress changes and, through them, the electrocapillary coupling strength during the potential scans of cyclic voltammetry. At typical potential scan rates, cycles take of the order of 1 min. The considerations on theory in Section 5.3 connect to such experiments by considering the electrode processes as quasi-static. Classic electrochemical approaches go beyond cyclic voltammetry by instigating potential cycles of small amplitude but high frequency, ω, around a given mean potential value, $\bar{E}$. This is the realm of electrochemical impedance spectroscopy, which yields important extra information on the kinetics of transport in the electrolyte and of electrosorption in the inner Helmholtz layer.

As compared to the signals of cyclic voltammetry, the electrocapillary coupling coefficients of the individual electrode processes provide separate information, since the ς-values emphasize the phenomena – for instance, changes in bond-strength – within the solid surface and are essentially independent of the processes in the solution. By analogy, it is conceivable that experiments exploring the electrocapillary response on short timescales might provide extra information that is over and above what can be learned from impedance spectroscopy. It is then of interest to set up measuring schemes that access the variation of the electrocapillary coupling with the frequency. First reports of two approaches towards frequency-dependent measurements of the electrocapillary response have recently been reported and will be addressed below.

The approach by Stafford and coworkers [78] explores, to what extent surface stress measurements by cantilever bending can be carried out at frequency. For small potential amplitudes around a given $\bar{E}$, the observed amplitude of the surface stress variation scales with ς. In an earlier work, Smetanin, Deng and the present author implemented a different approach [28,33], in which a small cyclic strain is imposed on the electrode while charge or potential are held constant. This approach accesses ς via the potential–strain response, the other side of the Maxwell relation, Eq. (5.29). As compared to cantilever bending, it brings the additional option of measuring a strain-induced modulation of the Faraday current at constant electrode potential. This enables characterization of the strain-dependence of nonequilibrium phenomena and, in particular, strain-dependent electrocatalytic reactivity (see Section 5.5).

While the experiments so far have indicated the feasibility of frequency-dependent studies of the electrocapillary coupling, systematic studies of the frequency dependence remain to be reported.

5.4.2
Cyclic Cantilever-Bending Experiments

Consider an electrochemical potential scan, for example, starting from E_0 and with scan rate $\dot{E}$, which has a cyclic potential modulation of frequency ω and amplitude $\hat{E}$ superimposed so that $E = E_0 + \dot{E}t + \hat{E}\sin(\omega t)$. If the electrode processes are fast compared to the cycle time $2\pi/\omega$, then the electrode charge and the surface stress will follow the modulation with amplitudes $\hat{q} = c\hat{E}$ and $\hat{f} = \varsigma\hat{q}$. Since the two amplitudes are readily measured in cantilever-bending experiments, the electrocapillary coupling parameter ς can be obtained as the ratio of these amplitudes, $\varsigma = \hat{f}/\hat{q}$. Figure 5.10 presents the result of such an experiment from Ref. [78]. Its general features agree well with the results of static cantilever-bending experiments and of dynamic electro-chemo-mechanical analysis, see Figure 5.13 below and the discussion there. The authors of Ref. [78] have found that centimeter-sized cantilevers enable such cyclic experiments with frequencies of up to 100 Hz.

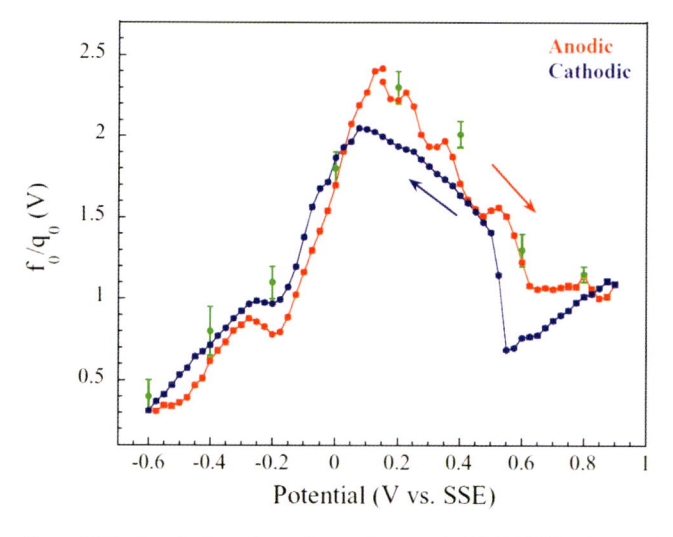

Figure 5.10 Results from the cyclic cantilever-bending study of Ref. [78]. The ratio, f_0/q_0, of surface stress variation with charge density variation as measured from the vibration of a cantilever excited at the frequency of 1 Hz by a small cyclic potential variation, superimposed to the much slower potential scan (abscissa) of a cyclic voltammogram. The surface is Au(111) in 0.1 M HClO$_4$. Arrows denote the potential scan direction. The parameter f_0/q_0 here agrees with the magnitude, $|\varsigma|$, of the electrocapillary response parameter. Note the agreement with results of dynamic electrochemical mechanical analysis, see Figure 5.13 below. Reproduced from Ref. [78].

5.4.3
Dynamic Electro-Chemo-Mechanical Analysis

Consider an experiment where the strain e varies cyclically with time, t, so that

$$e = \hat{e}\sin(\omega t) \qquad (5.62)$$

with frequency ω and amplitude $\hat{e}$. Consider further that the electrode is ideally polarizable (no Faraday processes) and that open circuit conditions apply. The potential will then vary in phase with the strain,

$$E = \hat{E}\sin(\omega t) = \varsigma\hat{e}\sin(\omega t) \qquad (5.63)$$

with $\hat{E}$ the potential amplitude. Therefore, the electrocapillary coupling coefficient ς can be obtained as the ratio of the experimental amplitudes of E (in-phase or real part) and e, that is, $\varsigma = \hat{E}_{re}/\hat{e}$.

Analogously to the strain cycles at constant charge, let us also consider an experiment where e is cycled at constant potential. In view of Eq. (5.35), the charge density will vary according to

$$q = -\varsigma c\hat{e}\sin(\omega t) \qquad (5.64)$$

which implies that the net current density, j, varies as

$$j = \frac{dq}{dt} = -\varsigma c\hat{e}\omega\cos(\omega t) \qquad (5.65)$$

It is seen that the current modulation is phase-shifted by 90°, so that the electrocapillary coupling coefficient is obtained from the out-of-phase (imaginary) part of the complex current amplitude via $\varsigma = -(c\omega)^{-1}\hat{j}_{im}/\hat{e}$

Equations (5.63) and (5.65) link the amplitudes of the potential and current modulations in response to cyclic strain to the electrocapillary coupling coefficient. DECMA is a technique which exploits that link for experimental studies of the electrocapillary coupling [28,98]. A cyclic strain is imposed on the electrode by a suitable loading device, while potential or charge are controlled. A lock-in technique detects the real and imaginary parts of the potential or current modulation. In this way, the technique affords the measurement of the coupling coefficient. Figure 5.11 shows a schematic representation of the DECMA experimental set-up, and Figure 5.12 depicts the electrode and the standing meniscus of the electrolyte by which it is wetted in an experimental set-up.

The experimental processes of interest are limited to surfaces with e near zero, and we can, therefore, ignore the strain-dependence of ς here. However, similar to other second derivatives of the surface free energy function, such as c, the coupling coefficient ς varies with E. As is explained in Ref. [28,98], DECMA can be set up to measure the modulation of either the potential (Eq. (5.63)) or the current (Eq. (5.65)) in response to strain *in situ* during the potential sweep of cyclic voltammetry. The technique thus provides a way of exploring the function $\varsigma(E)$ in the potential interval accessible to cyclic voltammetry.

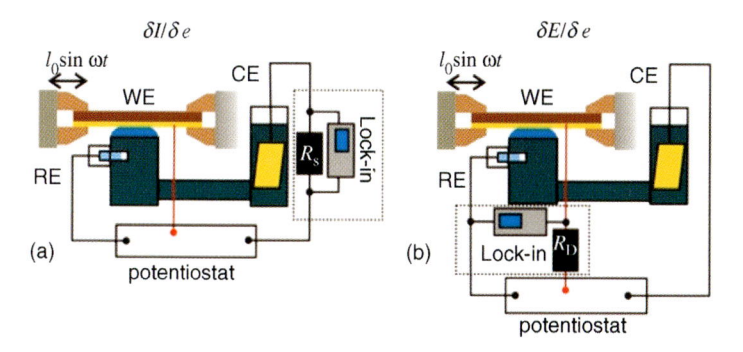

Figure 5.11 Schematic display of experimental set-ups for current–strain response and potential–current response during potentiostatic cyclic voltammetry using DECMA. Drawings showing counter, working, and reference electrodes (CE, WE, RE), potentiostat, as well as shunt resistance R_S for cyclic current measurement in (a) and delay resistance R_D acting as low-pass filter in (b). After Ref. [28].

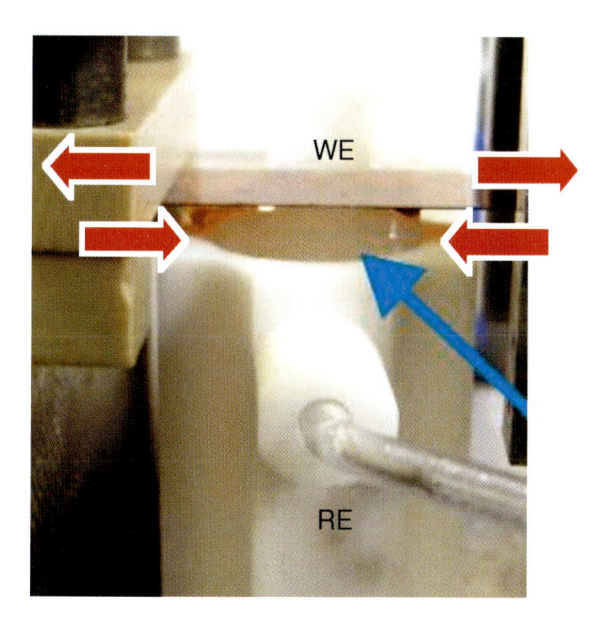

Figure 5.12 Detail of the experimental set-up for DECMA. The working electrode (WE) is a 50 nm thin metal film evaporated onto the bottom side of a polymer film substrate. The substrate is suspended between grips, one of which is seen on the left of the image. Grip displacement controlled by a piezoactuator allows the film to be cyclically strained (red arrows). A standing meniscus of electrolyte (blue arrow) wets the electrode from below. Part of the electrochemical cell (white polymer casing), including the reference electrode (RE) port can be seen. After Ref. [98].

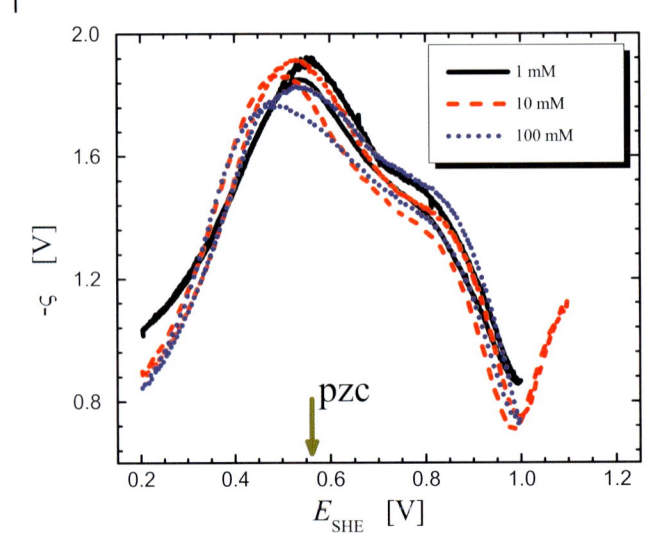

Figure 5.13 Experimental results of DECMA for a thin-film Au electrode in $HClO_4$ of different concentration as indicated in the figure. The negative, $-\varsigma$, of the electrocapillary coupling coefficient is plotted versus the electrode potential E. Note the extremum in $\varsigma(E)$ near the pzc, which is at ~560 mV here. After Ref. [28].

First DECMA results have recently been published [28,92]. As an example, Figure 5.13 shows the variation of the electrocapillary coupling coefficient with electrode potential for Au in $HClO_4$ as an electrolyte with weak adsorption [28]. Several observations are remarkable. First, it is seen that ς is practically independent of the electrolyte concentration, suggesting that processes which are characteristic of the electrode surface control the coupling. Second, it is noteworthy that $\varsigma(E)$ exhibits an extremum near the pzc, which is at ~560 mV here. As was already discussed in Section 5.3.2, thermodynamics does not require such an extremum. However, if similar features would empirically be confirmed at other metal surfaces, the measurement of $\varsigma(E)$ might evolve into an independent way of determining the pzc. Thirdly, the value of ς at the pzc, -1.9 ± 0.2 V [28], agrees with the value obtained from the cantilever-bending experiment described in Section 5.3.3 and in Ref. [41]. In other words, the Maxwell relation, Eq. (5.29), is confirmed by comparison of surface stress changes inferred from cantilever bending with potential changes observed in DECMA.

By inspection of Figure 5.10 it is seen that the data from cyclic cantilever bending of Ref. [78] is in good agreement with the data from DECMA, Ref. [28]. Specifically, the maximum of $|\varsigma|$ in the vicinity of the pzc is reproduced, and the numerical values of the coefficients are consistent.

As another example for DECMA results, Figure 5.14 shows $\varsigma(E)$ for Pd in dilute sulfuric acid [92]. The potential interval here includes several electrode processes, and it can be seen that the values of $\varsigma(E)$ provide signatures of these processes. At intermediate potentials, in the nominally capacitive regime of the voltammogram, ς

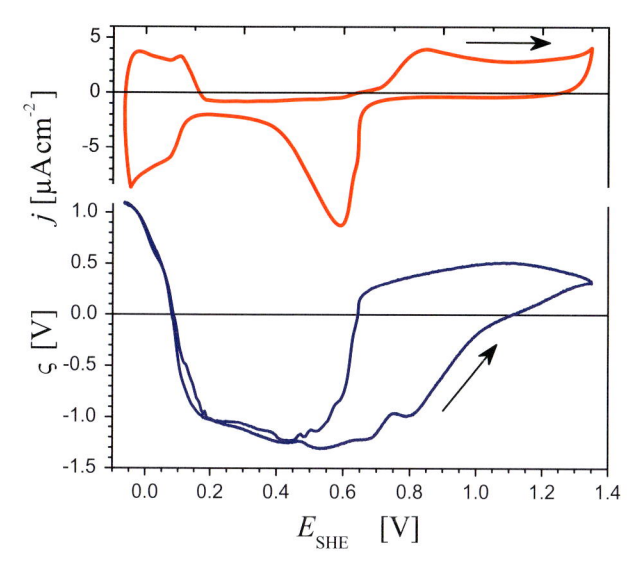

Figure 5.14 Experimental results of DECMA on a thin-film Pd electrode in 10 mM H_2SO_4. Top part, cyclic voltammogram of current density j versus electrode potential E. Bottom part, electrocapillary coupling coefficient ς versus E. Arrows denote scan direction. After Ref. [92].

assumes a negative value, as on Au. OH adsorption at more positive potentials is accompanied by a decrease in ς. Because oxygen species are rather stable on the Pd surface, the negative-going scan at positive potentials involves dominantly capacitive processes on an oxygen-covered Pd surface. This process is distinguished by a positive-valued ς, which prevails until the oxygen is desorbed and ς returns to negative. At the negative end of the potential scan, hydrogen adsorbs on the surface and is absorbed in the bulk of the electrode. This regime is again characterized by a positive-valued ς, in agreement with the surface stress data discussed in Section 5.3.8. The features of Figure 5.13 are also well compatible with observations of the surface stress variation with charge density during porous metal dilatometry experiments, for instance, on Pt [79].

5.5 Mechanically Modulated Catalysis

5.5.1 Outline

The considerations in Sections 5.2 and 5.3 refer to ideally polarizable electrodes near equilibrium. Interfacial reactions under non-equilibrium conditions will be inspected next, with attention to the impact of strain on the reaction rate. The analysis is based on recent work by Deng *et al.* [99], who propose a kinetic rate equation for electrocatalysis on strained surfaces and compare the predictions to first experimental data.

As a prerequisite for a phenomenological description of the mechanically modulated reactivity in heterogeneous catalysis we start out by inspecting the distinction between Faraday and capacitive current. We have in mind experiments by DECMA (see Section 5.4.3), using a small cyclic strain to impose a small modulation on the reaction rate, which can be detected based on the modulation of the reaction current. In view of the small variation of strain and reaction rate, the focus of the theory can be restricted to a linearized description of the interrelation between the two quantities. The results are coupling coefficients that are precisely defined and that can be quantified based on the DECMA data.

Subsequently, we shall explore the strain-dependence within the framework of simple microscopic models for electrocatalytic reactions, specifically the Butler–Volmer equation and a model for the Heyrowsky reaction. This will afford a connection of the phenomenological coupling coefficients to the microscopic parameters such as adsorption enthalpies and energy barriers for the various reaction steps.

5.5.2
Phenomenology; Distinguishing Capacitive from Faraday Current

So far, we have restricted attention to situations where the superficial charge density q is an appropriate state variable, in other words, to ideally polarizable electrodes (see Section 5.1 in Ref. [100]). Variations of q involve capacitive (charge-storage via ion accumulation in the diffuse layer) or pseudocapacitive (charge storage via ion accumulation in the inner Helmholtz layer) processes.

By definition, (see Section 5.2.3 and specifically Eq. (5.8)), q is linked to the superficial excess Γ_i of those species i which, in the bulk of the electrolyte, carry a charge. The definition of Γ_i does not distinguish between excess species residing in the inner and in the outer Helmholtz layer. When an ion – for instance, a proton –undergoes specific adsorption, then one may think of charge transfer between electrode and adsorbate that converts the ion effectively into a neutral atom (in the example, an adsorbed hydrogen atom). This local process is irrelevant for a phenomenological description, since it does not change the net charge on the electrode. The adsorbed atom still contributes to the excess and, thereby, to q. In a cyclic process, the local charge transfer can be undone, the hydrogen adatom desorbed and the proton in solution recovered along with the extra electron in the metal surface. In this sense the stored charge is recoverable. This holds even though specific adsorption/desorption processes typically involve hysteresis and are, therefore, not reversible in a thermodynamic sense. It is this more general type of process which we have in mind when speaking about pseudo-capacitive storage of charge.

With the above remarks in mind we introduce the pseudo-capacitive current density, j^{pc}, (current per referential area), which includes capacitive and pseudocapacitive processes and which is linked to q via the time (t) derivative

$$j^{pc} = \frac{dq}{dt} \tag{5.66}$$

Since q is linked to the excess of ions, this quantity remains well-defined, even when the electrode is away from equilibrium. It is thus possible to retain q and, consequently, the pseudo-capacitive current density j^{pc} as meaningful quantities, even in the presence of interfacial reactions.

Over and above the pseudo-capacitive current, we admit the presence of Faraday current, of density j^F, due to the interfacial reactions. This current relates to the transfer of charge that cannot be recovered and that is, therefore, inherently unrelated to q. We again restrict attention to isothermal processes at constant values of the chemical potential of all species in solution. All further analysis is then based on the constitutive assumption that the net current – which is the sum of the pseudo-capacitive variations of the interfacial charge density q as well as the reaction current – is given by the expression:

$$j = \frac{dq}{dt} + j^F(E, e). \tag{5.67}$$

Specifically, therefore, we assume that j^F is completely specified by the electrode potential and by the strain. Microscopic models for catalytic reactions will typically involve parameters such as the superficial excess, Γ_i, of various species i (reactants, reaction products, intermediate and spectator species). This is compatible with our assumption as long as the Γ_i themselves are functions of E and e.

In this section we restrict attention to a small-strain scenario with a linear response of the reaction current to strain. With this in mind we approximate j^F by

$$j^F(E, e) = j_0^F(E) + \iota(E)e \tag{5.68}$$

where j_0^F is the Faraday current density of the unstrained electrode and

$$\iota = \frac{dj^F}{de}\bigg|_E \tag{5.69}$$

at $e = 0$ is the current–strain coupling coefficient. As will become apparent below, it is also of interest to introduce an alternative measure for the reaction–strain coupling in the form of the coefficient λ, defined as

$$\lambda = \frac{RT}{F} \frac{1}{j^F} \frac{dj^F}{de}\bigg|_E \tag{5.70}$$

Let us now consider a cyclic strain as in Eq. (5.62), and examine an experiment at constant potential. With the description developed so far, the net current density is then readily found as

$$j = j_0^F + \iota \hat{e} \sin(\omega t) - \varsigma c \hat{e} \omega \cos(\omega t) \tag{5.71}$$

The first term on the right-hand side is the ordinary Faraday current in the absence of strain, the second is the modulated Faraday current, and the third represents the cyclic charge exchange associated with pseudocapacitive processes.

Equation (5.71) provides a basis for separating the effects of electrocapillary coupling (represented by ς) from the mechanical modulation of the Faraday current (represented by ι): The first effect is phase-shifted relative to the strain by 90° and it scales with the frequency of the mechanical excitation. By contrast, the second effect is in-phase as well as invariant when the frequency is varied. Ongoing experiments in our laboratory aim at exploiting this opportunity by means of DECMA experiments during electrode reactions [99].

We shall now discuss microscopic models that afford a link between λ and the underlying processes at the atomic scale.

5.5.3
Rate equations: Butler–Volmer kinetics

As a representation of the simplest type of reaction kinetics, let us consider the symmetric Butler–Volmer equation [100],

$$j^F = j^{ex}\left\{\exp\left(-\frac{nF\Delta E}{2RT}\right) - \exp\left(\frac{nF\Delta E}{2RT}\right)\right\} \tag{5.72}$$

Here

$$\Delta E = (E - E_{eq}) \tag{5.73}$$

is the overpotential for a reaction that is at equilibrium at potential E_{eq}, j^{ex} is the exchange current density, and n is number of electrons involved in the electrode reaction. The symmetry factor has been set to 1/2 by assumption. The current–potential response of the Butler–Volmer equation is characteristic for a reaction where the surface is saturated with the reactand species and the rate-limiting step is overcoming an energy barrier for combination or desorption of the product species.

The kernels of the exponentials in Eq. (5.72) contain the net free energy of reaction, $nF\Delta E$, as the driving force. In heterogeneous catalysis the reactants and products are all in solution, so that the free energy of reaction at any given value of E is not coupled to the strain of the electrode. For the Butler–Volmer model, the coupling of reactivity to strain is, therefore, entirely embodied in the exchange current density j^{ex}.

We may take j^{ex} to depend on the rate-limiting energy barrier via

$$j^{ex} = \text{const.} \exp\left(-\frac{\Delta h^{ex}}{RT}\right) \tag{5.74}$$

where the constant prefactor const. embodies the density of active sites and the attempt frequency. Δh^{ex} represents the activation enthalpy associated with the barrier. Admitting that Δh^{ex} depends on the strain, we may write in a linear approximation valid for small strain

$$\Delta h^{ex} = \Delta h_0^{ex} - F\varsigma^{ex}e \tag{5.75}$$

Here, the strain-response of Δh^{ex} is measured by the coupling coefficient ς^{ex}. The notation is thereby analogous to that introduced in the discussion of electrosorption, and specifically to Eq. (5.49). The resulting equation for the exchange current density is then

$$j^{ex} = j_0^{ex} \exp\left(\frac{F\varsigma^{ex}e}{RT}\right) \tag{5.76}$$

with j_0^{ex} the exchange current density at no strain. Analogously, the reaction current is obtained as

$$j^F = j_0^F \exp\left(\frac{F\varsigma^{ex}e}{RT}\right) \tag{5.77}$$

The reaction-current strain coupling coefficients ι and λ are readily obtained from the above expression for j^F:

$$\iota = j_0^F \varsigma^{ex} \frac{F}{RT} \tag{5.78}$$

and

$$\lambda = \varsigma^{ex} \tag{5.79}$$

It is seen that the current–strain response here scales with the current, and that the relative change, when normalized to F/RT, has a constant value that gives directly the coupling coefficient for the barrier-height variation with strain. For the special model considered here (namely Butler–Volmer behavior), λ is a constant in the entire potential range, both positive and negative of the equilibrium potential and independent of the sign of the reaction current.

5.5.4
Rate Equations: Heyrowsky Reaction

As a second example for a microscopic process, let us inspect the hydrogen evolution reaction (HER) with the ion + atom (Heyrowsky) [101,102] reaction,

$$H_{aq}^+ + H_{ad} + e^- \rightleftharpoons H_2 \tag{5.80}$$

as the rate-controlling step. Figure 5.15 illustrates the reaction pathway and its energetics. The H_2 desorption step as well as the discharge (Volmer) reaction, $H_{aq}^+ + e^- \rightleftharpoons H_{ad}$, are taken to be fast. Consistent with that assumption, the coverage θ with adsorbed atomic hydrogen at any given value of E is assumed to take on the equilibrium value of the Langmuir isotherm with molar free energy of adsorption Δg^{ad}. This latter parameter is defined as

$$\Delta g^{ad} = \Delta h^{ad} - T\Delta s_0^{ad} \tag{5.81}$$

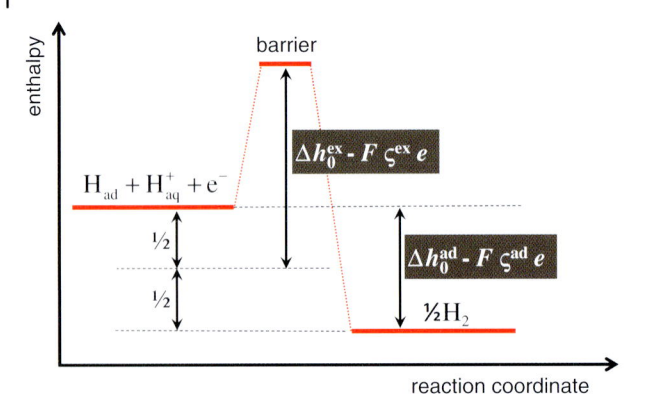

Figure 5.15 Schematic enthalpy diagram for the Heyrowsky reaction, showing enthalpy levels of the charge-neutral species (adsorbed atomic hydrogen, H^{ad}, and H_2 in adsorbed or gas form) as well as the energy of the transition state. The small strain e is taken to affect the values of the adsorption enthalpy, Δh^{ad}, and of the barrier height, Δh^{ex}, through the coupling parameters ς^{ad} and ς^{ex}.

where the adsorption enthalpy Δh^{ad} represents the change in molar enthalpy for the reaction $1/2 H_2^{gas} \rightarrow H^{ad}$, and Δs_0^{ad} refers to the corresponding entropy change, accounting for the entropies in the gas and for a θ-independent vibrational entropy of the adsorbed H, but excluding the configurational entropy of the adsorbate phase. Equation (5.45) along with the equilibrium condition of Eq. (5.11) gives the Langmuir isotherm for θ in its explicit form as the function

$$\theta^L = \left(1 + \exp\frac{\Delta g^{ad} + F\Delta E}{RT}\right)^{-1} \tag{5.82}$$

Figure 5.16 gives a schematic presentation of the various quantities involved in the present discussion of the HER within the Heyrowsky frame. Part (a) illustrates the potential-dependent coverage of the Langmuir isotherm. In the example, the electrosorption potential E^{ad} is at negative overpotential, as for metals with weak H adsorption (positive Δh^{ad}).

Parsons has analyzed the relevant reaction kinetics in detail [101]. For conciseness, we again restrict attention to energies that vary symmetrically with the reaction coordinate (symmetry parameter $\alpha = 1/2$). Parsons' expression for the Faraday current can be written in the form (see, for instance, Ref. [103])

$$j^F = j^{ex}\left\{\frac{(1-\theta)}{(1-\theta_{eq})}\exp\left(+\frac{F\Delta E}{2RT}\right) - \frac{\theta}{\theta_{eq}}\exp\left(-\frac{F\Delta E}{2RT}\right)\right\} \tag{5.83}$$

with θ_{eq} the coverage at the equilibrium potential of the HER. The number of electrons involved in the reaction was taken as $n = 1$. Figure 5.16b illustrates the reaction kinetics by means of a Tafel plot. One notes a transition between two different Tafel slopes at the H electrosorption potential.

We here take strain to affect the microscopic processes during the reaction in two fundamental ways: (i) a strain-dependent activation enthalpy for the ion + atom

5.5 Mechanically Modulated Catalysis

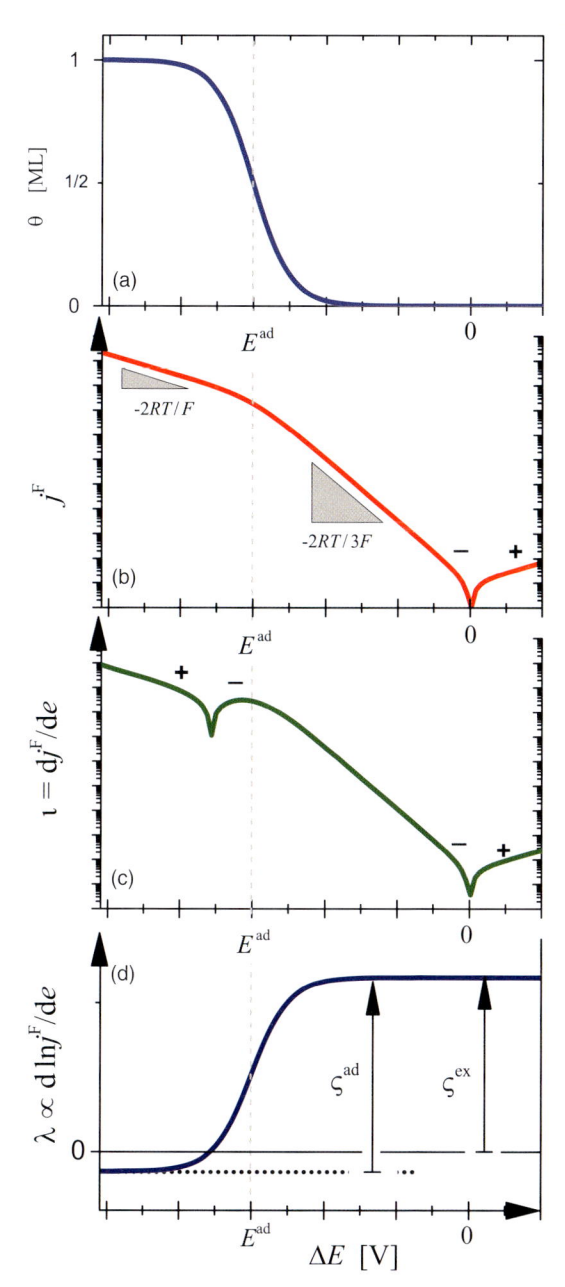

reaction step, analogous to that in Eq. (5.75) of the previous Section, and (ii) a strain-dependence of the hydrogen adsorption enthalpy.

As in Eq. (5.75), we express the strain response of the hydrogen adsorption enthalpy through

$$\Delta h^{ad} = \Delta h_0^{ad} - F\varsigma^{ad} e \tag{5.84}$$

where ς^{ad} is given by Eq. (5.49) with $z = 1$.

The strain-dependence of Δh^{ad} affects the Faraday current of Eq. (5.83) since the isotherm, Eq. (5.46), becomes strain-dependent. This propagates into the coverages, so that $\theta = \theta^L(E, e)$. It is important to note that the strain-dependence needs to be carried through to the coverage at the equilibrium potential. Otherwise, the principle of detailed balance would be violated for the strained surface and the reaction current at equilibrium would deviate from zero. Thus, we also have $\theta_{eq} = \theta^L(E_{eq}, e)$.

The reaction-strain coupling coefficient λ is obtained by evaluating Eq. (5.83) in terms of the definition of λ. Specifically, Eq. (5.76) is used for the exchange current density in Eq. (5.83), and the coverage θ is obtained from Eq. (5.82) with Eq. (5.84) for the adsorption enthalpy. By taking the derivative with respect to the strain e, the result for the current-strain response is obtained as

$$\lambda = \left(\frac{1}{1 + \exp\frac{\Delta g^{ad}}{RT}} - \frac{1}{1 + \exp\frac{\Delta g^{ad} + F\Delta E}{RT}}\right)\varsigma^{ad} + \varsigma^{ex} \tag{5.85}$$

In view of Eq. (5.82), this expression can be transformed to the simpler result

$$\lambda = \varsigma^{ex} + (\theta_{eq} - \theta)\varsigma^{ad} \tag{5.86}$$

That result expresses the current-strain coupling coefficient λ as the weighted sum of the two relevant electrocapillary coupling parameters ς^{ex} and ς^{ad}. The weighting factor emerges as linearly dependent on the H coverage θ. That implies that λ may vary significantly near the potential of electrosorption, while λ takes on approximately constant values at more negative or more positive potentials. Table 5.3 displays values of λ for important limiting cases.

It is remarkable that the coupling remains well-behaved near the equilibrium potential of the reaction. Even when the current inverts its sign along with the overpotential, λ retains the constant value ς^{ex}. In other words, λ near equilibrium

Figure 5.16 Schematic representation of the various quantities involved in the discussion of strain-dependent Heyrowsky kinetics. Example is for weakly adsorbing H, as for instance on Au; furthermore, coupling parameters for adsorption (ς^{ad}) and activation (ς^{ex}) have been linked – arbitrarily – by $\varsigma^{ex} = 0.9\varsigma^{ad}$. (a) H coverage θ versus overpotential ΔE, with electrosorption potential ΔE^{ad} marked on abscissa. (b) Magnitude of reaction current density, $|j^F|$, versus E. Shaded triangles denote Tafel slopes, changing from $-2RT/3F$ while $\theta \ll 1$ at small overpotential to $-2RT/F$ when $\theta \approx 1$ at large overpotential. Note logarithmic ordinate scale and labels indicating sign of j^F, for example, $j^F < 0$ in the hydrogen evolution regime at $\Delta E < 0$. (c) Response parameter, $\iota = dj^F/de$, for the reaction current change with strain. Note logarithmic ordinate scale and labels indicating sign as in (b). Since $j^F < 0$, negative sign of response here implies enhanced reactivity for tensile strain. (d) Reduced response parameter, $\lambda = RTF^{-1}d\ln j^F/de$, for the reaction current change with strain. Positive value means reactivity increases with tensile strain.

Table 5.3 The values of the reaction rate–strain coupling coefficient λ for the Heyrowsky reaction on surfaces with different values of the adsorption enthalpy, Δh^{ad}, for hydrogen.

	$\Delta E \ll \Delta h^{ad}/F$ $\theta \approx 1$	$\Delta E = 0$ θ arbitrary	$\Delta E \gg \Delta h^{ad}/F$ $\theta \ll 1$
$\Delta h^{ad} \gg 0$ (Au)	$\varsigma^{ex} - \varsigma^{ad}$	ς^{ex}	ς^{ex}
$\Delta h^{ad} \approx 0$ (Pt)	$\varsigma^{ex} - \varsigma^{ad}/2$	ς^{ex}	$\varsigma^{ex} + \varsigma^{ad}/2$
$\Delta h^{ad} \ll 0$ (Ni)	ς^{ex}	ς^{ex}	$\varsigma^{ex} + \varsigma^{ad}$

Examples for relevant materials are indicated (in brackets) for illustration. The three limiting cases of the overpotential, ΔE, refer to concentrated and dilute hydrogen adsorbate (coverage $\theta \approx 1$ and $\theta \ll 1$, respectively) and to the equilibrium of the hydrogen evolution reaction, $\Delta E = 0$, where the results shown apply irrespective of the value of θ. The parameters ς^{ex} and ς^{ad} refer to the electrocapillary coupling parameters for the activation barrier of the Heyrowsky reaction and for the hydrogen adsorption enthalpy, respectively. It is seen that λ jumps by $-\varsigma^{ad}$ when the surface undergoes the transition from dilute to concentrated adsorbate during hydrogen underpotential deposition.

measures the strain-dependence of the exchange current density. Depending on the sign of Δh^{ad}, the transition from dilute to concentrated adsorbate population at the hydrogen electrosorption potential occurs at negative or positive overpotential. That transition is accompanied by a jump in λ, of magnitude $-\varsigma^{ad}$.

The present approach to the strain dependence treats the coupling coefficients ς^{ad} and ς^{ex} as independent phenomenological parameters. In this respect it is of interest that empirical findings suggest that adsorption and activation energies for catalytic reactions may be linked. The Brønsted–Evans–Polanyi relation [104,105] states that, for a given reaction on different surfaces, the activation energy scales with the adsorption energy. Scaling factors near to but slightly less than unity have been reported [106]. Conceivably, this concept may be transferred to mechanically modulated catalysis. In that sense one might suspect a similar scaling when the reaction takes place on the same crystal surface in different states of strain. This would imply that the effect of strain – in, say, the tensile direction – on Δh^{ex} is of same sign as that on Δh^{ad}. In other words, if the Brønsted–Evans–Polanyi relation was applicable, the coefficients ς^{ad} and ς^{ex} could be expected to be of same sign and similar magnitude, $\varsigma^{ex}/\varsigma^{ad} \lesssim 1$. However, the applicability of the Brønsted–Evans–Polanyi relation to strain-induced variations in the energies is be no means obvious.

Figure 5.16c and d show the reactivity–strain coupling coefficients ι and λ, staying with the example of weakly adsorbing H (as, for instance, on Au). Consistent with the we above considerations on electrocapillary coupling for H on transition metal surfaces and on Brønsted–Evans–Polanyi scaling, the example considers positive-valued ς^{ad} along with $\varsigma^{ex}/\varsigma^{ad} = 0.9$. As a consequence, $\varsigma^{ex} - \varsigma^{ad} < 0$. This negative value is remarkable in view of the transition between $\lambda = \varsigma^{ex}$ and $\lambda = \varsigma^{ex} - \varsigma^{ad}$ at small and large overpotential magnitude, respectively (see Table 5.3). The implication is that the scaling results in a sign-inversion of the reactivity–strain coupling at large overpotential, near the electrosorption potential. Tensile strain *enhances* the exchange current density and the reactivity for hydrogen evolution under small forward bias. By contrast, under large forward bias the impact of strain is inverted and tensile strain *inhibits* the reaction. This transition is clearly illustrated by the graphs in Figure 5.16c and d.

So far, experiments exploring the variation of the reactivity–strain coupling with overpotential have not been reported. However, ongoing work in our laboratory suggests that DECMA may be adapted to investigate the issue [99].

5.6
Summary and Outlook

Experimental studies of the electrocapillary coupling go back to Gokhshtein's work almost half a century ago. The past decade has seen the field move into a state where reproducible, quantitative data are reported. Results of experiment and theory by several groups now agree accurately. In parallel, several instances have emerged where the electrocapillary coupling parameters are the key descriptors of materials behavior. This relates in particular to electrochemical actuation and sensing. Within the context of the present book, electrocapillary coupling provides the link between mechanics and electrosorption that governs the mechanical modulation of reactivity. This chapter outlines the theory of capillarity of solid electrodes, and quotes selected experimental data for illustration.

Section 5.2 introduces the distinction between surface tension and surface stress as two capillary parameters for solid surfaces. It is emphasized that the two parameters are separately relevant for different phenomena. For instance, the surface tension governs the wetting angle of fluids on solid electrode surfaces, whereas the surface stress measures the mechanical interaction of the surface with the underlying bulk solid. Fundamentally different capillary equations govern the respective equilibria. Confusing the capillary parameters or the capillary equations may lead to grossly erroneous predictions.

Section 5.3 introduces the notion of electrocapillary coupling. The coupling parameter ς describes the impact of changes of state of solid electrodes on the capillary forces. It is relevant in situations as different as the surface stress variation during capacitive charging or during adsorption, the electron emission from surfaces in vacuum, or the shift of electrosorption potentials when an electrode is strained. A considerable body of evidence has accumulated specifically for the electrocapillary coupling of charge neutral metal surfaces, either in vacuum or in electrolytes near the pzc. This evidence comes from experiment – using cantilever bending or porous metal expansion schemes – as well as *ab inito* computation. Both approaches typically reach excellent agreement.

Section 5.4 introduces recent approaches toward a direct measurement of the electrocapillary coupling in experiment. Both approaches are based on fast excitations at small amplitude, superimposed on the much slower potential cycles of the voltammetry. Dynamic electro-chemo-mechanical analysis measures ς from the response of the electrode potential to cyclic strain, whereas the more conventional approaches explore the response of the surface stress to charge. Both schemes achieve excellent mutual agreement. The results so far are the first data for $\varsigma(E)$ in wide potential intervals. There is also the prospect of extending those studies to explore the frequency-dependence of electrocapillary coupling.

Section 5.5 explores how the electrocapillary coupling affects the reactivity of the surface. As a simple constitutive assumption, it allows for a strain-dependence of the adsorption enthalpies as well as the energy of the transition state that defines the barrier for the reaction. The relevant electrocapillary coupling coefficients are allowed to be independent. The results indicate that the magnitude and sign of the coupling between reactivity and strain depend not only on the strain dependence of the adsorption enthalpy of the reactants, but also on the overpotential and on the regime – dilute or concentrated – of the electrosorption process.

In conclusion, the following is a random selection of current issues in the field of electrocapillarity of solid electrodes, as they emerge from the discussion in this chapter:

- *Quantifiable parameters.* The field of electrocapillary coupling has progressed to a state where experiments as well as theory produce data that are sufficiently detailed, accurate and reproducible to warrant a quantitative comparison. As a prerequisite for linking experiment and theory, quantifiable parameters must be defined so that they can be evaluated by both approaches, and their values must be reported. The parameters ς of this work, in their application to various electrode processes, afford an example for such a quantifiable basis.
- *Strength for adsorption–strain coupling.* Whereas studies of the coupling of work-function or pzc to strain have seen good progress over the past decade, much less is known about the coupling of adsorption enthalpies to strain. Part of the theory, such as Refs. [107,108], has focused on adsorption on expitaxial monolayers. This is relevant for experiments, but the impact of ligand effects prevents understanding the impact of strain in isolation. Experimental values for electrocapillary coupling coefficients associated with adsorption, for instance measured using cantilever bending or DECMA, would be valuable as a basis for understanding the impact of mechanics on the reactivity of electrodes. The same applies for additional DFT studies of adsorption on strained substrates.
- *Quantitative experiments on mechanical modulation of catalysis.* Theory has predicted that, qualitatively, it is possible to modulate the catalytic reactivity of surfaces mechanically. That prediction is confirmed by experiments, some of which are described elsewhere in this book. Most of those experimental findings have so far also remained qualitative. The field stands to benefit from quantitative data. As outlined above, this will – first and foremost – require the definition of quantifiable parameters. The quantity λ, which was introduced in Section 5.5.2, provides an example of such a parameter that can be linked to the underlying electrocapillary coupling terms. It is noteworthy that even simple kinetic models, such as that of the Heyrowsky reaction, show transitions as the function of the overpotential, where λ switches between values that are in fact representative of ς of distinctly different electrode processes. Since the ς values can be studied separately – for instance, through surface stress changes at equilibrium – it may be expected that studies of the mechanical modulation of catalysis can contribute to the identification of the reaction mechanisms.

- *Microscopic origin of the work-function–strain response.* Even though several computer studies have explored the coupling between the electron work function and strain by *ab initio* simulation, no theory has so far emerged that would afford a prediction of magnitude or even the sign of the coupling strength. Since this is arguably the simplest instance of electrocapillary coupling, its understanding might be a starting point for a more general theory of coupling effects, including adsorption processes or charging of adsorbate-covered surfaces. Furthermore, while the Maxwell relation that equates the work-function–strain response with the response of surface stress to charging is well established on a phenomenological level, the link between the two separate coupling parameters has not been explained in terms of microscopic phenomena. This is puzzling and should prompt further studies.
- *Is there a relation between adsorption–strain coupling and work function–strain coupling?* The electron chemical potential is coupled to the strain via the work-function strain response. Furthermore, strain affects the electrosorption. Since the electrosorption of ions involves the transfer of electronic charge, one might wonder if the impact of strain on the work function and on adsorption are interrelated, and what is the nature of that interrelation?
- *Rough electrodes.* Many experiments in the field of electrocapillary coupling use electrodes with surfaces that are stepped, faceted, or otherwise not perfectly planar. General results for the mechanics of such surfaces are available [86–88] and allow the local capillary forces to be backed out from experiment even when the surface is rough. Yet, the relevant corrections are not yet routinely applied in studies of the capillarity of solid electrodes. The accuracy of data in the field could be further enhanced if roughness correction algorithms were established and routinely applied in experimental studies.
- *Electrocapillary forces in energy storage?* In principle, the electrocapillary coupling phenomena might be expected to have an impact on energy storage strategies based on lithium ion or hydrogen absorption in solid aprticles. First, the capillary forces induce large stresses in the bulk of nanoparticles, as they are used in energy storage devices. These surface-induced stresses impact the energetics of dissolution, changing the alloy phase diagrams and the electric or chemical potentials. Secondly, the incorporation of Li or H into a solid storage material typically involves interfacial reactions with an energy barrier. Similar to electrocatalysis, the energy of the barrier can be expected to depend on the strain. Storage materials typically undergo large strain when their composition changes during the absorption of lithium or hydrogen. These considerations suggest that capillary forces and electrocapillary coupling phenomena may conceivably have a significant impact on the performance of energy storage materials. This impact appears not to have found due attention so far and may deserve further studies.

Acknowledgements

The author acknowledges the close cooperation with Qibo Deng, Dominik Kramer, Maxim Smetanin and Raghavan N. Viswanath, and many stimulating discussions over the past decade with Ludwig A. Kibler, Dieter M. Kolb, Karl Sieradzki and Gery R. Stafford. Special thanks to Qibo Deng and Maxim Smetanin for permission to present, in Section 5.5.4, results of joint work on mechanically modulated catalysis prior to publication. Thanks also to Maxim Smetanin for a critical reading and for translation of selected passages from publications, in Russian, by AY Gokhshtein. Permission to reproduce figures was kindly granted by Elsevier Global rights (Figure 5.5), by The PCCP Owner Society (Figures 5.7, 5.8, 5.11 and 5.13) and by The Electrochemical Society (Figures 5.10 and 5.14).

References

1 Gibbs, J.W. (1928) *The Collected Works of J. W. Gibbs*, volume **1**, Longmans, Green & Co., text and footnotes on p. 315.
2 Weissmüller, J. and Cahn, J.W. (1997) Mean stresses in microstructures due to interface stresses: A generalization of a capillary equation for solids. *Acta Mater.*, **45** (5), 1899–1906.
3 Shuttleworth, R. (1950) The surface tension of solids. *Proc. Phys. Soc. London A*, **63** (365), 444–457.
4 Herring, C. (1951) *Surface Tension as a Motivation for Sintering*, McGraw-Hill, New York, pp. 143–179.
5 Gurtin, M.E. and Murdoch, A.I. (1975) A continuum theory of elastic material surfaces. *Arch. Ration. Mech. An.*, **57**, 291–323.
6 Gurtin, M.E., Weissmüller, J., and Larche, F. (1998) A general theory of curved deformable interfaces in solids at equilibrium. *Philos. Mag. A*, **78** (5), 1093–1109.
7 Inglesfield, J.E. (1985) Reconstructions and relaxations on metal-surfaces. *Prog. Surf. Sci.*, **20** (2), 105–164.
8 Lemier, C. and Weissmüller, J. (2007) Grain boundary segregation, stress and stretch: Effects on hydrogen absorption in nanocrystalline palladium. *Acta Mater.*, **55** (4), 1241–1254.
9 Nichols, R.J., Nouar, T., Lucas, C.A., Haiss, W., and Hofer, W.A. (2002) Surface relaxation and surface stress of Au(111). *Surf. Sci.*, **513** (2), 263–271.
10 Umeno, Y., Elsässer, C., Meyer, B., Gumbsch, P., and Weimueller, J. (2008) Reversible relaxation at charged metal surfaces: An ab initio study. *EPL (Europhysics Letters)*, **84** (1), 13002.
11 Nozières, P. and Wolf, D.E. (1988) Interfacial properties of elastically strained materials .1. thermodynamics of a planar interface. *Z. Phys. B Condens. Matter*, **70** (3), 399–407.
12 Weissmüller, J. and Kramer, D. (2005) Balance of force at curved solid metal-liquid electrolyte interfaces. *Langmuir*, **21** (10), 4592–4603.
13 Lipkowski, J., Schmickler, W., Kolb, D.M., and Parsons, R. (1998) Comments on the thermodynamics of solid electrodes. *J. Electroanal. Chem.*, **452** (2), 193–197.
14 Shreiber, D. and Jesser, W.A. (2006) Size dependence of lattice parameter for Si_xGe_{1-x} nanoparticles. *Surf. Sci.*, **600** (19), 4584–4590.
15 Nozières, P. (1991) *Solid Surfaces: Capillarity vs. Elasticity, Lecture Notes in Physics*, Springer, pp. 113–128.
16 Cammarata, R.C. and Sieradzki, K. (1994) Surface and interface stresses. *Annu. Rev. Mater. Sci.*, **24**, 215–234.
17 Ibach, H. (1997) The role of surface stress in reconstruction, epitaxial growth and stabilization of mesoscopic structures. *Surf. Sci. Rep.*, **29** (5–6), 195–263.
18 Haiss, W. (2001) Surface stress of clean and adsorbate-covered solids. *Rep. Prog. Phys.*, **64** (5), 591–648.
19 Müller, P. and Saúl, A. (2004) Elastic effects on surface physics. *Surf. Sci. Rep.*, **54** (5–8), 157–258.
20 Kramer, D. and Weissmüller, J. (2007) A note on surface stress and surface tension and their interrelation via Shuttleworth's

equation and the Lippmann equation. *Surf. Sci.*, **601** (14), 3042–3051.
21 Weißmüller, J. (1998) Grenzflächen in Nanostrukturierten Materialien - Lokale Eigenschaften und Makroskopische Mittelwerte. Habilitationsschrift. Mathematisch-Naturwissenschaftliche Fakultät der Universität des Saarlandes, Saarbrücken, Germany.
22 Weissmüller, J., Kramer, D., Viswanath, R.N., and Gleiter, H. (2003) Durchstimmbare Dehnung in platin. *Physik in Unserer Zeit*, **34** (4), 155–156.
23 Cahn, J.W. and Hilliard, J.E. (1958) Free energy of a nonuniform system. 1. Interfacial free energy. *J. Chem. Phys.*, **28** (2), 258–267.
24 Cahn, J.W. (1961) On spinodal decomposition. *Acta Met.*, **9** (9), 795–801.
25 Cahn, J.W. (1977) Critical-point wetting. *J. Chem. Phys.*, **66** (8), 3667–3672.
26 Cahn, J.W. (1978) *Thermodynamics of Solid and Fluid Surfaces*, ASM, Metals Park, Ohio, pp. 3–23.
27 Young, Thomas. (1805) An essay on the cohesion of fluids. *Philos. Trans. R. Soc. London*, **95**, 65–87.
28 Smetanin, M., Deng, Q.B., and Weissmüller, J. (2011) Dynamic electro-chemo-mechanical analysis during cyclic voltammetry. *Phys. Chem. Chem. Phys.*, **13** (38), 17313–17322.
29 Stoney, G.G. (1909) The tension of metallic films deposited by electrolysis. *Proc. Roy. Soc. London, Ser. A*, **82** (553), 172–175.
30 Laplace, P.S. (1805) *Mécaniqe Céleste*, vol. **4**, Courcier, Paris.
31 Gibbs, J.W. (1928) *The Collected Works of J. W. Gibbs*, volume **1**, Longmans, Green & Co., subsection 'The Stability of Surfaces of Discontinuity' in section 'Surfaces of Discontinuity between Fluid Masses'.
32 Weissmüller, J., Duan, H.L., and Farkas, D. (2010) Deformation of solids with nanoscale pores by the action of capillary forces. *Acta Mater.*, **58** (1), 1–13.
33 Smetanin, M., Kramer, D., Mohanan, S., Herr, U., and Weissmüller, J. (2009) Response of the potential of a gold electrode to elastic strain. *Phys. Chem. Chem. Phys.*, **11** (40), 9008–9012.

34 Smetanin, M., Deng, Q.B., Kramer, D., Mohanan, S., Herr, U., and Weissmüller, J. (2010) Reply to the 'comment on "Response of the potential of a gold electrode to elastic strain"' by A. Horvath, G. Nagy and R. Schiller. *Phys. Chem. Chem. Phys.*, **12** (26), 7291–7292.
35 Pannikkat, A.K. and Raj, R. (1999) Measurement of an electrical potential induced by normal stress applied to the interface of an ionic material at elevated temperatures. *Acta Mater.*, **47** (12), 3423–3431.
36 Lippmann, G. (1875) Relations entre les phénomènes électriques et capillaries. *Ann. Chim. Phys.*, **5**, 494.
37 Couchman, P.R. and Davidson, C.R. (1977) Lippmann relation and surface thermodynamics. *J. Electroanal. Chem.*, **85** (2), 407–409.
38 Gokhshtein, A.Ya. (1969) Charge acquirement by an elastically strained electrode. *Dokl. Akad. Nauk SSSR*, **187** (3), 601.
39 Gokhshtein, A.Ya. (1975) The estance method. *Russ. Chem. Rev.*, **44** (11), 921–932.
40 Gokhshtein, A.Ya. (2012) On the simple check of electrocapillarity. *J. Solid State Electrochem.* pages doi: 10.1007/s10008-012-1843-z.
41 Smetanin, M., Viswanath, R.N., Kramer, D., Beckmann, D., Koch, T., Kibler, L.A., Kolb, D.M., and Weissmüller, J. (2008) Surface stress-charge response of a (111)-textured gold electrode under conditions of weak ion adsorption. *Langmuir*, **24** (16), 8561–8567.
42 Nørskov, J.K., Bligaard, T., Logadottir, A., Kitchin, J.R., Chen, J.G., and Pandelov, S. (2005) Trends in the exchange current for hydrogen evolution. *J. Electrochem. Soc.*, **152** (3), J23–J26.
43 Weissmüller, J., Viswanath, R.N., Kibler, L.A., and Kolb, D.M. (2011) Impact of surface mechanics on the reactivity of electrodes. *Phys. Chem. Chem. Phys.*, **13** (6), 2114–2117.
44 Kibler, L.A., El-Aziz, A.M., Hoyer, R., and Kolb, D.M. (2005) Tuning reaction rates by lateral strain in a palladium monolayer. *Angew. Chem. Int. Ed.*, **44** (14), 2080–2084.

45 Ruban, A., Hammer, B., Stoltze, P., Skriver, H.L., and Nørskov, J.K. (1997) Surface electronic structure and reactivity of transition and noble metals. *J. Mol. Catal. A-Chem.*, **115** (3), 421–429.

46 Alefeld, G. (1978) *Hydrogen in Metals I, Topics in Applied Physics*, vol. **28**, Springer, Berlin.

47 Mott, N.F. and Nabarro, F.R.N. (1940) An attempt to estimate the degree of precipitation hardening, with a simple model. *Proc. Phys. Soc.*, **52** (1), 86.

48 Viswanath, R.N. and Weissmüller, J. (2013) Electrocapillary coupling coefficients for hydrogen electrosorption on palladium. *Acta Mater.* doi: 10.1016/j.actamat.2013.07.013.

49 Trasatti, S. (1971) Work function, electronegativity, and electrochemical behaviour of metals 2. Potentials of zero charge and electrochemical work functions. *J. Electroanal. Chem.*, **33** (2), 351.

50 Rath, D.L. and Kolb, D.M. (1981) Continuous work function monitoring for electrode emersion. *Surf. Sci.*, **109** (3), 641–647.

51 Bohnen, K.P. and Kolb, D.M. (1998) Charge- versus adsorbate-induced lifting of the Au(100)-(hex) reconstruction in an electrochemical environment. *Surf. Sci.*, **407** (1–3), L629–L632.

52 Lozovoi, A.Y. and Alavi, A. (2003) Reconstruction of charged surfaces: General trends and a case study of Pt(110) and Au(110). *Phys. Rev. B*, **68** (24), 245416.

53 Weigend, F., Weissmüller, J., and Evers, F. (2006) Structural relaxation in charged metal surfaces and cluster ions. *Small*, **2** (12), 1497–1503.

54 Smoluchowski, R. (1941) Anisotropy of the electronic work function of metals. *Phys. Rev.*, **60** (9), 661.

55 Lang, N.D. and Kohn, W. (1971) Theory of metal surfaces - work function. *Phys. Rev. B*, **3** (4), 1215–1223.

56 Umeno, Y., Elsässer, C., Meyer, B., Gumbsch, P., Nothacker, M., Weissmüller, J., and Evers, F. (2007) Ab initio study of surface stress response to charging. *Europhys. Lett.*, **78** (1), 13001.

57 Gokhshtein, A.Ya. (1970) Investigation of surface tension of solid electrodes at several frequencies simultaneously. *Electrochim. Acta*, **15** (1), 219–223.

58 Gokhshtein, A.Ya. (1976) *Surface Tension of Solids and Adsorption*, Nauka, Moscow.

59 Beck, T.R. (1969) Electrocapillary curves of solid metals measured by extensometer instrument. *J. Phys. Chem.*, **73** (2), 466.

60 Lin, K.F. and Beck, T.R. (1976) Surface stress curves for gold. *J. Electrochem. Soc.*, **123** (8), 1145–1151.

61 Seo, M., Makino, T., and Sato, N. (1986) Piezoelectric response to surface stress change of platinum-electrode. *J. Electrochem. Soc.*, **133** (6), 1138–1142.

62 Seo, M., Jiang, X.C., and Sato, N. (1987) Piezoelectric response to surface stress change of gold electrode in sulfate aqueous-solutions. *J. Electrochem. Soc.*, **134** (12), 3094–3098.

63 Haiss, W. and Sass, J.K. (1995) Adsorbate-induced surface stress at the solid–electrolyte interface measured with an STM. *J. Electroanal. Chem.*, **386** (1–2), 267–270.

64 Raiteri, R. and Butt, H.J. (1995) Measuring electrochemically induced surface stress with an atomic-force microscope. *J. Phys. Chem.*, **99** (43), 15728–15732.

65 Haiss, W. and Sass, J.K. (1996) Quantitative surface stress measurements on Au(111) electrodes by scanning tunneling microscopy. *Langmuir*, **12** (18), 4311–4313.

66 Brunt, T.A., Rayment, T., Oshea, S.J., and Welland, M.E. (1996) Measuring the surface stresses in an electrochemically deposited metal monolayer: Pb on Au (111). *Langmuir*, **12** (24), 5942–5946.

67 Haiss, W., Nichols, R.J., Sass, J.K., and Charle, K.P. (1998) Linear correlation between surface stress and surface charge in anion adsorption on Au(111). *J. Electroanal. Chem.*, **452** (2), 199–202.

68 Ibach, H., Bach, C.E., Giesen, M., and Grossmann, A. (1997) Potential-induced stress in the solid-liquid interface: Au (111) and Au(100) in an $HClO_4$ electrolyte. *Surf. Sci.*, **375** (1), 107–119.

69 Ueno, K. and Seo, M. (1999) Study of adsorption of iodide ions on gold electrode by a laser-beam deflection method compared with a piezoelectric

technique. *J. Electrochem. Soc.*, **146** (4), 1496–1499.

70 Lang, G.G., Ueno, K., Ujvari, M., and Seo, M. (2000) Simultaneous oscillations of surface stress and potential in the course of galvanostatic oxidation of formic acid. *J. Phys. Chem. B*, **104** (13), 2785–2789.

71 Kongstein, O.E., Bertocci, U., and Stafford, G.R. (2005) In situ stress measurements during copper electrodeposition on (111)-textured Au. *J. Electrochem. Soc.*, **152** (3), C116–C123.

72 Stafford, G.R. and Bertocci, U. (2006) In situ stress and nanogravimetric measurements during underpotential deposition of bismuth on (111)-textured Au. *J. Phys. Chem. B*, **110** (31), 15493–15498.

73 Friesen, C., Dimitrov, N., Cammarata, R.C., and Sieradzki, K. (2001) Surface stress and electrocapillarity of solid electrodes. *Langmuir*, **17** (3), 807–815.

74 Vasiljevic, N., Trimble, T., Dimitrov, N., and Sieradzki, K. (2004) Electrocapillarity behavior of au(111) in SO_4^{2-} and F^-. *Langmuir*, **20** (16), 6639–6643.

75 Tabard-Cossa, V., Godin, M., Beaulieu, L.Y., and Grütter, P. (2005) A differential microcantilever-based system for measuring surface stress changes induced by electrochemical reactions. *Sens. Actuat. B-Chem.*, **107** (1), 233–241.

76 Tabard-Cossa, V., Godin, M., Burgess, I.J., Monga, T., Lennox, R.B., and Grütter, P. (2007) Microcantilever-based sensors: Effect of morphology, adhesion, and cleanliness of the sensing surface on surface stress. *Anal. Chem.*, **79** (21), 8136–8143.

77 Floro, J.A., Chason, E., Lee, S.R., Twesten, R.D., Hwang, R.Q., and Freund, L.B. (1997) Real-time stress evolution during $Si_{1-x}Ge_x$ heteroepitaxy: Dislocations, islanding, and segregation. *J. Electron. Mater.*, **26** (9), 969–979.

78 Lafouresse, M.C., Bertocci, U., Beauchamp, C.R., and Stafford, G.R. (2012) Simultaneous electrochemical and mechanical impedance spectroscopy using cantilever curvature. *J. Electrochem. Soc.*, **159** (10), H816–H822.

79 Viswanath, R.N., Kramer, D., and Weissmüller, J. (2008) Adsorbate effects on the surface stress-charge response of platinum electrodes. *Electrochim. Acta*, **53** (6), 2757–2767.

80 Lavrik, N.V., Tipple, C.A., Sepaniak, M.J., and Datskos, P.G. (2001) Gold nanostructures for transduction of biomolecular interactions into micrometer scale movements. *Biomed. Microdevices*, **3**, 35–44.

81 Weissmüller, J., Viswanath, R.N., Kramer, D., Zimmer, P., Würschum, R., and Gleiter, H. (2003) Charge-induced reversible strain in a metal. *Science*, **300** (5617), 312–315.

82 Kramer, D., Viswanath, R.N., and Weissmüller, J. (2004) Surface-stress induced macroscopic bending of nanoporous gold cantilevers. *Nano Lett.*, **4** (5), 793–796.

83 Jin, H.J., Wang, X.L., Parida, S., Wang, K., Seo, M., and Weissmüller, J. (2010) Nanoporous Au-Pt alloys as large strain electrochemical actuators. *Nano Lett.*, **10** (1), 187–194.

84 Detsi, E., Chen, Z.G., Vellinga, W.P., Onck, P.R., and De Hosson, J.T.M. (2012) Actuating and sensing properties of nanoporous gold. *J. Nanosci. Nanotechnol.*, **12** (6), 4951–4955.

85 Dahmen, K., Lehwald, S., and Ibach, H. (2000) Bending of crystalline plates under the influence of surface stress - a finite element analysis. *Surf. Sci.*, **446** (1–2), 161–173.

86 Duan, H.L., Weissmüller, J., and Wang, Y. (2008) Instabilities of core-shell heterostructured cylinders due to diffusions and epitaxy: Spheroidization and blossom of nanowires. *J. Mech. Phys. Solids*, **56** (5), 1831–1851.

87 Wang, Y., Weissmüller, J., and Duan, H.L. (2010) Mechanics of corrugated surfaces. *J. Mech. Phys. Solids*, **58** (10), 1552–1566.

88 Li, W.N., Duan, H.L., Albe, K., and Weissmüller, J. (2011) Line stress of step edges at crystal surfaces. *Surf. Sci.*, **605** (9–10), 947–957.

89 Jin, H.J., Parida, S., Kramer, D., and Weissmüller, J. (2008) Sign-inverted surface stress-charge response in nanoporous gold. *Surf. Sci.*, **602** (23), 3588–3594.

90 Soliman, K.A. and Kibler, L.A. (2007) Variation of the potential of zero charge

for a silver monolayer deposited onto various noble metal single crystal surfaces. *Electrochim. Acta*, **52** (18), 5654–5658.

91 Albina, J.M., Elsässer, C., Weissmüller, J., Gumbsch, P., and Umeno, Y. (2012) Ab initio investigation of surface stress response to charging of transition and noble metals. *Phys. Rev. B*, **85** (12), 125118.

92 Deng, Q., Smetanin, M., and Weissmüller, J. (2012) Dynamic electro-chemo-mechanical analysis. *Meeting Abstracts*, **MA2012-01** (23), 959.

93 Wang, X.F., Li, W., Lin, J.G., and Xiao, Y. (2010) Electronic work function of the Cu(100) surface under different strain states. *EPL*, **89**, 66004 doi: 10.1209/0295-5075/89/66004.

94 Feibelman, P.J. (1997) First-principles calculations of stress induced by gas adsorption on Pt(111). *Phys. Rev. B*, **56** (4), 2175–2182.

95 Mavrikakis, M., Stoltze, P., and Nørskov, J.K. (2000) Making gold less noble. *Catal. Lett.*, **64** (2–4), 101–106.

96 Gsell, M., Jakob, P., and Menzel, D. (1998) Effect of substrate strain on adsorption. *Science*, **280** (5364), 717–720.

97 Miyatani, T. and Fujihira, M. (1997) Calibration of surface stress measurements with atomic force microscopy. *J. Appl. Phys.*, **81** (11), 7099–7115.

98 Smetanin, M. (2010) Mechanics of Electrified Interfaces in Diluted Electrolytes. PhD thesis, http://scidok.sulb.uni-saarland.de/volltexte/2010/3380/.

99 Deng, Q., Smetanin, M., and Weissmüller, J. (2013) Mechanical modulation of the electrocatalytic hydrogen evolution reaction. *In preparation*.

100 Bard, A.J. and Faulkner, L.R. (1980) *Electrochemical Methods, Fundamentals and Applications*, John Wiley and Sons, Inc., New York.

101 Parsons, R. (1958) The rate of electrolytic hydrogen evolution and the heat of adsorption of hydrogen. *Trans. Faraday Soc.*, **54** (7), 1053–1063.

102 Kibler, L.A. (2006) Hydrogen electrocatalysis. *ChemPhysChem*, **7** (5), 985–991.

103 Gasteiger, H.A., Sheng, W., and Shao-Horn, Y. (2010) Hydrogen oxidation and evolution reaction kinetics on platinum: Acid vs alkaline electrolytes. *J. Electrochem. Soc.*, **157** (11), B1529–B1536.

104 Brønsted, J.N. (1928) Acid and basic catalysis. *Chem. Rev.*, **5** (3), 231–338.

105 Evans, M.G. and Polanyi, M. (1938) Inertia and driving force of chemical reactions. *Trans. Faraday Soc.*, **34** (0), 11–24.

106 Nørskov, J.K., Bligaard, T., Logadottir, A., Bahn, S., Hansen, L.B., Bollinger, M., Bengaard, H., Hammer, B., Sljivancanin, Z., Mavrikakis, M., Xu, Y., Dahl, S., and Jacobsen, C.J.H. (2002) Universality in heterogeneous catalysis. *J. Catal.*, **209** (2), 275–278.

107 Kitchin, J.R., Nørskov, J.K., Barteau, M.A., and Chen, J.G. (2004) Role of strain and ligand effects in the modification of the electronic and chemical properties of bimetallic surfaces. *Phys. Rev. Lett.*, **93** (15), 156801.

108 Greeley, J., Nørskov, J.K., Kibler, L.A., El-Aziz, A.M., and Kolb, D.M. (2006) Hydrogen evolution over bimetallic systems: Understanding the trends. *ChemPhysChem*, **7** (5), 1032–1035.

6
Synthesis of Precious Metal Nanoparticles with High Surface Energy and High Electrocatalytic Activity

Long Huang, Zhi-You Zhou, Na Tian, and Shi-Gang Sun

6.1
Introduction

Catalysts are vital in modern society. To support a world population of more than 7 billion population, large amounts of crops are needed, which is far beyond our nature's ability. Fortunately, the Haber–Bosch process, based on an iron catalyst, makes it possible to fix N_2 artificially to produce ammonia, reaching 160 million tons per year, most of which is consumed as ammonium sulfate, to fertilize the soil for crop growth [1]. In order to make sufficient petroleum, a critical resource in the modern world, platinum-based catalysts are indispensable. To enhance the efficiency, one would expect to make the catalytic surface as large as possible to interact with the reactant. For a heterogeneous catalyst, this means that we should maximize the surface of a catalyst with a given mass since it is generally considered that the heterogeneous catalytic process is only influenced by the topmost layers of atoms of the catalysts. Nanoparticles, with extremely high surface-to-volume ratio, fulfill the requirement. Moreover, the synergistic size effect and quantum confinement effect make the catalyst even more powerful. Nanoparticles-based catalysts, therefore, are widely used in catalytic processes. Though great progress in catalyst design has been achieved, the efficiency and selectivity of most of the man-made catalysts are still far inferior to natural ones. Just take the ammonium synthesis as an example, the natural catalyst – nitrogenase, whose catalytic active center is believed to be related to the iron, can catalyze the nitrogen immobilization reaction using N_2 from the air as substrate at room temperature; in contrast, the ammonia produced industrially is usually done so under extreme conditions, such as temperatures of 400–600 °C and pressure greater than 20 atm [2]. In order to design a proper nanoscale catalyst and further improve its performance, it is crucial to understand the nanoscale topography of surface sites, such as terraces, steps, kinks, adatoms, and vacancies, and their effects on catalytic and other physicochemical properties.

Electrocatalysis: Theoretical Foundations and Model Experiments, First Edition.
Edited by Richard C. Alkire, Ludwig A. Kibler, Dieter M. Kolb, and Jacek Lipkowski.
© 2013 Wiley-VCH Verlag GmbH & Co. KGaA. Published 2013 by Wiley-VCH Verlag GmbH & Co. KGaA.

Among all the nanocatalysts, the platinum-group and iron-group metals are the most important industrial catalyst materials, such as the previously mentioned iron for the Haber–Bosch process in ammonia synthesis [2], platinum–rhenium for petroleum refining in the petrochemical industry [3], platinum–palladium–rhodium for three way convertors in the automobile industry, and platinum for the oxygen reduction reaction (ORR) and direct oxidation of fuel molecules in fuel cells, as well as for nitric acid production in the chemical industry. In all these related applications, the platinum metals are unsubstitutable due to their unique reactivity and stability. However, the platinum group metals are rare resources, their reserve on the earth is quite limited. As a consequence, to develop metal nanoparticle catalysts with excellent activity, stability, selectivity and to increase significantly the utilization efficiency of precious metals is always a focal key issue.

Taking the electrocatalysts of proton exchange membrane fuel cells (PEMFC) as an example, much scientific work has been focused on the non-noble metal catalysts, some reports have even claimed that they have synthesized the non-noble metal-based catalyst with activity comparable to platinum. However, platinum is still the irreplaceable material in a real PEMFC device. Long-term efforts of research and development of the catalyst for PEMFC over the past decades have consisted in increasing the catalyst activity while minimizing the loading of platinum on the carbon support. Investigations have been mainly devoted to decreasing the size of Pt particles yet maintaining their stability, as well as uniformly dispersing the Pt nanoparticles on substrate while keeping their loading as low as possible. The results are evident: the platinum loading has been decreased by about two orders of magnitude during the last half century, that is, from a few tens mg cm^{-2} in the 1960s to a few hundred µg cm^{-2} nowadays [4]. Even though the high price is still one of the bottle necks of commercialization of the FEMFC, in which the most expensive component is the membrane electrode assembly (MEA) the Pt electrocatalysts contribute the major cost. It is worth noting that the effect of decreasing the size of the Pt nanoparticles not only increases the surface-to-volume ratio of the particle, which reduces the Pt loading effectively, but also enhances the reactivity of Pt particles. When the size of a particle decreases, the ratio of active atoms, which have a low coordination number (CN) and locate on the corners and edges of the particle, over the total surface atoms will drastically increase, especially with the particles below a few nanometers in size [5]. Nevertheless, the normally synthesized Pt nanoparticles have often a cuboctahedral shape in order to reach the lowest total surface energy. On such nanoparticles, the active atoms are only 28–19% of the total surface atoms when their size ranges from 3 to 5 nm as in actual commercial Pt nanoparticle catalysts. Only the corners and edges of the nanoparticles are low-coordinated atoms, the faces consist of {111} or {100} low-index facets, whose CN is relatively high and which are catalytically less reactive. It is evident that to significantly increase the ratio of active atoms on metal nanoparticles would be an effective direction for further enhancing the catalytic activity of nanoparticles.

The current platinum loading of a few hundred µg cm^{-2} is still far too expensive to be attractive for a consumer, which makes driving a PEMFC-powered automobile unrealistic. Besides decreasing the size of nanoparticles, which is almost reaching an optimum value, other aspects, such as surface structure and composition, should be optimized to further increase the activity and stability and lower the cost of the catalyst, mainly noble metals such as platinum, to a level that is competitive to the gasoline engine. It has been demonstrated, from fundamental studies using metal single crystal planes as model catalysts, that platinum high-index planes with open surface structure exhibit much higher reactivity than low-index planes like {111} or {100}, because high-index planes have a large density of low-coordinated atoms situated on steps and kinks with high reactivity [6–8]. More importantly, on high-index planes, there exist short-ranged steric sites (such as "chair" sites) that consist of the combination of several (typically 5~6) step and terrace atoms [9,10]. Due to the synergistic effect between step and terrace atoms, the steric sites usually serve as active centers, and such steric sites of short-ranged ordered domain also display a high stability [8,9].

Therefore, if the surface atoms of a metal nanoparticle are organized in such a way that they all (not only the atoms of low coordination number) form active sites, its catalytic activity and stability will be greatly enhanced. Unfortunately, possessing such surface structure usually means a high surface energy and instability. From the thermodynamics point of view, the metal nanoparticles of high surface energy could never be obtained, because the thermodynamics of crystallization require a minimization of total surface energy of the crystal and the nanoparticles of high surface energy will all be eliminated during their growth process [9,11–13]. Therefore, the synthesis of metal nanocatalysts of high surface energy and thus high activity presents a big challenge in the fields of both catalysis and nanoscience and technology [14].

A breakthrough in synthesizing metal nanoparticles of high surface energy was made in 2007 by Tian et al. [15] They developed an electrochemical square wave method. Through the periodic oxidation/dissolution and reduction/crystallization, Pt nanoparticles of tetrahexahedral shape enclosed with {730} and vicinal high-index facets were synthesized in high yield. Stimulated by this work, much progress has been made using three main strategies: (i) periodic oxidation and reduction, such as the electrochemical square wave method and the etching/growth process in wet chemistry methods [16]; (ii) preference-face-blocking adsorption or face-promoting adsorption using functional molecules of surfactants and additives in wet chemistry methods [17–19]; and (iii) controlling the work function of formation of the two-dimensional {hkl} nuclei by fine-tuning the growth potential in electrochemical methods [20,21].

This chapter reviews recent progress in shape-controlled synthesis of metal nanoparticles with high-energy facets and their applications in electrocatalysis, through both electrochemical methods and wet chemistry synthesis. The discussions are focused first on the monometallic nanoparticles, and then on the bi-/tri-metallic nanoparticles. The chapter concludes by considering the future challenges and exciting perspectives of the shape-controlled synthesis of metal nanoparticles with high surface energy and high reactivity in electrocatalytic applications.

6.2
Shape-Controlled Synthesis of Monometallic Nanocrystals with High Surface Energy

6.2.1
Electrochemical Route

Since the first Pt nanocrystal with high surface energy was made by an electrochemical method, that is, the Pt tetrahexahedron nanocrystal by the electrochemical square wave method [15], we will begin this section with the superb and powerful electrochemical route for controlling the shape and thus the surface structure of metal nanoparticles.

6.2.1.1 Platinum
Bearing both high activity and good reactive stability, platinum, without any doubt, becomes the star element of catalysts. Much more attention, not only fundamental but also practical, has been focused on platinum than any other noble element.

It is well known that nanocrystals (NCs) enclosed by high-index facets with high surface energy cannot be fabricated through the thermodynamic control method, since the growth rate in the direction perpendicular to high-index facets is faster than that along low-index ones, resulting in the disappearance of high-index facets during the growth of the nanocrystals [22–24]. We have, therefore, developed an electrochemical method for synthesis of tetrahexahedral(THH) Pt NCs with {hk0} high-index facets [15]. In brief, Pt nanospheres of size ~750 nm were first electrochemically deposited on a glassy carbon substrate, and then subjected to a treatment with a square-wave potential at 10 Hz, with upper potential of 1.20 V (vs SCE) and lower potential between −0.10 and −0.20 V, in 0.1 M H_2SO_4 containing 30 mM ascorbic acid for 5–60 min. The SEM image of the as-prepared THH Pt NCs is shown in Figure 6.1a, which is quite similar to the geometrical model of the ideal tetrahexahedron presented in Figure 6.1b. The surface structure of the THH Pt NCs has been further identified as mainly {730} facets by high-resolution transmission electron microscopy (HRTEM) and selected-area electron diffraction (SAED). The crucial factor for the formation of high-index facets was proposed to be the repetitive adsorption/desorption of oxygen generated by the square-wave potential, namely: (i) the dynamic oxygen adsorption/desorption mediated by the square-wave potential, oxygen adsorption occurring at high potential and desorption at low potential; (ii) the significant impediment to place-exchange between oxygen and Pt surface atoms on high-index surfaces; and (iii) the decrease in surface energy of the Pt NCs by electrochemical adsorption of hydrogen at low potential and oxygen species at high potential. All those factors result in the growth of THH Pt NCs during a treatment of square-wave potential. Guided by the proposed mechanism, we further developed a one-step method to synthesize Pt THH instead of the two-step one mentioned above. The one-step method involves quick nucleation under the modified square wave potential, the THH Pt NCs are formed by direct growth along the surfaces of the Pt nucleus [15]. This straightforward one-step method avoids the coexistence of THH NCs with residual polycrystalline nanospheres, in contrast to with the previous two-step method.

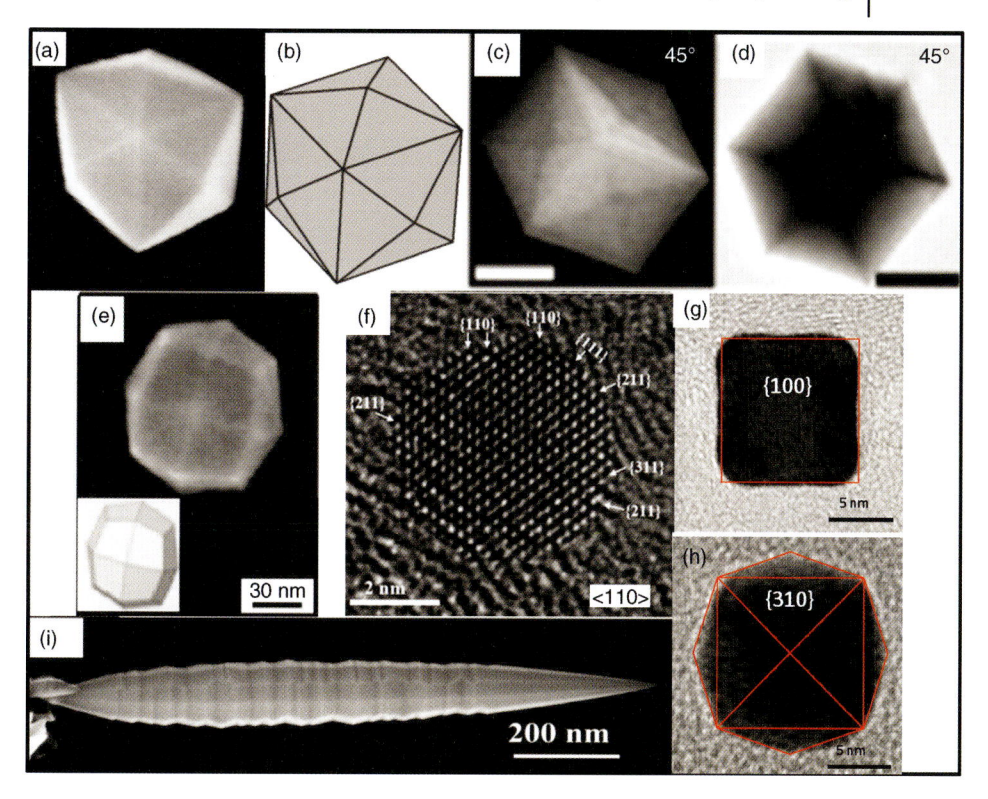

Figure 6.1 (a) High-magnification SEM image of Pt THH, scale bar: 100 nm. (b) Geometrical model of an ideal THH [15]. (c) SEM and (d) TEM images of concave THH Pt NCs synthesis in DESs, scale bar: 100 nm [25]. (e) SEM image of the TPH Pt NCs, the insets show an ideal geometrical model of the TPH NCs with the same orientation [26]. (f) Aberration-corrected HRTEM images of HIF-Pt/C catalysts [27]. HRTEM images of Pt nanocubes/CNTs (g) and THH Pt NCs/CNTs (h), the nanocrystals are oriented along the [001] direction [28]. (i) HRSEM image of Pt fivefold twinned nanorods [29].

Using the one-step method while replacing the water with deep eutectic solvents (DESs), other NCs enclosed with high-index {hk0} facets can be synthesized, as the SEM and TEM image in Figure 6.1c and d show [25]. The concave THH Pt NCs were mainly enclosed by {910} and {10, 1, 0} facets, determined by AFM. The formation mechanism of the convex and concave THH Pt nanocrystals should be similar since both of them are enclosed with {hk0} facets. However, the two cases are to some extent different: the shape of the NCs synthesized in H_2SO_4/H_2O is convex while those synthesized in DES are concave. The reason is still unclear, it is certainly a possibility that the solvent–metal interaction could affect the shape of the nanocrystals.

By simply modifying the potential and the precursor, we can use the square-wave method to synthesize Pt trapezohedral (TPH) NCs in the [01$\bar{1}$1] crystallographic zones, whose surfaces are enclosed by {hkk} facets (Figure 6.1e) [26]. The precursor is changed from K_2PtCl_6 for THH NCs synthesis to H_2PtCl_6 for TPH NCs growth, the concentration is also varied. At the same time, the potential has been optimized for fabricating TPH Pt NCs. The surface structure of the TPH Pt NCs has been identified by HRTEM and SAED as mainly {522} facets.

Besides single crystalline Pt NCs, fivefold twinned Pt nanorods with {hk0} high-index facets have also been synthesized by the square-wave method (Figure 6.1i) [29]. The preparation processes are very similar to those for the THH Pt NCs, except that after Pt nanospheres were deposited on the glass carbon electrode, the electrode was exposed to air for 3–5 h prior to the square-wave treatment. Such a procedure could cause the glass carbon surface to be hydrophobic or inert so that new Pt nuclei are hardly generated, instead, the nuclei form preferentially on Pt nanospheres and grow into nanorods during the square-wave potential treatment. The length of the Pt nanorods is about 1 μm, and their diameter is not uniform along the growth direction, being broadest at the middle and gradually tapering to both ends. Further analysis confirmed the surface of the fivefold twinned nanorods consists of a series of {hk0} facets, like{210} and {310}.

The periodic oxygen adsorption/desorption induced by the square wave is a highly valid route to construct different kinds of Pt NCs with high-index facets, as displayed before. However, the relatively large size (greater than 20 nm) of the NCs and their being deposited on glassy carbon obstructs the potential applications such as a catalyst in fuel cells owing to the low utilization efficiency of noble metals. In practise, Pt electrocatalysts are usually supported on carbon black, which can effectively reduce the size of Pt nanoparticles and disperse the Pt nanoparticles evenly. In 2010, we reported a new result concerning the synthesis of high-index faceted Pt NCs supported on carbon black (HIF-Pt/C) with a size (2–10 nm) comparable to that of commercial catalysts by using a square-wave potential method [27]. The key for decreasing the size is the employment of insoluble Cs_2PtCl_6 dispersed on carbon black as precursor. The typical TEM image of the as-synthesized Pt NPs is shown in Figure 6.1f, in which the border atoms are clearly resolved. The shape of the nanoparticle is not spherical, as some small facets can be clearly observed, which indicate that the surface of the nanoparticle is composed of a large number of low-coordinated Pt atoms.

Since nanoparticles surrounded by low-energy facets are facile to synthesize, we have developed another new method to synthesize nanoparticles enclosed with high-index faces, based on the shape transformation. This method could transfer Pt cubic nanoparticles with a {100} facet into THH Pt NCs with a high-index facet and has the potential to prepare sub-10 nm Pt THH NCs with high-index facets [28]. It was found that after subjecting to square-wave potential treatment for 2 min, Pt nanocubes of 10 nm in edge length will be converted to THH Pt NCs with sizes varying from 6 to 20 nm. The HRTEM image (Figure 6.1g) along the [001] orientation shows a nearly perfect cubic shape, verifying the nature of {100} facets while the TEM image of the THH Pt NCs (Figure 6.1h) along the same direction exhibits an octagonal shape, which is in good agreement with a THH model enclosed by {310} facets (red line in Figure 6.1h). This result demonstrates that Pt nanocubes bounded by {100} facets have been successfully converted into THH Pt NCs with high-index facets by square-wave treatment.

All the nanoparticles synthesized above are enclosed with high-index facets. As commonly accepted, high-index facets of fcc metals, bearing a high density of low-coordinated surface atoms, could be active for most catalytic reactions. Electrooxidation of small molecules is the anodic reaction of direct fuel cells and is also widely used to probe the surface activity of nanoparticles. THH Pt NCs (81 nm), as shown in Figure 6.2a, possess a catalytic activity superior to that of spherical Pt nanoparticles of a similar size (115 nm) and the commercial 3.2 nm Pt/C catalyst toward electrooxidation of formic acid and ethanol [15]. The steady-state current density of formic acid oxidation on the THH Pt NCs is 1.6–4.0 times higher than that on the Pt nanospheres, and 2.0–3.1 times higher than on the commercial Pt/C catalyst. For ethanol electrooxidation, the enhancement factor varies from 2.0 to 4.3 and 2.5 to 4.6 compared to that of Pt nanospheres and the commercial catalyst, respectively (Figure 6.2b). In addition, at a given oxidation current density of

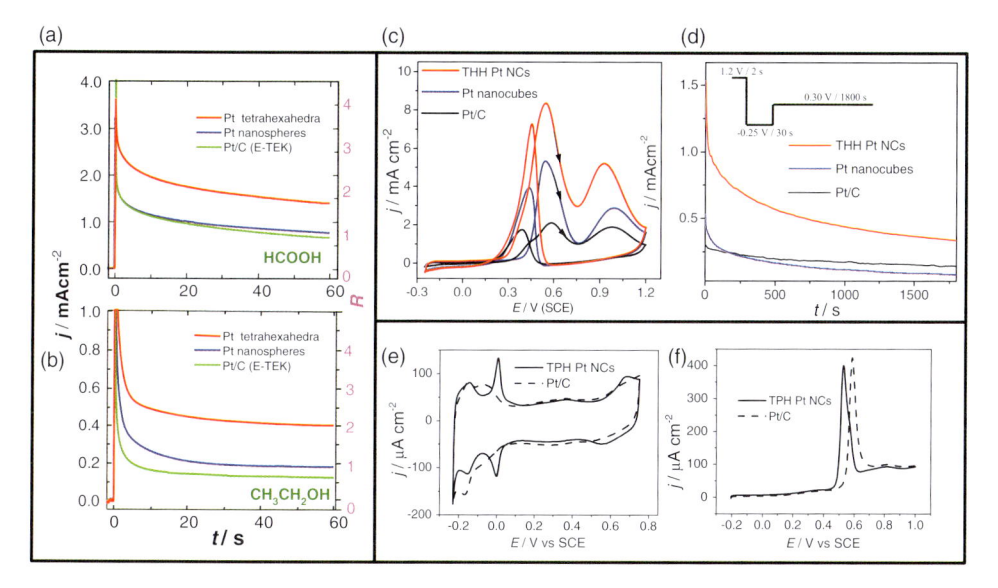

Figure 6.2 (a) Comparison of catalytic activity among THH Pt NCs, polycrystalline Pt nanospheres, and 3.2-nm Pt/C catalyst. Potential-dependent steady-state current density (left, recorded at 60 s) of formic acid oxidation, and the ratios R (right) between that of THH with the latter two [15]. The inset in (a) is an SEM image of a THH Pt NC after reaction, indicating the preservation of shape. (b) Potential dependent steady-state current density (left) of ethanol oxidation, and the ratios R (right) [15]. (c) Cyclic voltammograms (100 mV s^{-1}) and (d) current–time curves measured at 0.3 V (vs SCE) of ethanol oxidation on THH Pt NCs/CNTs (red line), Pt nanocubes/CNTs (blue line), and commercial Pt/C (black line) in 0.1 M ethanol + 0.1 M HClO$_4$ solution at 60 °C [28]. Cyclic voltammograms of the electrooxidation of (e) formic acid in 0.25 M HCOOH + 0.1 M HClO$_4$ solution and (f) methanol in 1 M CH$_3$OH + 0.1 M HClO$_4$ solution by TPH Pt NCs and commercial Pt/C (JM) at a scan rate of 50 mV s^{-1} [26].

technical significance in fuel cells, the corresponding potential on the THH Pt NCs is much lower than that on the Pt nanospheres or the commercial catalyst. For formic acid oxidation, the potential on THH Pt NCs is shifted negatively by about 60 mV as compared to that on Pt nanospheres at a current density of 0.5 mA cm^{-2}; while for ethanol oxidation, the negative shift is about 80 mV at a current density of 0.2 mA cm^{-2}. The negative shift was even larger in comparison with the commercial catalyst. These results demonstrate that the THH Pt NCs exhibit much enhanced catalytic activity for the electrooxidation of small molecules.

Both the convex and concave THH Pt crystals show excellent activity toward ethanol electrooxidation [15,25]. When the temperature is further raised to 60 °C, the catalytic enhancement of high-index facets toward ethanol electrooxidation is more prominent, as illustrated in Figure 6.2c and d by comparing the catalytic activity of THH NCs, nanocubes and Pt/C [28]. The peak oxidation current density on THH Pt NCs is almost 4 times as high as that of Pt/C, while considerably high current density can be maintained after 1800 s. Considering 60 °C is comparable to the real temperature at which a fuel cell works, the nanoparticles with high-index facets bearing much higher catalytic activity than the commercial Pt/C are promising materials for future fuel cell applications. Further, *in situ* FTIR spectroscopic studies show that high-index facet Pt nanoparticles can facilitate C—C bond cleavage, as a consequence, they show a higher tendency for complete oxidation of ethanol to CO_2 [27].

As for the TPH Pt NCs, formic acid and methanol are selected as probe molecule to test the surface activity [27]. Figure 6.2e and 6.2f compare CVs of electrooxidation of formic acid and methanol on TPH Pt NCs and commercial Pt/C catalysts, respectively. The peak current density of formic acid electrooxidation at 0.60 V (vs SCE) in the positive going scan is 4.1 mA cm^{-2} on the TPH Pt NCs, 2.9 times as high as that on commercial Pt/C (1.4 mA cm^{-2}), while for methanol the value is 5.1 (8.1 mA cm^{-2} for TPH Pt NCs vs 1.6 mA cm^{-2} for commercial Pt/C). The high density of step sites on TPH Pt NCs is believed to be responsible for the superior catalytic activity for the electrooxidation of small molecules such as formic acid and methanol.

6.2.1.2 Palladium

Besides platinum, palladium is another very important element of catalysts in different applications, such as electrooxidation of formic acid [30,31], the Suzuki coupling reaction, and the Heck coupling reaction [32,33]. Therefore, shape-controlled synthesis of Pd nanocrystals has also attracted great attention. By using the electrochemical square-wave method, we have successfully synthesized different kinds of Pd NCs bounded by high-index facets.

THH Pd NCs were electrodeposited directly on a glass carbon electrode in 0.2 mM $PdCl_2$ + 0.1 M $HClO_4$ solution by a one-step square-wave method, an overview SEM image of the THH Pd NCs and a corresponding THH polyhedron model are display in Figure 6.3a and b, respectively [34]. In the high magnification SEM image (Figure 6.3a inset), three perfect square-based pyramids can be clearly observed, confirming that the Pd nanocrystal has a THH shape. The dominant facets of the Pd

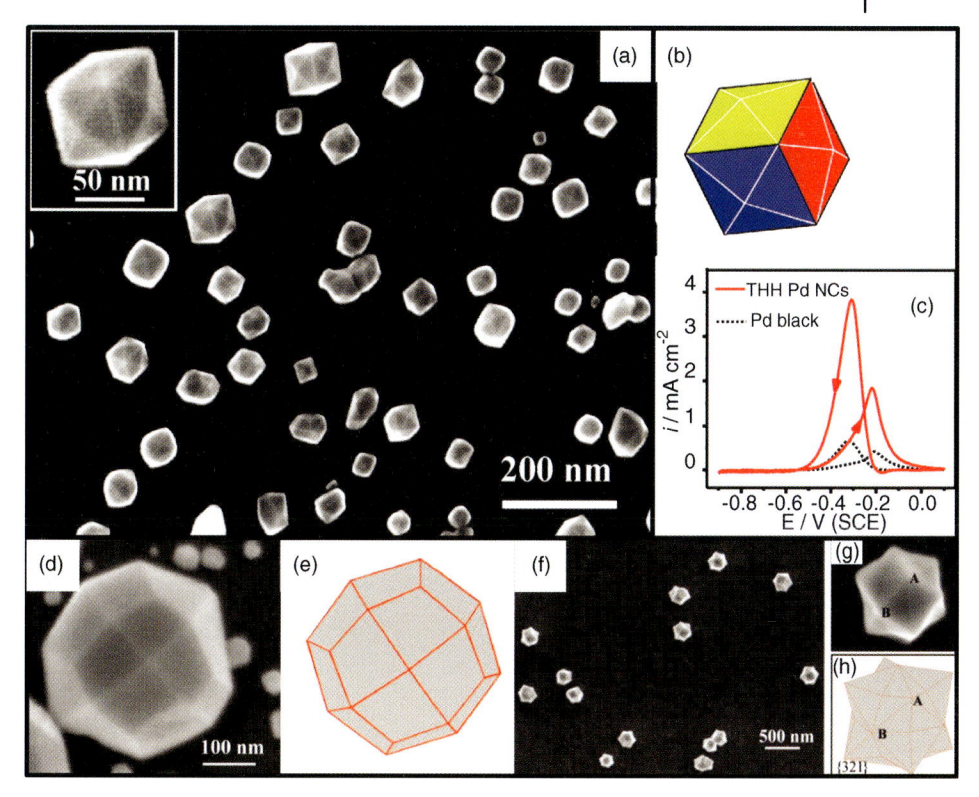

Figure 6.3 (a) SEM image of THH Pd NCs. The inset is a high-magnification SEM image. (b) Geometrical model of a tetrahexahedron [34]. (c) Cyclic voltammograms of THH Pd NCs (solid line) and Pd black catalyst (dashed line) in 0.1 M ethanol + 0.1 M NaOH solution at 10 mV s^{-1}. SEM images of trapezohedral Pd NCs with {hkk} facets (d) and concave hexoctahedral Pd NCs (f and g). The models of trapezohedron with {311} facets (e) and concave hexoctahedron with {321} facets (h) are also shown [29].

THH NC are {730}, determined by HRTEM and SAED. Figure 6.3c depicts cyclic voltammograms of THH Pd NCs and commercial Pd black recorded in 0.1 M ethanol + 0.1 M NaOH solution. In the positive-going potential scan, the peak current densities of ethanol oxidation on the Pd THH NCs and Pd black catalyst were 1.84 and 0.42 mA cm^{-2}, respectively; in the negative-going potential scan, the corresponding values were 3.83 and 0.65 mA cm^{-2}. The THH Pd NCs exhibited 4–6 times higher catalytic activity than commercial Pd black catalyst, which was attributed to the high density of surface atomic steps on THH Pd NCs. In addition, the THH Pd NCs also exhibited high stability. After 1000 potential cycles, 95.0 and 75.0% of the initial catalytic activity in the positive- and negative-going potential scans was maintained.

By changing the synthesis conditions, Pd nanocrystals of different shape can be fabricated, like trapezohedral Pd NCs with {hkk} facets and concave hexoctahedral

Pd NCs with {hkl} facets [29]. Figure 6.3d illustrates a SEM image of a trapezohedral Pd NC, on which four facets surrounded by eight facets can be seen clearly. This feature is well consistent with the model of a trapezohedron bounded by 24 {hkk} facets (Figure 6.3e). The trapezohedral Pd NC was most likely to be bounded by {311} facets through comparing the trapezohedral models with the Miller indices (i.e., h, k) with the as-synthesized trapezohedral Pd NC. Figure 6.3g and h depict the SEM images of concave hexoctahedral Pd NCs, whose shape is fairly similar to the model of a concave hexoctahedron with {321} facets shown in Figure 6.3h.

6.2.2
Wet-Chemical Route

Since the first report the synthesis of Pt NCs with high-index facets in 2007 [15], reports concerning metal nanoparticles with a high energy surface synthesized through the wet-chemical route have boomed. Functional molecules, like organic capping agents [35], inorganic ions [36,37], polymers [38], and even peptide and DNA [39] have been used to control the shape of the nanoparticles. Here we will review some published work relating to the shape-controlled synthesized nanoparticles with high-index facets.

6.2.2.1 Platinum

Zheng *et al.* reported the synthesis of concave Pt NCs having high-index {411} facets by use of amine as capping agent via a facile wet-chemical method [35]. As shown by the SEM images in Figure 6.4a, the product consisted of uniform four-armed starlike particles. A structural model was proposed (inset in Figure 6.4b) for the as-prepared Pt NCs. Further, HRTEM and SAED analysis determined that the exposed facets of the trigonal pyramidal arms are {411} facets. Since the thermodynamics drives the formation of nanoparticles with the lowest surface energy, in the synthesis of the concave Pt NCs with high-index facets the surface energy must be modified. As a capping agent, the use of the amine was essential. Perfect Pt octapod NCs were obtained only by supplying a certain amount of methylamine. When the amount of methylamine was reduced, the degree of concavity of the obtained NCs was significantly reduced. The essential role of methylamine in the formation of the octapod NCs originates from their selective binding on the Pt high-index {411} facets to stabilize the low-coordinated Pt sites during growth, which was confirmed by FTIR spectroscopy. Interestingly, when methylamine was substituted with other amines (i.e., ethylamine, butylamine, 4-methylpiperidine, trimethylamine), the reactions also yielded concave NCs. The concave Pt nanoparticles with high-index facets show their structure advantage in electrocatalysis, in comparison with commercial Pt black and Pt/C (E-TEK), as shown in Figure 6.4c and d. Formic acid and ethanol are selected as probe molecules. The peak current densities of formic acid oxidation at 0.61 V (vs SCE) in the forward potential scan were 3.9, 1.7, and 0.7 mA cm^{-2} on the concave Pt NCs, commercial Pt black, and Pt/C catalyst, respectively. The activity of the concave Pt NCs was accordingly 2.3 and 5.6 times as

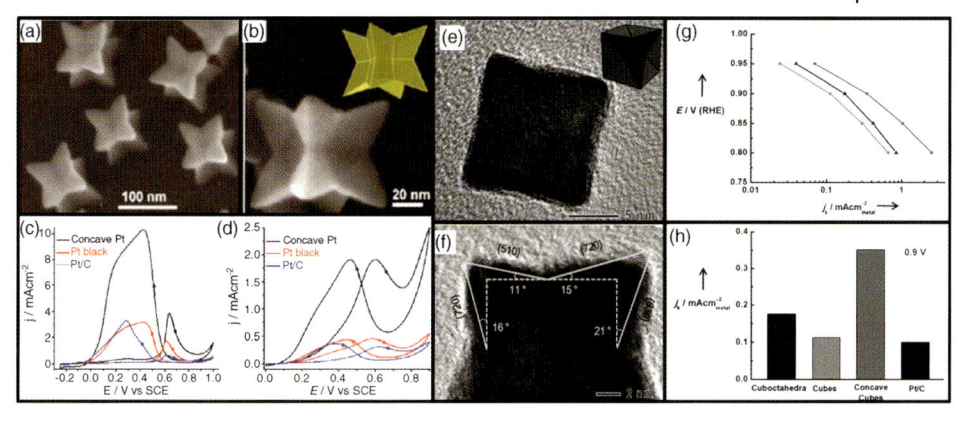

Figure 6.4 (a) Typical SEM images of the concave Pt nanocrystals. Reproduced with permission from Ref. [35]. (b) High-magnification SEM image of a single concave Pt nanocrystal. The top-right inset shows an ideal geometrical model of the concave Pt nanocrystal with the same orientation as the nanocrystal in the SEM image. CV curves for electrooxidation of (c) formic acid and (d) ethanol by the concave Pt nanocrystals, commercial Pt black, and Pt/C (E-TEK). The formic acid oxidation was recorded in 0.5 M $H_2SO_4 + 0.25$ M HCOOH solution at a scan rate of 50 mV s^{-1}. The ethanol oxidation was recorded in 0.1 M $HClO_4 + 0.1$ M CH_3CH_2OH solution at a scan rate of 50 mV s^{-1} [35]. (e), (f) HRTEM images of Pt concave nanocubes. Reproduced with permission from Ref. [40] Comparison of the electrocatalytic properties of the Pt concave cubes (■), cubes (●), and cuboctahedra (▲). (g) Specific activities given as kinetic current densities (j_k) normalized against the ECSA of the catalyst. (h) Specific activities of these catalysts at 0.9 V versus RHE. The ORR measurements performed at room temperature in O_2-saturated 0.1 mol L^{-1} $HClO_4$ solutions, with a sweep rate of 10 mVs^{-1} and a rotation rate of 1600 rpm [40].

high as those of commercial Pt black and Pt/C, respectively. Similar behavior was also found in the electrooxidation of ethanol, for which the concave Pt NCs demonstrated the best activity, 4.2 and 6.0 times as high as those of commercial Pt black and Pt/C, respectively, at 0.60 V (vs SCE). These results illustrate that the presence of the {411} facets endows the concave Pt NCs with outstanding electrocatalytic activity.

Another case of synthesis of concave Pt cubic nanoparticles with high-index facets was reported by Xia's group [40]. The concave Pt THH nanoparticles enclosed by a high-index facet, such as {510}, {720}, {830}, were synthesized by reduction of K_2PtCl_4 with $NaBH_4$ in the presence of $Na_2H_2P_2O_7$ and KBr. The as-synthesized concave Pt THH nanoparticles were characterized by HRTEM as shown in Figure 6.4e and f. The key to synthesizing concave Pt nanoparticles in this case is the use of a Pt pyrophosphato complex as the precursor, which is formed by reaction between K_2PtCl_4 and $Na_2H_2P_2O_7$, while Br$^-$ serves as a capping agent to block the {100} facets as in many cases. The Pt pyrophosphato complex was more difficult to reduce than the initial Pt^{II} ions, which resulted in a reduced reduction rate. The sluggish reduction rate is vital for the kinetically controlled growth of concave Pt nanocrystals. For the electrocatalysis toward ORR, as depicted in

Figure 6.4g, the specific activity of the Pt concave cubes was 3.1–4.1 and 1.9–2.8 times greater than that of the Pt cubes and cuboctahedra, respectively, in the potential region between 0.8 and 0.95 V. In comparison with the commercial Pt/C catalyst, the Pt concave cubes exhibited a specific activity that was 3.6 times higher than the 3.2 nm Pt particles on Pt/C catalyst at 0.9 V (Figure 6.4h). Both the results demonstrated that the Pt concave cubes exhibited greatly enhanced catalytic activity in the ORR because of their high-index facets.

Unfortunately, although numerous efforts have been devoted to synthesizing the Pt THH nanoparticles through a wet-chemical method from a Pt precursor like H_2PtCl_6, to the best of our knowledge, so far none has been successful. Some examples by growing epitaxially in the presence of external-added seed as core will be discussed in Section 6.3.

6.2.2.2 Palladium

Xia's group reported a simple wet-chemical route based on seeded growth to the synthesis of Pd concave nanocubes bounded by high-index {730} facets, where Pd nanocubes were used as seeds for the reduction of a Pd precursor in an aqueous solution [41]. The reduction kinetics was controlled by manipulating the concentrations of reagents, including Na_2PdCl_4, KBr, and ascorbic acid (AA). Figure 6.5a, b shows SEM and TEM images of the as-prepared Pd concave nanocubes. It is clear that each face of the nanocube in the product was excavated by a square pyramid in the center. The Miller indices of the as-synthesized concave nanocube were determined by HRTEM and SAED to be {730}. Further experiments demonstrated that, in general, reducing the concentrations of Na_2PdCl_4 and KBr or increasing the concentration of AA is beneficial to the formation of Pd concave nanocubes. Since Br^- selectively adsorbs onto the {100} facets, the new Pd atoms will be preferentially added to the corners and edges of a Pd cubic seed, thus resulting in a concave nanocube. This observation is different from the cases previously reported for the synthesis of Pt concave nanocubes. The difference can be attributed to the use of different reducing agents. For the Pt system, Pt atoms were immediately formed in the solution phase by reduction of the Pt precursor with a strong reducing agent, $NaBH_4$, and then added to the surface of a growing Pt seed. In the case of Pd, the Na_2PdCl_4 precursor had to diffuse to the surface of a growing Pd seed and then be reduced to Pd atoms owing to the use of the relatively weak reducing agent ascorbic acid. As a result, the dependence of growth behavior on the concentration of reducing agent showed opposite trends in these two systems. The catalytic activity of as-synthesized concave nanocubes towards formic acid electrooxidation was tested. The performance of these concave nanocubes was compared with that of Pd nanocubes enclosed by {100} facets with the same average size of 37 nm. Figure 6.5c shows cyclic voltammograms of these two catalysts for electrooxidation of formic acid. It is clear that the Pd concave nanocubes exhibited superior electrocatalytic activity to conventional nanocubes, with a 1.9 times increase for the peak current. Again, the high-index {730} facets with a higher density of low-coordinate atomic steps and kinks should be responsible for the enhanced activity for formic acid oxidation.

Figure 6.5 (a) SEM and (b) TEM images of Pd concave nanocubes. Reproduced with permission from Ref. [41] (c) Cyclic voltammograms of the Pd nanocubes enclosed by concave or flat faces in a solution containing 2 M HCOOH and 0.1 M HClO$_4$ at a sweep rate of 50 mVs^{-1} at room temperature. The current density was normalized against the corresponding ECSA [41]. (d) A typical SEM image of the concave Pd NCs. Reproduced with permission from Ref. [42] (e) High-magnification SEM images and (f) models of the concave Pd NCs, viewed from different directions. (g) CV curves measured on Pd NCs with different degrees of concavity: a, concave Pd NCs-4:1; b, concave Pd NCs-1:1; c, concave Pd NCs-1:4; and d, the commercial Pd-black catalyst in 0.1 M ethanol +0.1 M NaOH solution (scan rate: 10 mV s^{-1}). The arrows in the CV curves indicate the direction of the scan [42].

A similar work has been reported by Zheng's group, concerning directly synthesized concave Pd cubic nanoparticles with {310} facets by reducing H$_2$PdCl$_4$ with ascorbic acid while carefully controlling the ratio of CTAB/CTAC = 4 : 1 [42]. The corresponding SEM images are displayed in Figure 6.5d and e. Based on the measured morphologic information, the concave facets can be determined as {hk0}. The following formation mechanism is proposed: at the beginning of the reaction, PdCl$_4^{2-}$ partially transforms to the [PdCl$_4$ (CTA)$_2$] and [PdBr$_4$ (CTA)$_2$] complexes in the presence of CTAC and CTAB, which have been detected in UV/Vis absorption spectra. These complexes are rapidly reduced to Pd NCs by ascorbic acid. The rapid reduction process leads to the formation of lots of interstices in the Pd nanocrystals. As the reaction continues, the deposition of Pd in the interstices and the dissolution of surface Pd atoms reach equilibrium. Finally, the interstices are filled with Pd atoms during the aging period. In the last stage, the additional $PdCl_4^{2-}$ ions are slowly reduced and deposited onto the concave Pd NCs. The formation of a high-index surface could be due to the coadsorption of Br^- and Cl^- ions; this thermodynamically lowers the surface energy of the $\{hk0\}$ facets. The angle of the concave structure of the cubic Pd NCs can also be tuned, by simply changing the CTAC/CTAB ratio. The catalytic activities of Pd NCs with different degrees of concavity and the commercial Pd-black catalyst toward ethanol electrooxidation were compared, as shown in Figure 6.5g. The electrocatalytic activities of all concave, cubic, Pd NCs are much higher than for the commercial Pd-black catalyst and the normal Pd nanocubes. Furthermore, the electrocatalytic activity of Pd NCs increases with increase in the concavity. In addition, comparing with Pd black, the onset potential of the ethanol electrooxidation of the concave Pd NCs-4:1 shifts from -0.59 to -0.67 V, and the peak potential shifts from -0.16 to -0.19 V. All these results indicate that the high-index surface reduces the reaction barrier of ethanol oxidation.

6.2.2.3 Gold

An interesting case concerning Au high-index facets nanoparticle synthesis is illustrated in Figure 6.6a [43]. As can be seen, by simply varying the capping agent while keeping other conditions almost the same, two different kinds of nanoparticles can be synthesized. When using CTAB as capping agent, Au THH nanoparticles are formed; while replacing the CTAB with CTAC, Au concave cube nanoparticles will be the product. The anion species, Br^- and Cl^-, respectively, accounted for the different facet direction of the two kinds of nanoparticles. It seems that CTAC is widely used in controlling the morphology of Au nanoparticles, both CTA^+ and Cl^- play important roles for the shape-controlled synthesis. Ma et al. reported the synthesis of trisoctahedral Au NCs with high-index facets by reduction of an aqueous solution of $HAuCl_4$ with ascorbic acid in the presence of CTAC at room temperature [44]. The SEM images are shown in Figure 6.6b. Further HRTEM and SAED experiments indicate that the exposed surfaces of the NCs correspond to $\{221\}$ planes. This is the first report concerning the synthesis of well-shaped metal NCs with high-index surfaces by a wet-chemical method. The electrochemical behavior of the Au NCs was different from that of a surface of polycrystalline gold, as demonstrated by the CV curve displayed in Figure 6.6d, and also different from that of the $\{111\}$, $\{110\}$, and $\{100\}$ surfaces (the low-index surfaces) of a single crystal of gold [45]. This result suggests that the as-prepared well-faceted gold nanocrystals are not enclosed by low-index surfaces, or by a polycrystalline surface, consistent with the SEM and TEM findings.

Another study using CTAC as capping agent was reported by Yu et al., forming the concave trisoctahedral (TOH) Au NCs as illustrated in Figure 6.6c [46]. Further characterization demonstrates that the surface of the TOH NCs was bounded by high-index $\{hhl\}$ facets in the $[1\bar{1}0]$ zone, such as $\{221\}$, $\{331\}$, and/or $\{441\}$.

Others capping agent, like PDDA, a kind of organic quaternary ammonium salt, and inorganic ions, such as Ag^+, Pd^{2+}, have all been reported to control the shape of the Au nanoparticles with high-index facets [38,43,47–49], we do not have space to discuss it in detail here.

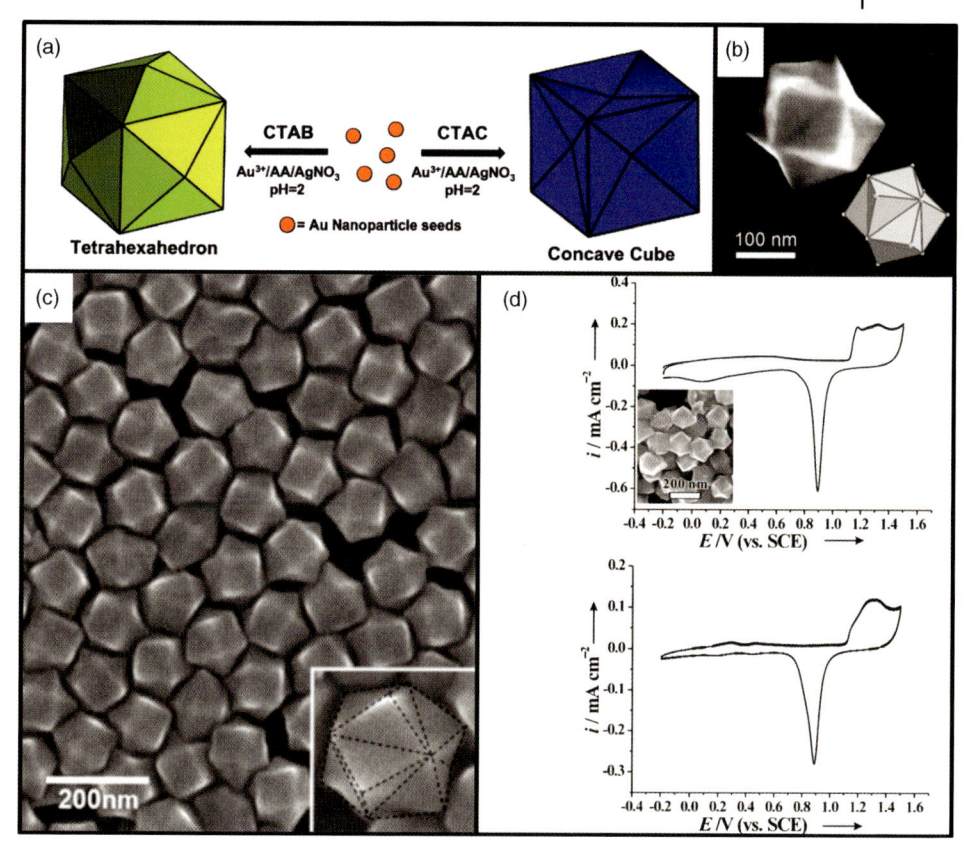

Figure 6.6 (a) Schematic illustration showing the effect of the counterion on product morphology in the seed-mediated synthesis: CTAC leads to the formation of concave cubes, and CTAB leads to the formation of tetrahexahedra (convex cubes) [43]. (b) HRSEM image of a single NC. Reproduced with permission from Ref. [44] inset: model of an ideal trisoctahedron enclosed by {221} surfaces, in the same orientation as the NC in the SEM image [44]. (c) Representative FESEM images of TOH Au NCs at high magnifications. Reproduced with permission from Ref. [46] (d) Upper: CV trace of the trisoctahedral gold NCs loaded onto a glassy carbon electrode; inset: SEM image of the NCs used in the experiment; Lower: CV trace of a polycrystalline gold wire electrode [46].

6.3
Shape-Controlled Synthesis of Bimetallic NCs with High Surface Energy

As for bimetallic nanoparticles, there exist several kinds of structure depending on the spatial distribution of the elements [50]. In this section, we focus on three kinds of structure of bimetallic NCs with high-index facets, namely, surface (sub)monolayer modification, solid solution (alloy), and core–shell structure, as illustrated in Figure 6.7.

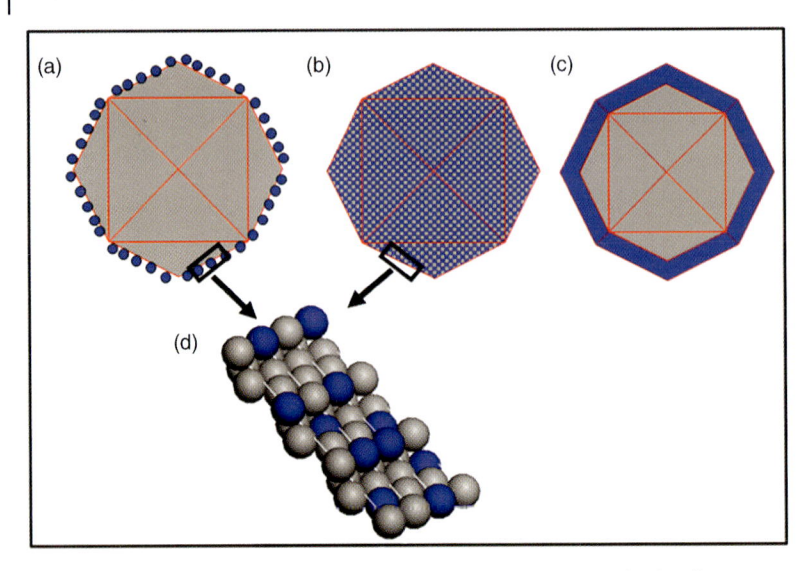

Figure 6.7 Illustration composition of (a) surface-modified nanoparticle, (b) alloy nanoparticle and (c) core–shell structure nanoparticle, (d) surface atoms arrangement in (a), (b).

6.3.1
Surface Modification

Surface modification is an efficient way to tune the surface structure of nanoparticles, thus improving their catalytic properties. It is also a convenient method to prepare bimetallic nanocatalysts with high-index facets through surface modification of the synthesized nanoparticles with high-index facets by using foreign metal atoms.

Different from the bimetallic alloy, in surface modification it is believed that the ligand effect and the so-called "third-body" effect account for the changing properties of the NCs, without the size effect since the foreign atoms only stay on the surface and do not go into the bulk lattice. There are two main strategies to modify a given metal surface through electrochemistry, as illustrated in Figure 6.8. One is direct electrodeposition of the target atoms, which is usually done by applying a potential scanning or a constant potential in a solution containing the foreign metal ions. The coverage of the foreign metal, θ, can be controlled by simply changing parameters such as the number of scanning cycles, the concentration of the metal ion, the polarization time and so on. The second involves first carrying out the so-called under potential deposition (UPD) of a monolayer or sub-monolayer of foreign metal, usually Cu, and then, through galvanic displacement, replacing the UPD Cu with target metal M following the reaction below:

$$n\text{Cu} + 2\text{M}^{n+} \rightarrow n\text{Cu}^{2+} + 2\text{M}$$

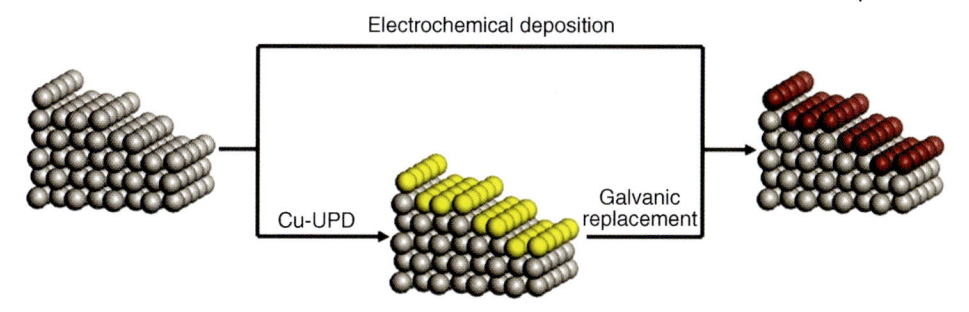

Figure 6.8 Illustration of two different electrochemical approaches to modify a metal surface [51].

From the equation we can see that the second strategy works only when the foreign metals have higher standard redox potential than the Cu^{2+}/Cu. The unique UPD makes the Cu only be deposited on the surface of NCs, which guarantees the formation of monolayer or sub-monolayer foreign atoms instead of possibly several layers of foreign atoms as in the first strategy. In this section, we will first review electrochemical surface modification of THH Pt NCs with Bi, Ru and Au, then describe Pt-modified Au prisms with high-index facets.

6.3.1.1 Bi-Modified THH Pt NCs

It is well known that formic acid electrooxidation on Pt is easily poisoned by adsorbed CO, a reaction intermediate generated from spontaneous dissociative adsorption of formic acid on Pt [52–54]. However, the oxidation via the CO pathway can be blocked if the Pt surfaces are modified with some foreign atoms (such as Bi, Pb, and Sb) [55–58]. These foreign atoms provide steric hindrance for formic acid to decompose into CO_{ad}, this is the so-called "third-body" effect [59,60]. Chen et al. prepared Bi-modified THH Pt NCs through Bi irreversible adsorption, which drastically improved the catalytic activity for formic acid oxidation [61]. The preparation of THH Pt NCs was described previously. Figure 6.9a illustrates the as-synthesized THH Pt nanocrystal. The modification of the THH Pt NCs was carried out by means of potential cycling between 0.06 and 0.80 V (vs. RHE) in H_2SO_4 solution containing $\sim 10^{-5}$ M of Bi^{3+}. Figure 6.9b compares the voltammograms of THH Pt NCs with increasing amounts of Bi on the surface. The Bi coverage (θ_{Bi}) varied from 0.23 to 0.90 by changing the sweeping time. The modification of Bi has significantly enhanced the catalytic activity, as shown in Figure 6.9c. On the bare THH Pt NCs, the surface is very active for formic acid dissociative adsorption to form CO species [62], leading to surface poisoning at low potential. As a consequence, the peak current density in the positive-going potential scan (0.21 mA cm^{-2}) is much smaller than that in the negative-going one (1.22 mA cm^{-2}). When the THH Pt NCs are modified with Bi, the oxidation current increases significantly, accompanied by a negative shift of onset oxidation potential from about 0.30 to 0.08 V. The current keeps increasing as the Bi coverage increases until the

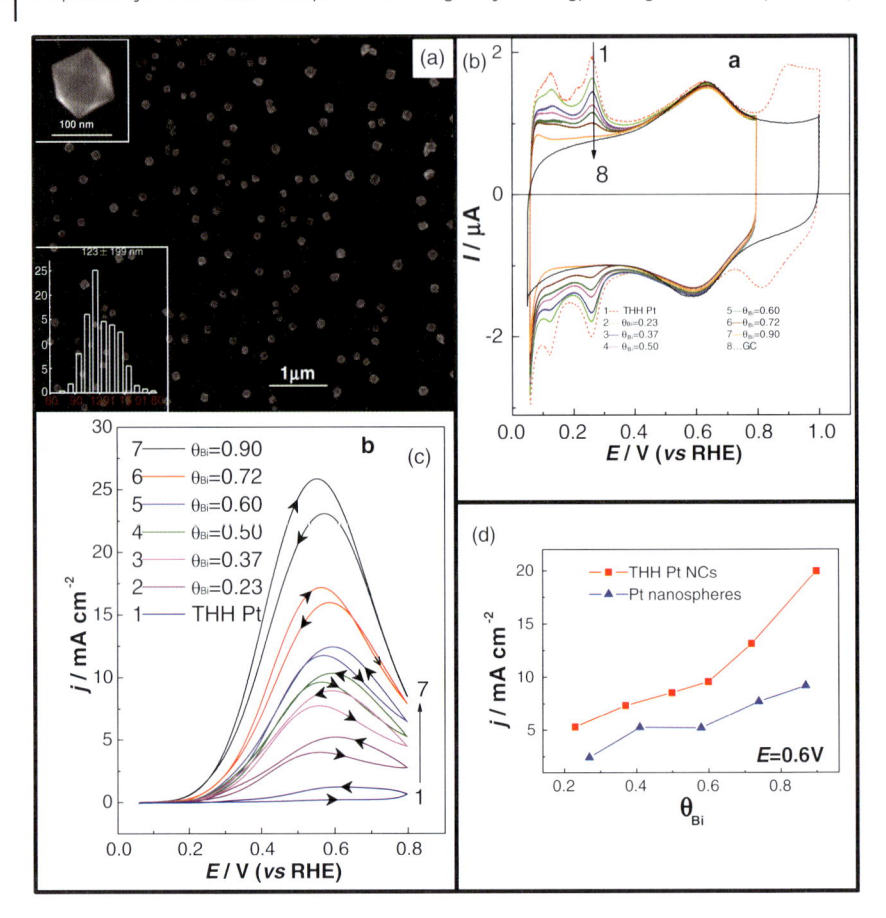

Figure 6.9 (a) SEM image of THH Pt NCs. The insets are a high-magnification SEM image and size distribution histogram, respectively; (b) CVs of THH Pt NCs with different Bi coverage as indicated in the figure, in 0.5 M H_2SO_4 solution (scan rate: 50 mV s^{-1}); (c) CVs of THH Pt NCs with different Bi coverage in 0.25 M HCOOH + 0.5 M H_2SO_4 solution (scan rate: 20 mV s^{-1}); (d) Comparison of steady-state current density of formic acid oxidation at 0.6 V on Bi-decorated THH Pt NCs with that of Bi-decorated Pt nanospheres as a function of Bi coverage in 0.25 M HCOOH + 0.5 M H_2SO_4 solution [61].

highest peak current density of 25.8 mA cm^{-2} is achieved at the largest Bi coverage of 0.90. The oxidation peak current density is 21 times larger than that obtained on bare THH Pt NCs. In addition, the hysteresis of CV profiles in the positive- and negative-going scans is greatly reduced, indicating the absence of CO poisoning at high Bi coverage. It is worthwhile noting that although Bi-decorated Pt nanospheres of polycrystalline structure can also enhance the catalytic activity, the steady-state current densities measured on such surfaces are always smaller than those on Bi-decorated THH Pt NCs. A comparison is illustrated in Figure 6.9d with oxidation potential at 0.6 V. The current densities on THH Pt NCs are almost double those on Pt nanospheres for all Bi coverages investigated, in accordance with the enhancement of bare THH Pt NCs compared with Pt nanospheres for formic acid electrooxidation reported before. This result further confirms that the Bi decoration can boost significantly the catalytic activity of THH Pt NCs, on which the main surface sites are of $\{hk0\}$ structure together with active sites at the corners and edges of the nanocrystals.

6.3.1.2 Ru-Modified THH Pt NCs

PtRu alloy is widely used in direct methanol fuel cells as an anode material due to the so-called bi-functional effects: Pt facilitates the cleavage of the C—H bond in methanol to form some adsorbed species, including CO_{ads}; while Ru promotes the formation of oxygen-containing species at low potential, which can oxidize adsorbed CO to thus free the catalyst from poisoning. As a result, PtRu alloy exhibits significant enhanced catalytic activity over pure Pt in the lower overpotential region (<0.65 V vs. RHE). The Ru-decorated Pt surfaces have similar enhancement effects. However, the degree of enhancement has been shown to be strongly influenced by the surface structure of underlying Pt. On the basal planes of Pt single crystal, both the activity and the Ru-induced enhancement factor follow the sequence Pt (111) Ru > Pt (110) Ru > Pt (100) Ru [63].

THH Pt NCs provide a model substrate for studying Ru-decorated Pt nanoparticles with high-index facets. The Ru decoration can also be carried out by potential cycling between -0.24 and 0.65 V (vs SCE) in 5×10^{-5} M $RuCl_3 + 0.1$ M $HClO_4$ solution [64]. Figure 6.10a shows the CVs of Ru-modified THH Pt NC electrode with various Ru coverage (θ_{Ru}) in H_2SO_4 solution. The θ_{Ru} was evaluated by the degree of blockage of hydrogen adsorption sites, as mentioned in the Bi-modified case. Along with increasing the θ_{Ru}, the oxygen adsorption/desorption currents between 0.50 and 0.65 V decline rapidly, suggesting that Ru atoms preferentially adsorb at step sites on Pt high index planes since the oxygen adsorption/desorption at such low potential mainly occurs at Pt step atoms [65–67]. Increasing Ru coverage is also accompanied by the broadening of the double layer region, a process well known for Pt–Ru materials and associated with the onset of OH adsorption by the adatoms [65–67]. The decoration of Ru promotes the electrooxidation of adsorbed CO, as shown in Figure 6.10b. On the clean THH Pt NC electrode, the onset potential for CO oxidation is 0.41 V, with the peak observed at about 0.53 V. As θ_{Ru} increases, both the oxidation onset potentials and peak potentials are shifted to more negative values. The ability to remove CO on Ru-modified THH Pt NCs should increase the tolerance to CO poisoning for methanol oxidation. As expected, the onset potential of methanol oxidation was negatively shifted by over 100 mV, and the oxidation current in the low potential range was greatly boosted with Ru decoration (Figure 6.10c). It is interesting to observe that the bare THH Pt NCs have a catalytic activity much higher than commercial Pt black, and the Ru-modified THH Pt NCs are even superior to commercial PtRu alloy catalyst. The enhancement effect of Ru decoration on THH Pt NCs is more significant than that on Pt black (Figure 6.10d). For example, the catalytic current

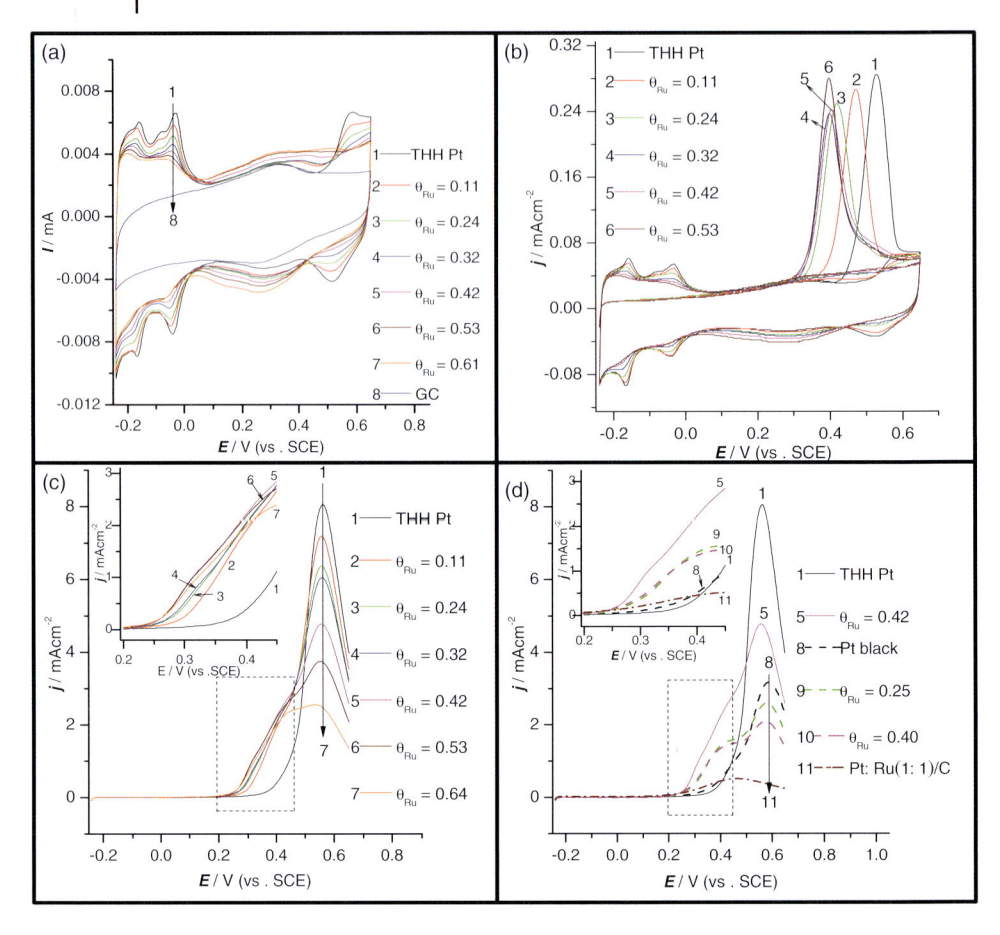

Figure 6.10 (a) CVs of the THH Pt NC electrodes with various coverages of Ru adatoms, recorded in 0.1 M H_2SO_4 solution at a scan rate of 50 mV s^{-1}. (b) CO oxidative stripping CVs for the bare and Ru-decorated THH Pt NC electrodes. The electrolyte solution was 0.1 M H_2SO_4, and the scan rate was 50 mV s^{-1}. (c) Positive segments of CVs characterizing the methanol electrooxidation performance of Ru-decorated THH Pt NCs. (d) Comparison of positive segments of CVs for methanol oxidation on Ru-decorated THH Pt NCs (solid line), Ru-decorated Pt black (dashed line), and PtRu/C catalyst (dashed-dotted line). The test solution was 1.0 M CH_3OH + 0.1 M $HClO_4$, and the scan rate was 50 mV s^{-1} [64].

density at 0.40 V on Ru-modified ($\theta_{Ru} = 0.42$) THH Pt NCs is about 66% greater than that of the Ru-modified Pt black catalyst ($\theta_{Ru} = 0.40$). These facts indicate that THH Pt NCs with high density of surface step atoms can promote methanol oxidation, and also serve as a better substrate for surface decoration of Ru that preferentially blocks the step sites. In contrast, Ru decoration on a normal Pt nanoparticle will consume most of the high active step sites, which is not in favor of methanol oxidation.

6.3.1.3 Au-Modified Pt THH NCs

Besides Bi and Ru, other foreign atoms have also been incorporated into Pt electrode surfaces for studies of formic acid electrooxidation. However, the instability of the introduced foreign atoms hampers their use as a stable catalyst. Due to the identical stability of Au and Pt, the Pt–Au bimetallic system has received much attention recently [68]. Liu et al. reported Au-modified Pt THH nanoparticles with high-index facets and evaluated the effect of these modifications on the formic acid electrooxidation performance [69].

The gold decoration was accomplished through the initial UPD of a sub-monolayer of Cu atop the THH Pt surfaces, followed by the galvanic charge displacement of the Cu with Au adatoms [70]. The coverage of Au was varied by changing the time of the initial Cu deposition process. The variation, with Au coverage, of the voltammetric profile of the Pt THH NC electrode is shown in Figure 6.11a. As the coverage is increased there is a progressive diminishing of the charge associated with hydrogen adsorption/desorption peaks at -0.02 and -0.16 V, owing to the blockage of Pt surface sites by Au adatoms. The same explanation accounts for the decline in Pt oxide formation (above 0.46 V) and reduction (about 0.30–0.75 V) currents. Figure 6.11b features the forward scans of cyclic voltammograms recorded for unmodified, and various Au-decorated THH Pt NC electrodes for formic acid electrooxidation. The formic acid electrooxidation current via the direct pathway has an onset just above the hydrogen region and it rises with potential up to ~ 0.25 V (Peak I). CO_{ads} adsorption occurs in parallel to the dehydrogenation process of HCOOH and its increasing coverage limits the number of sites available to the direct reaction, causing an arrest in the current above 0.25 V. The current increases again beyond ~ 0.45 V owing to the oxidative removal of CO_{ads} by oxygenated species generated on Pt sites in this potential range. A peak is observed just above 0.61 V (Peak II) with the current decreasing at higher potentials due to oxide formation on the Pt surface [68]. The ratio of the charge associated with peak I compared with peak II is an indication of the extent of the direct path relative to the indirect path for a given electrocatalyst. The peak I and peak II current density in Figure 6.11b and the peak current of the backward segments is presented in Figure 6.11c. The liberation of the active Pt sites from CO_{ads} at the upper end of the forward sweep permits formic acid oxidation to occur at a much greater rate, and predominantly by the direct pathway, on the backward scan. The introduction of Au adatoms onto the stepped Pt surface has a profound effect on the voltammetric profiles for formic acid electrooxidation. It can be seen from Figure 6.11b that, with the increase in Au coverage the current associated with Peak I enhances at the expense of Peak II as the surface becomes progressively more selective towards the direct pathway. Accordingly, the decorated electrodes offer higher activities at lower potentials. In the case of $\theta_{Au} = 0.72$, the peak current density 4.6 mA cm^{-2} at 0.28 V, is 20 times greater than the plateau current density for the clean electrode. The activity decreases for $\theta_{Au} > 0.72$, since above this threshold the promoting effect of the Au atoms is beginning to be offset by the loss of active Pt catalytic centers. The peak potential of Peak I and Peak II is almost unaltered at different θ_{Au}, indicating that the promoting effect of the Au atoms may not be due to the bi-functional mechanism. Figure 6.11c

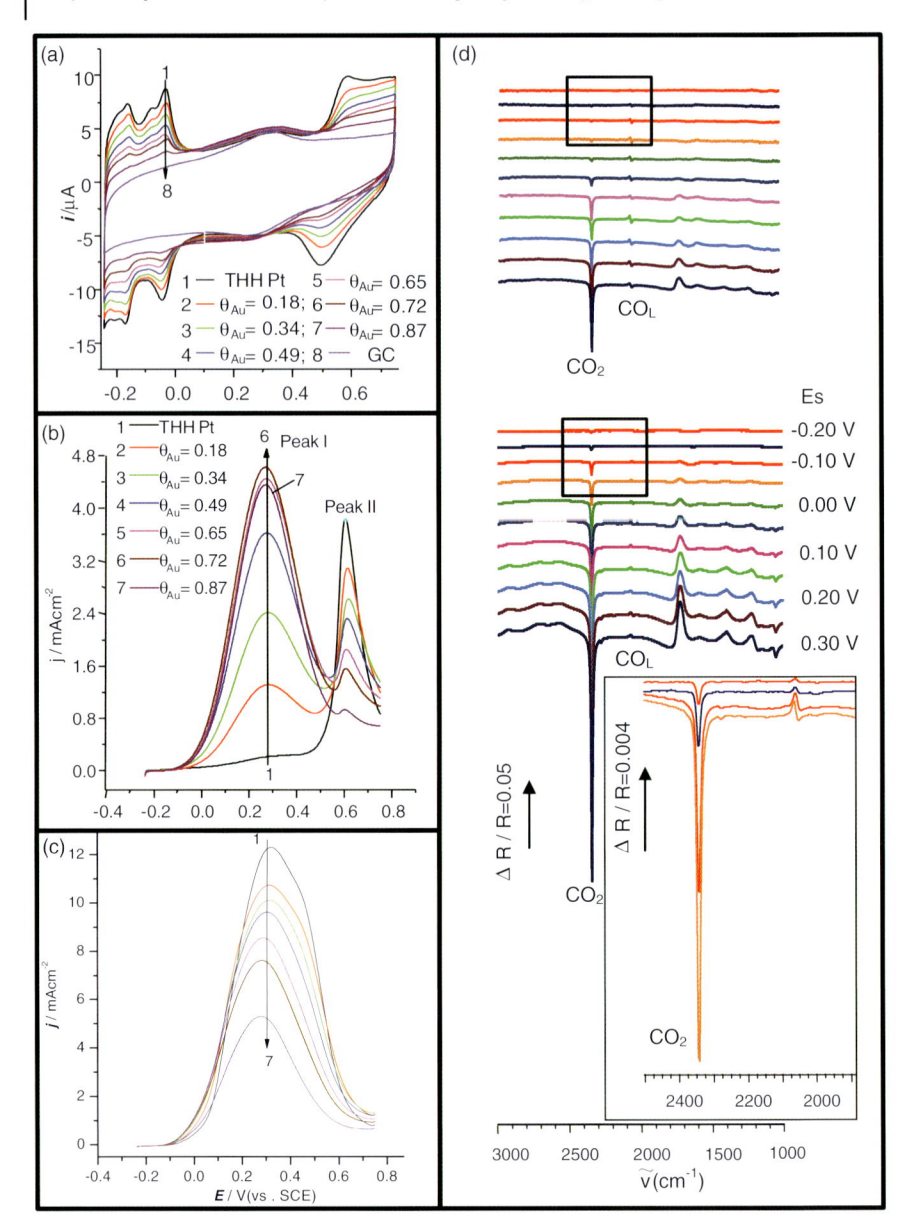

Figure 6.11 (a) Cyclic voltammograms of THH Pt NCs decorated by various coverage of Au adatoms recorded in 0.1 M H_2SO_4 solution at 50 mV s^{-1}. (b) Forward segments of CVs characterizing formic acid electrooxidation on THH Pt NC electrodes with different Au coverages. (c) Backward segments of CVs characterizing formic acid on the various electrodes. The test solution was 0.25 M HCOOH + 0.1 M H_2SO_4 and the scan rate was 50 mV s^{-1}. (d) In situ FTIR spectra for formic acid electrooxidation in 0.25 M HCOOH + 0.1 M H_2SO_4 solution on an unmodified THH Pt NC electrode (upper), and the same electrode with $\theta_{Au} = 0.74$ (lower). The reference spectra were recorded at $E_R = -0.25$ V [69].

shows the backward potential scan of the Au-decorated THH Pt NCs for formic acid electrooxidation. In this case, the HCOOH oxidation is free from CO poisoning, and the observed activity presents intrinsic catalytic activity of unmodified surfaces. The peak current decreases with the increase in θ_{Au}, demonstrating that the increase in peak I in Figure 6.11b is due to the "third body" effect, that is, reducing CO poisoning by foreign adatoms. As for the backward scan, the surface is free from CO poisoning, while decoration with Au atom will decrease the active Pt site for electrocatalysis. As a consequence, the backward current decreases with the increase in Au coverage. This is based on the fact that CO_{ads} formation via HCOOH dehydration requires at least three adjacent Pt atoms [71].

In situ FTIR spectra recorded at various potentials in 0.25 M HCOOH + 0.1 M H_2SO_4 solution are presented in Figure 6.11d for both unmodified and Au-decorated ($\theta_{Au}=0.74$) THH Pt NCs. The positive bands at ~1720, 1400 and 1225 cm^{-1} arise from the consumption of formic acid [72,73]. The negative band at 2341 cm^{-1} was ascribed to the formation of CO_2 [72]. Linearly bonded CO (CO_L), adsorbed on Pt, gives rise to the band observed at ~2070 cm^{-1} [72,74]. The bipolar appearance of the latter feature is due to a Stark effect in IR absorption frequency, which in turn indicates the presence of CO_L at both the sampling and reference potentials. This implies that dissociative adsorption of formic acid to form CO_L occurs by the displacement of adsorbed H atoms from THH Pt NCs surfaces, even at very low potentials (−0.25 V). The detection of CO_L also verifies the CO poisoning for formic acid oxidation on Pt. The effect of Au decoration on the catalytic performance of the THH Pt NCs is most obvious when comparing the CO_L (around 2070 cm^{-1}) and the CO_2 (2341 cm^{-1}) signals for the unmodified and decorated surfaces. For THH Pt NCs, the CO_2 band appears at $E = -0.15$ V, while for the Au-decorated one it starts at $E = -0.2$ V. The intensity of the CO_2 band is also much larger over the entire range of potentials studied for the Au-decorated THH Pt NCs. In this regard there is close agreement between the in situ FTIR measurements and the electrochemical activity data described above, with the Au-decorated anodes yielding enhanced products of CO_2 in the former, and enhanced oxidation currents in the latter. Since the CO stripping experiments indicated that Au decoration does not lower the oxidation onset potential of CO_{ads}, the enhanced yield of CO_2 at the decorated THH Pt NCs must arise from Au adatoms induced switch in the HCOOH oxidation mechanism towards the dehydrogenation pathway. The improved catalytic activity displayed by the Au decoration is therefore due, not to a bi-functional effect, but rather to an increase in the importance of the direct relative to the indirect HCOOH oxidation pathway, by the third-body effect.

6.3.1.4 Pt-Modified Au Prisms with High-Index Facets

Lu et al. reported that they used truncated ditetragonal Au prisms as nanofacet activators for platinum-catalyzed reactions [75]. The truncated ditetragonal Au prisms (TDP) are enclosed by 12 (310) high-index facets, as illustrated in Figure 6.12. In a typical synthesis, Au seeds were injected into a growth solution containing $HAuCl_4$, $AgNO_3$, HCl, ascorbic acid, and cetylpyridinium chloride (CPC) under magnetic stirring. Ag^+, Cl^-, and CPC played a synergetic effect in controlling the

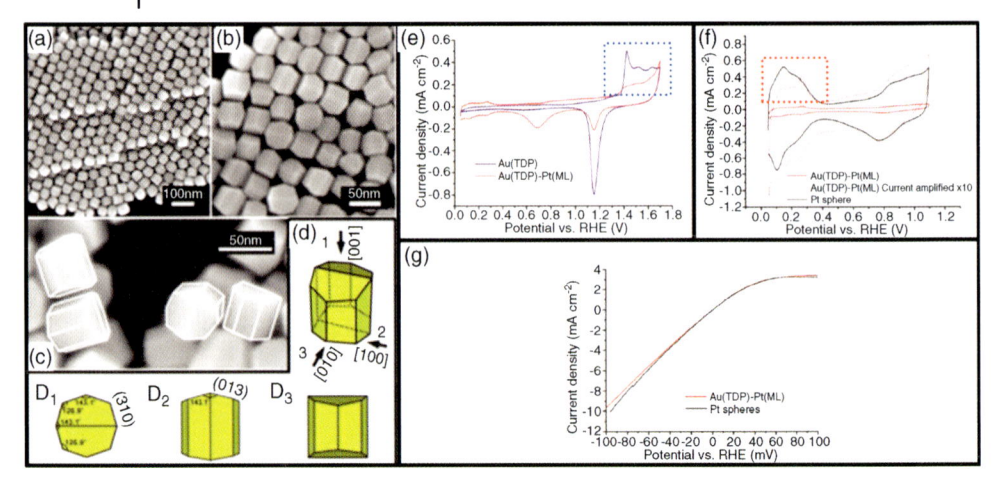

Figure 6.12 (a) Typical large-area and (b) locally magnified SEM images of the as-prepared Au TDP NPs, with an edge length of 45 ± 4 nm. (c) High-magnification SEM image stressing typical TDP profiles in random orientations. (d) Model of an ideal TDP bound by (310) facets and projections along three viewing directions indicated with arrows, as shown in (d1–3). (e, f) CV curves in deaerated 0.1 M H_2SO_4 solutions, 50 mVs^{-1}. (g) Polarization curve in hydrogen-saturated 0.1 M H_2SO_4 solutions, 5 mV s^{-1}, and 2500 rpm rotation rate. The current densities were normalized by the geometry area of the 0.5-cm-diameter electrode surface [75].

shape of truncated ditetragonal Au prisms. The effect of Ag UPD becomes dominant since the binding affinity of CPC for Au surfaces is relatively low. Ag^+ can be considered as a selective face-blocking adsorbate through the UPD mechanism [37]. The difference in the onset of Ag^+ UPD for various crystalline facets of Au, namely {110} > {100} > {111}, leads to an Ag monolayer being formed more favorably on the Au {110} facets. An Ag monolayer or sub-monolayer on the Au {110} facets acts as a strongly binding surfactant that results in a slower growth of Au {110} facets. The repeated galvanic reaction of an Ag monolayer by Au ions would significantly retard the total growth of the Au {110} facets [37]. The type of surfactant is also crucial to determining the proportion of {110} and {100} sub-facets present in the high-index facets. A Pt monolayer was placed on the surface of the Au TDP NPs via galvanic replacement of a UPD Cu monolayer [76]. The smaller Au reduction peak at 1.15 V and an additional reduction peak at 0.7 V in the voltammetry curve for the Au (TDP)–Pt(monolayer) sample (red curve in Figure 6.12e) are consistent with the presence of a Pt monolayer that partly shifts the reduction of surface oxide below 0.9 V. For the Au(TDP)–Pt(monolayer) sample, the hydrogen desorption peak at 0.3 V is higher than that at 0.15 V, distinctly differing from the ratio of the two peaks in the CV curve for sphere-like Pt NPs (45% Pt/C was purchased from E-TEK), as shown in Figure 6.12f. This feature suggests that the Pt monolayer lattice is not a hexagonal close-packed {111} surface, but mimics the underlying Au NPs containing rich (100) sub-facets [77]. From the integrated hydrogen desorption charges, the ratio between the electrochemical surface areas of the Au(TDP)–Pt(monolayer) and the Pt NPs samples is estimated to be 1 : 20. Yet, the polarization curves for hydrogen evolution and oxidation reactions are similar (Figure 6.12g), which indicates that the Pt monolayer on the high-index facets is much more active per Pt surface area than the close-packed surface of spherical Pt NPs. In other words, Au TDPs serve as nanofacet-activating substrates by translating their high-index-facet features to the supported materials; this results in a high electrochemical catalysis activity of the added Pt monolayer. The results illustrate a feasible way to study facet-dependent catalytic behavior of reactive metals using well-shaped Au NPs as facet-specific supports, as well as creating new opportunities for an enhancement of catalytic properties.

6.3.2
Alloy NCs

Alloying is another strategy commonly used to modify the catalytic activity of nanoparticles, through the combination of electronic effect, steric effect and/or "third body" effect. It is expected that alloy NCs with high-index facets may exhibit exceptionally high catalytic activity. However, synthesizing alloy nanoparticles with high-index facets is extremely challenging, not only because of the high surface energy of the monometallic high-index faceted nanoparticles, but also because of the need to match at least two elements both in size and thermodynamic stability. In the following section, two cases are reviewed for the successful synthesis of alloy nanoparticles with high-index facets, through an electrochemical strategy and a chemical method, respectively.

6.3.2.1 THH PdPt Alloy

We proposed a method of programmed potential electrodeposition previously, by which Pd tetrahexahedral NCs (THH Pd NCs) enclosed by {730} high-index facet can been directly deposited on a glassy carbon electrode from a dilute $PdCl_2$ solution as in Figure 6.3a. This method can be further extended to the preparation of alloy Pd–Pt THH NCs, by using Pd as the main element while a small amount of Pt is incorporated [78]. In a typical procedure, the alloy THH $Pd_{0.90}Pt_{0.10}$ NCs can be electrodeposited directly onto the glass carbon substrate electrode in 200 μM $PdCl_2$ + 20 μM K_2PtCl_6 + 0.1 M $HClO_4$ solution with a programed potential stepping to control the nucleation and surface structure of Pd–Pt NCs. Figure 6.13a shows SEM images of the THH Pd–Pt NCs. The atomic composition, determined by EDS, was 90.47% Pd and 9.53% Pt. By changing the K_2PtCl_6 concentration in the plating solution, THH Pd–Pt NCs with different alloy composition, for example, $Pd_{0.94}Pt_{0.06}$, $Pd_{0.92}Pt_{0.08}$, $Pd_{0.86}Pt_{0.14}$, and $Pd_{0.82}Pt_{0.18}$, have also been prepared. The dominant surfaces of the THH $Pd_{0.90}Pt_{0.10}$ NCs are {10,3,0} facets, as determined by HRTEM and SAED. Figure 6.13b displays the STEM image of a THH $Pd_{0.90}$–$Pt_{0.10}$ NC and corresponding EDS elemental mapping of Pd and Pt. Clearly, the elemental distribution profiles of Pd and Pt are very similar except for slight enrichment of Pt at the edge sites. This result indicates that Pd and Pt are evenly distributed through

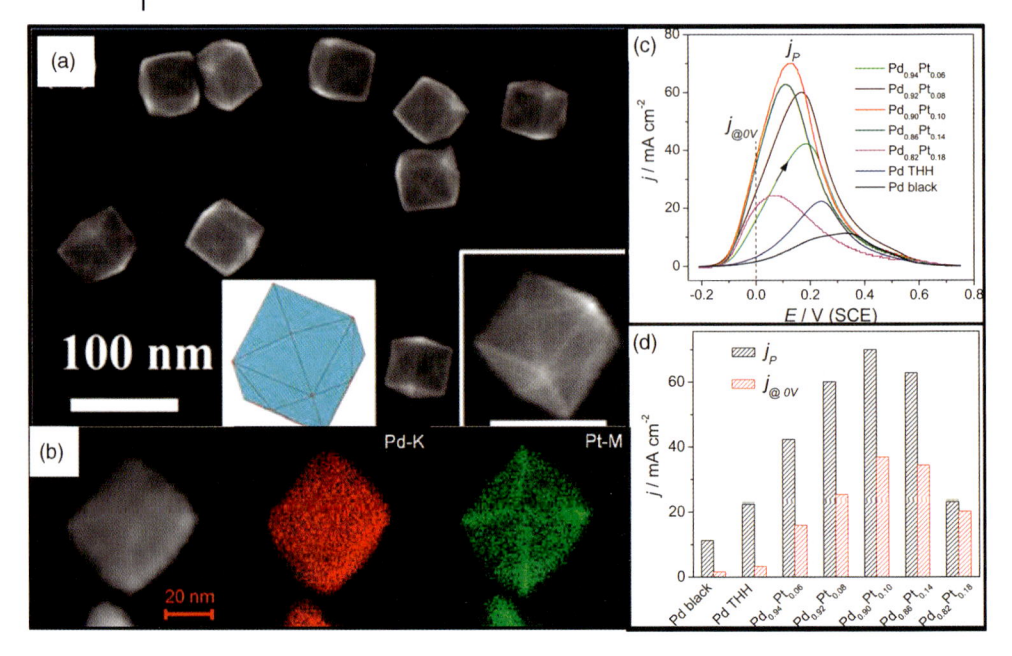

Figure 6.13 (a) High-magnification SEM image of THH Pd–Pt NCs. The scale bar is 50 nm in the inset. (b) STEM image and EDS elemental mapping of Pd and Pt in a THH Pd$_{0.90}$Pt$_{0.10}$ NC. Comparison of electrocatalytic activities of THH Pd–Pt NCs, Pd THH and commercial Pd black towards formic acid oxidation. (c) Current–potential curves recorded at 50 mV s^{-1} in 0.25 M HCOOH + 0.25 M HClO$_4$, The currents were normalized to the electrochemical surface area of the catalysts. (d) Comparison of oxidation current density at the peak (j_P) and at 0 V ($j = 0$ V) [78].

the alloy particle. The XPS test of THH Pd$_{0.90}$Pt$_{0.10}$ NCs indicated that the atomic percentage of Pd and Pt in the near-surface region was 87 : 13, close to that obtained from EDS data. Due to the high-index facet and alloy effect, as well as freedom from surfactant, the as-synthesized alloy THH Pd–Pt NCs exhibit extremely high catalytic activity for formic acid electrooxidation. Figure 6.13c compares the current–potential curves of formic acid electrooxidation on a series of alloy THH Pd–Pt NCs, THH Pd NCs and commercial Pd black. On the alloy THH Pd–Pt NCs, the peak current densities (j_P) increase with increasing Pt content, and reach a maximum at a Pt content of 10%. The j_P on THH Pd$_{0.90}$Pt$_{0.10}$ NCs is as high as 70 mA cm^{-2}, which is 3.1 times that measured on THH Pd NCs (22.4 mA cm^{-2}), and 6.2 times that on commercial Pd black (11.3 mA cm^{-2}). In addition, the peak potentials (E_P) for formic acid oxidation on alloy THH Pd–Pt NCs are shifted negatively as compared with pure Pd catalysts. The negative shift in the E_P results in a large enhancement in catalytic activity at low potential. For example, the oxidation current densities at 0.0 V (vs. SCE) are 36.8, 3.25, and 1.64 mA cm^{-2} on THH Pd$_{0.90}$Pt$_{0.10}$ NCs, THH Pd NCs, and commercial Pd black, respectively. The catalytic activity has been enhanced by one order of magnitude on the THH Pd$_{0.90}$Pt$_{0.10}$ NCs at this potential.

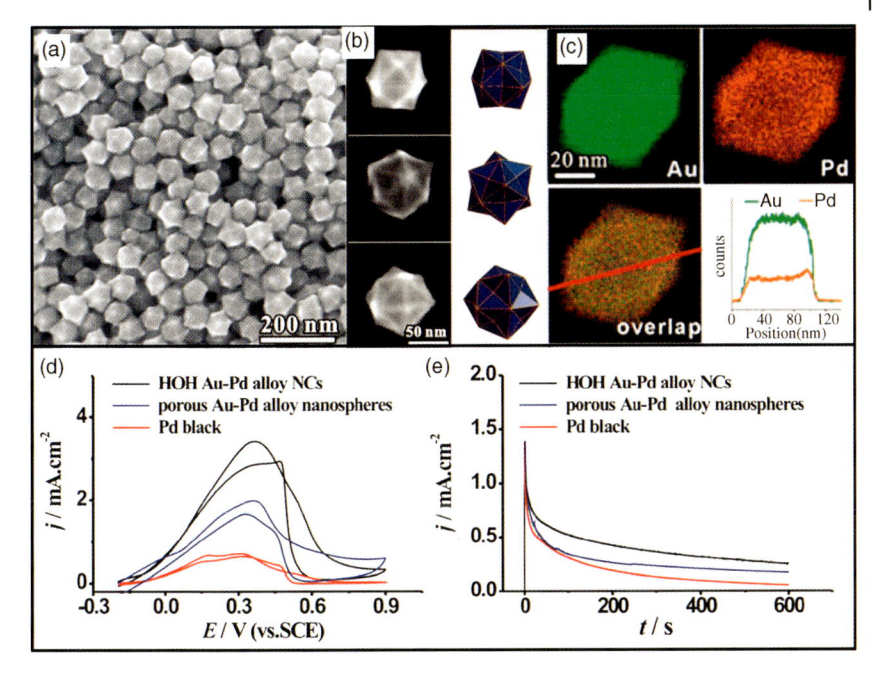

Figure 6.14 (a) SEM image of the as-prepared HOH AuPd alloy NCs. (b) A series of high-magnification SEM images and corresponding models of the HOH NCs with exposed {431} facets in different orientations. (c) Elemental scanning images and the cross-sectional compositional line-scanning profile of the Au–Pd alloy NCs. (d) CV curves measured on the as-prepared HOH Au–Pd alloy NCs, porous AuPd alloy nanospheres, and Pd black in 0.5 M H_2SO_4 + 0.25 M HCOOH solution, Scan rate: 50 mV s^{-1}. (e) Current–time curves of formic acid oxidation measured on the three kinds of nanocatalysts in 0.5 M H_2SO_4 + 0.25 M HCOOH solution at 0.40 V [79].

6.3.2.2 HOH PdAu Alloy

Xie's group first reported the chemical synthesis of alloy nanocrystals with high-index facet [79]. They employed UPD of Cu on Au as a bridge to realize simultaneous reduction of Au and Pd, and successfully prepared hexoctahedral (HOH) Au–Pd alloy NCs, which were enclosed by 48 {431} high-index facets. Figure 6.14a shows a representative SEM image of the HOH Au–Pd NCs. The crystal shape can be seen more clearly from the high-magnification SEM images of HOH Au–Pd NCs in different orientations, as well as from the corresponding models (Figure 6.14b). This polyhedron is a concave 48-facet polyhedron enclosed by 48 {hkl} high-index facets, similar to HOH NCs reported by Tian et al. [15] (Figure 6.4g). The exposed facets were determined mainly to be {431} facets. The element mapping analysis (Figure 6.14c) confirmed that the HOH NCs were definitely Au–Pd alloy. Both octadecyl trimethyl ammonium chloride (OTAC) and ethylene glycol act as capping agents to facilitate the formation of the special HOH shape. The presence of Cu^{2+} ions greatly improves the alloying of Pd into the Au lattice, which prevents the phase separation resulting from the different standard reduction potential between Au and Pd. In addition, the Cu^{2+}

ions play a key role in the formation of the HOH shape with the assistance of OTAC. The UPD process on the Au–Pd alloy particles' surface should affect the surface energy and slow down the growth rate along high-index facets, which leads to the formation of (hkl) high-index facets. It is interesting that they detected trace (0.2%) metallic Cu on the HOH Au–Pd NCs by XPS and inductively coupled plasma atomic emission spectroscopy (ICP-AES), which further confirms that Cu plays a role in the formation of HOH Au–Pd NCs. The as-prepared hexoctahedral Au–Pd alloy NCs exhibited high catalytic activity for formic acid electrooxidation due to synergism between the high-index facets and the alloy. As shown in Figure 6.14d, the peak current density of formic acid on alloy HOH Au–Pd NCs is double that of Au–Pd nanospheres, and four times that of Pd black. This result means that the {hkl} high-index facets consisting of high-density atomic steps and kinks indeed exhibit very high electrocatalytic activity. What is more, the HOH AuPd alloy NCs also showed good stability (Figure 6.14e) during the electrochemical tests.

6.3.3
Core–Shell Structured NCs

Core–shell structure NCs have unique advantages in catalytic processes, such as controlling the electronic and structure property of the shell-component by using different core materials, stabilizing the inner structure by coating a relatively stable shell while maintaining the activity of the inner component, and so on. From the preparation aspects, epitaxial growth has been widely used for the synthesis of core–shell nanoparticles with a high-index facet since the Au nanoparticles of different shapes are facile to prepare by chemical methods and are widely used as the core component.

Lu *et al.* reported the synthesis of Au–Pd core@shell heterostructures with THH morphology by using Au nanocubes as the structure-directing cores [80]. SEM and TEM characterizations of the resulting Au–Pd core–shell NCs generated using these Au nanocubes as cores for Pd shell growth are given in Figure 6.15. The images clearly show that the Au–Pd core–shell NCs possess a tetrahexahedral structure. Both the TEM image and high-angle annular dark-field scanning transmission electron microscopy (HAADF-STEM) image giving enhanced elemental contrast reveal a nanocube residing at the center of each THH nanocrystal. The EDS line scan and elemental mapping analyses on a single nanocrystal confirmed its Au core and Pd shell composition. The facet angles determined from the TEM image suggest that the THH Au–Pd core–shell NCs are bounded by the {730} facets. Four key factors were identified to facilitate the formation of the THH core–shell NCs: (i) A substantial lattice mismatch between Au and Pd; (ii) oxidative etching in the presence of chloride and oxygen; (iii) the use of cetyltrimethylammonium chloride (CTAC) as surfactant; (iv) moderate reaction temperature (30–60 °C) [81]. In comparison with the electrochemical square-wave potential method, the respective oxidation–reduction reactions for growth of high-index facets are quite similar. Electrooxidation of ethanol in KOH solution was performed to probe the relative electrocatalytic activities of the three Au–Pd core–shell structures

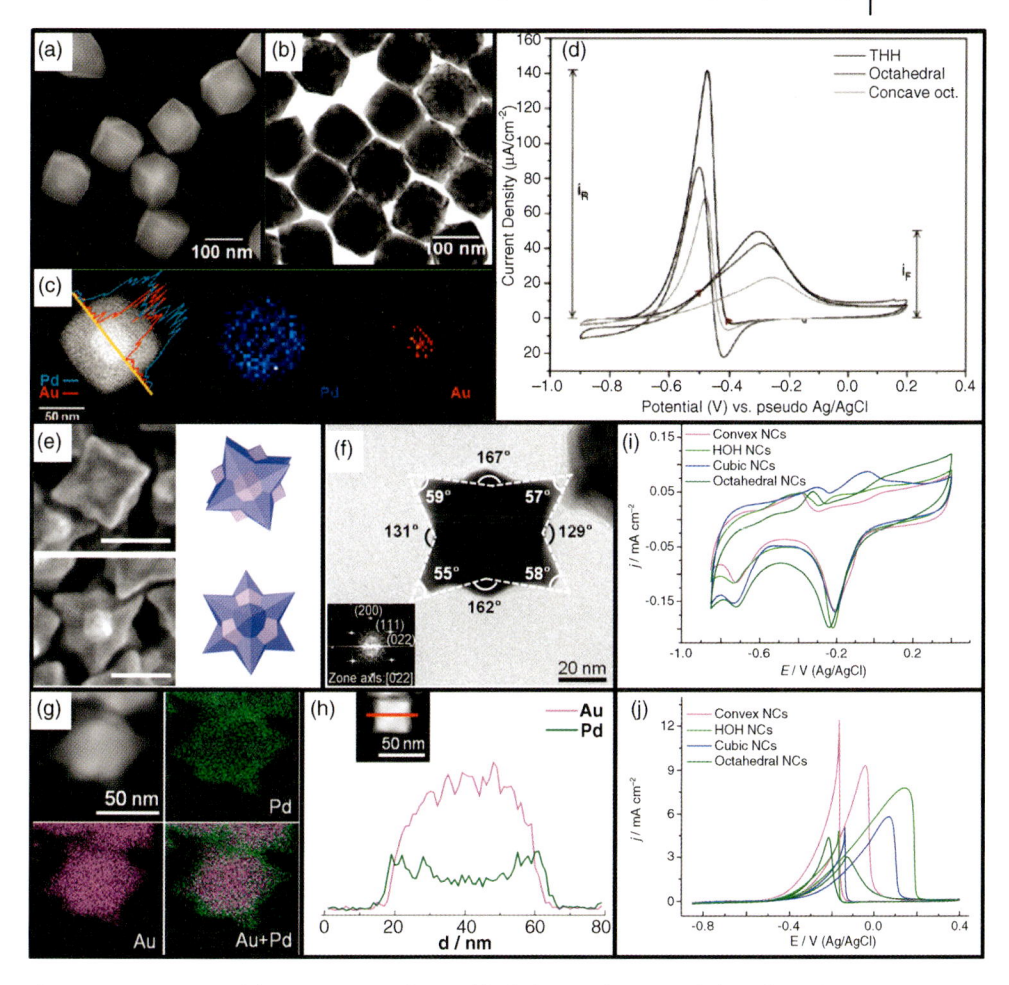

Figure 6.15 (a) SEM and (b) TEM images of the THH Au–Pd core–shell nanocrystals. (c) EDS line scan and elemental mapping image of a THH Au–Pd core–shell nanocrystal. (d) Cyclic voltammograms of different Au–Pd nanocrystal-modified electrodes in 0.2 M KOH solution containing 1.62 M ethanol with a scan rate of 50 mV s^{-1} [80]. (e) SEM images and corresponding geometric model of NCs in different orientations. The scale bars indicate 50 nm. (f) TEM image of a single convex polyhedral Au@Pd NC viewed along the [110] direction. The corresponding FFT pattern is shown in the bottom-left inset. (g) HAADF-STEM-EDS mapping images of the convex Au@Pd NCs. (h) HAADF-STEM image and cross-sectional compositional line profiles of a convex Au@Pd NC (d = distance). CVs of glassy carbon electrode modified with four different Au@Pd NCs obtained in (i) 0.1 M KOH and (j) 0.1 M KOH + 0.5 M ethanol solution at a scan rate of 50 mVs^{-1} [82].

synthesized in this study. It is clear that the THH NCs exhibit the highest electrocatalytic activity with a forward oxidation current (i_F) and a reverse oxidation current (i_R) value of 49.5 and 141.7 μA, respectively, as compared to those for the concave octahedra (i_F, 23.5 μA; i_R, 69.3 μA) and octahedra (i_F, 42.8 μA; i_R, 86.3 μA). The i_F peak current for the THH NCs is around 2.1 and 1.2 times as high as those recorded for the concave octahedra and octahedra. The high electrocatalytic activity of the THH Au–Pd core–shell NCs is attributed to the existence of all high-index {730} facets.

Kim et al. synthesized convex polyhedral Au@Pd core–shell NCs predominantly enclosed by high-index {12,5,3} facets through co-reduction of Au and Pd precursors in the presence of octahedral Au NC seeds, as shown in Figure 6.15e [82]. Elemental mapping of Au and Pd (Figure 6.15g) and the compositional line profiles of a nanocrystal (Figure 6.15h) obtained by high-angle annular dark-field scanning TEM-energy dispersive X-ray spectroscopy (HAADF-STEM-EDS) demonstrated that the NCs had a core–shell structure consisting of an Au core and a Pd shell. Since the reduction potential of Au^{3+} is higher than that of Pd^{2+}, the formation of Au@Pd core–shell NCs under the experimental conditions was most likely initiated by the preferential reduction of Au ions on the octahedral Au NC seeds, followed by the deposition of a Pd layer. The situation is different from the previously reported AuPd alloy, the reason is that in that case, the added Cu^{2+} ions formed a Cu UPD layer on the Au surface to modify the reaction rate, while in this case, no Cu^{2+} ions are present, so the $AuCl_4^-$ will be reduced first according to the standard reduction potential. The co-reduction of both metal precursors in a suitable molar ratio was indispensable for the successful formation of convex Au@Pd NCs. This can be attributed to the competitive reduction between Au and Pd precursors, which can modulate the growth kinetics of NCs [83–85]. Moreover, an adequate amount of reducing agent, ascorbic acid, was required to realize convex Au@Pd NCs, only from solutions containing ascorbic acid in a concentration higher than 50 mM were convex Au@Pd NCs formed.

To show a morphologic advantage of the convex Au@Pd NCs in catalysis, their electrocatalytic performance toward ethanol oxidation was investigated and the results were compared with the performance of other types of high-index-faceted Au@Pd core–shell NCs, such as HOH Au@Pd NCs enclosed by high-index {431} facets as well those of cubic and octahedral Au@Pd NCs that had low-index {100} and {111} facets on their surfaces, respectively (Figure 6.15i). The HOH, cubic, and octahedral Au@Pd NCs were prepared according to previously reported protocols [86,87]. Figure 6.15i shows the CVs of the NCs on a glassy carbon electrodes recorded in alkaline media. Typical current peaks associated with the oxidation/reduction of Pd were observed. During the cathodic sweep, the peak for the reduction of Pd oxide appeared around −0.2 V (vs Ag/AgCl) for convex Au@Pd NCs, the peak position being similar to those of other Au@Pd NCs, and even identical to that of cubic Au@Pd NCs. This further indicates that the shell of convex Au@Pd NCs consists of pure Pd. Figure 6.15j depicts the ethanol oxidation activities of the four different NCs. Well-separated anodic peaks in the forward and reverse sweeps, associated with the oxidation of ethanol, were identified. As shown in Figure 6.15j, the convex polyhedral Au@Pd NCs yielded a peak current density of 9.3 mA cm^{-2} in the forward scan, higher than those of HOH (7.7 mA cm^{-2}), cubic (5.8 mA cm^{-2}), and octahedral (3.0 mA cm^{-2}) Au@Pd NCs, indicating that high-index facets of convex Au@Pd NCs promote catalytic activity in the context of electrooxidation reactions.

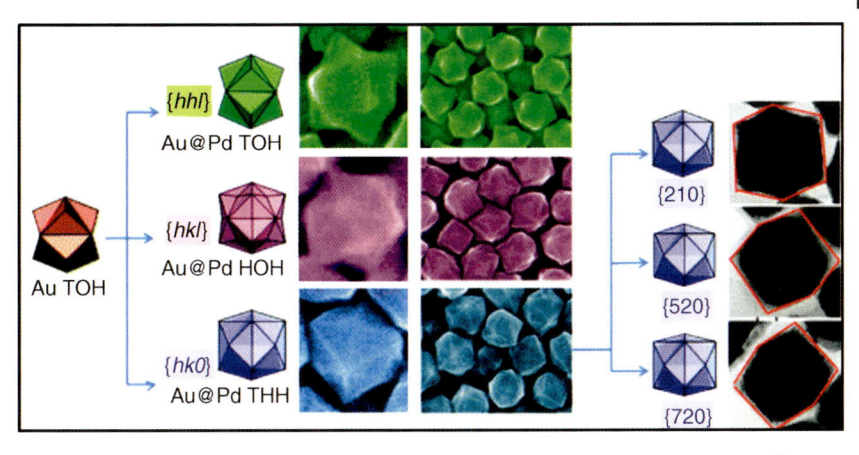

Figure 6.16 Schematic illustration of the reaction regions that form Au@Pd NCs with different polyhedral shapes and different high-index facets [88].

Yu et al. prepared polyhedral Au@Pd NCs with three different classes of high-index facets, including concave trisoctahedral (TSH) NCs with {hhl} facets, concave HOH NCs with {hkl} facets, and THH NCs with {hk0} facets (Figure 6.16) [88]. The key for the preparation is heteroepitaxial growth of Pd layers on concave TSH gold NC seeds while carefully controlling the Pd/Au ratio and NaBr concentration. The Miller indices of NCs are also modifiable, for example, from {210} to {720} for THH NCs. They found that Au@Pd NCs with high-index facets were more active for formic acid oxidation at low potentials than those with low-index facets with an order of cubes < octahedra < concave trisoctahedra < concave hexoctahedra < THH$_{\{720\}}$ < THH$_{\{210\}}$ < THH$_{\{520\}}$ [88]. Using a similar approach, Wang et al. [89] have also prepared Pd nanoshells with high-index {730} and {221} facets through heteroepitaxial growth on corresponding high-index-faceted Au NCs (i.e., tetrahexahedron and trisoctahedron) [89]. The catalytic activity for the Suzuki coupling reaction of the high-index-faceted Pd nanoshells is 3–7 times those of Pd and Au–Pd core–shell nanocubes, which possess only {100} facets. It is worth pointing out that the surfaces of Pd nanoshells prepared by Wang et al. [89] are much smoother (especially for THH) than the other two cases above [81,88].

As the previous several examples all take Au as the core and Pd as the shell, the following case concerns using Pd as the core and Au as the shell, in contrast with most of the above mentioned examples. The concave TOH Pd@Au core–shell NCs bound by {331} high-index facets were synthesized using Pd nanocubes as seed and CTAC as the structure-directing cores and capping agents, respectively [90]. Figure 6.17a shows the typical SEM image of the concave TOH Pd@Au core–shell NCs with uniform sizes and shapes. The TOH nanocrystal comprises eight trigonal pyramids, which can be generated from an octahedron by "pulling out" the centers of the eight triangular faces. The model of a concave TOH enclosed by {hhl} facets is shown in Figure 6.17c. The Miller indices of the edge-on facets of a TOH are {331}, determined through analyzing the projection angles from HRTEM. Figure 6.17d

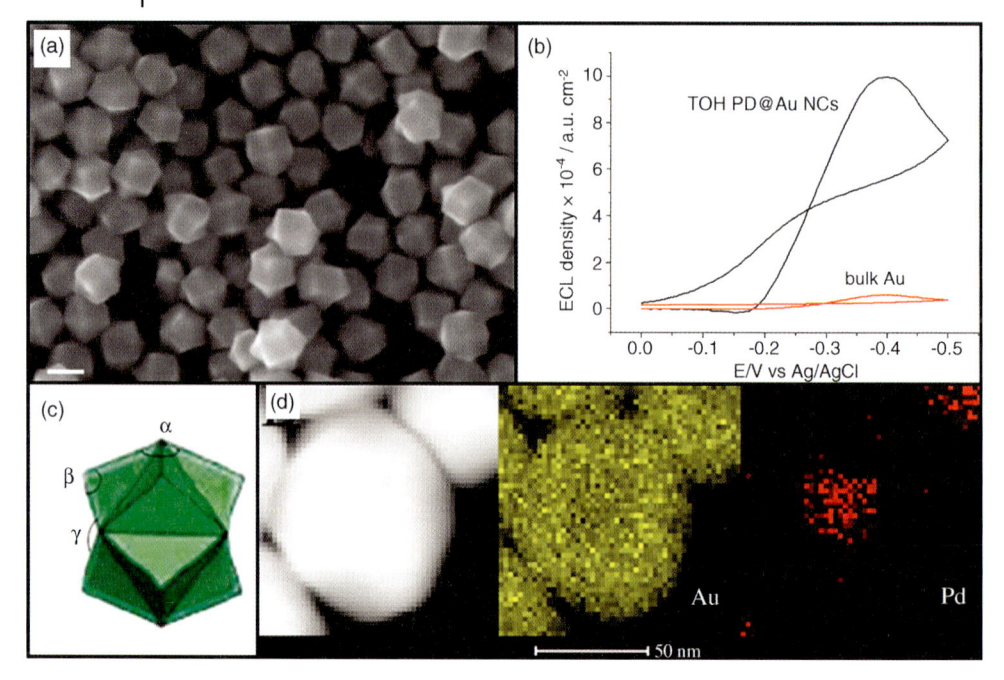

Figure 6.17 (a) SEM image of TOH Pd@Au core–shell NCs (scale bar: 200 nm). (b) ECL density on bulk Au electrode and TOH Pd@Au core–shell NCs modified glassy carbon electrode in 0.1 M pH 7.4 phosphate buffer solution containing 1 mM H_2O_2 and 0.2 mM luminol. PMT: −500 V. (c) A model TOH nanocrystal viewed in the [110] direction. (d) HAADF-STEM and elemental mapping images of an individual TOH Pd@Au core–shell nanocrystal [90].

illustrates the HAADF-STEM image of an individual Pd@Au core–shell nanocrystal and the elemental mapping analyses, confirming its Pd core and Au shell composition. The electrocatalytic activity of TOH Pd@Au core–shell NCs on the electrochemiluminescence (ECL) of luminal and H_2O_2 was investigated. As far as we know, it is the first study of the electrocatalytic properties of a polyhedron on ECL [91]. As shown in Figure 6.17b, the ECL density on TOH Pd@Au NCs is much higher than that of the bulk Au electrode, indicating the high electrocatalytic activity of TOH Pd@Au NCs toward ECL reactions of luminol and H_2O_2.

Zhang *et al.* synthesized sub-30 nm Au@Pd concave nanocubes and Pt-on-(Au@Pd) trimetallic nanostructures [92]. A multi-component heterogeneous seed-mediated growth route was employed to prepare the Au@Pd and Pt^(Au@Pd) samples, as illustrated schematically in Figure 6.18a. Further analysis determined the surface of Au@Pd nanoparticles are composed of {hk0} facets, such as {610}, {520}, and {730}. Figure 6.18b displays the HAADF-STEM image and STEM-EDS elemental mapping results for the trimetallic Pt^(Au@Pd) sample. Every particle in the HAADF-STEM image (Figure 6.18b) appears to be composed of a bright core particle and a less bright shell with rough surfaces. This bright core particle must be

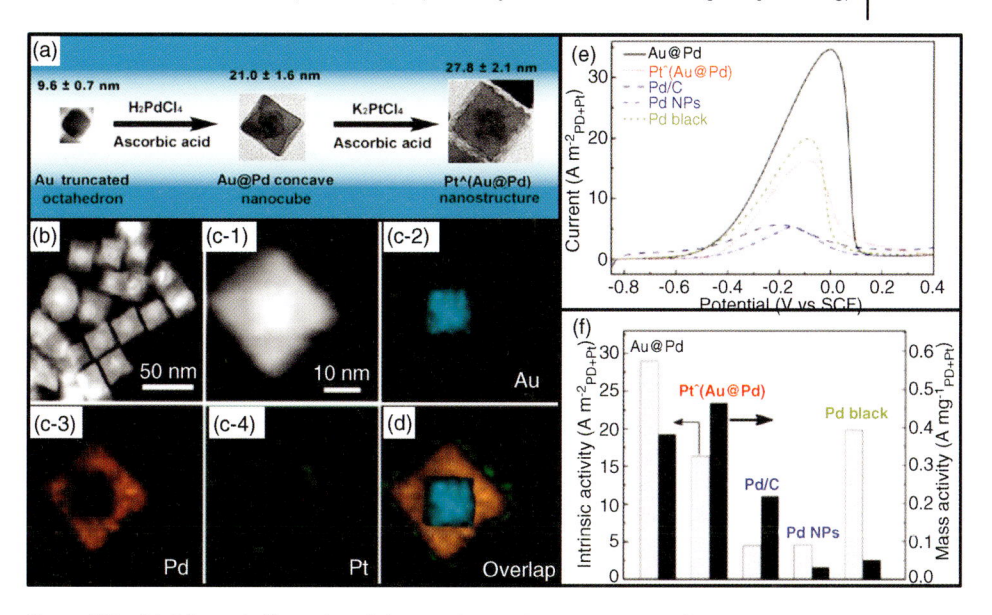

Figure 6.18 (a) Schematic illustration of the heterogeneous seed-mediated syntheses of Au@Pd concave nanocubes and Pt$^\wedge$(Au@Pd) nanostructures. The numbers above the particle images show the particle size distribution statistics obtained from TEM measurements. (c) HAADF-STEM image of the trimetallic Pt$^\wedge$(Au@Pd) nanostructures. (c) STEM-EDS elemental mapping images for a representative Pt$^\wedge$(Au@Pd) particle. (d) An adduct image of the Au (c-2), Pd (c-3), and Pt (c-4) elemental mappings. (e) Anodic-scan CV curves, and (f) comparison of intrinsic activity and mass activity data for ethanol electrooxidation on the indicated samples [92].

Au NP since Au would show a stronger atomic number contrast (Z-contrast) than Pd in the shell layers. Figure 6.18c shows the HAADF-STEM image of a representative Pt$^\wedge$(Au@Pd) particle and its corresponding elemental mapping images. Figure 6.18d is an adduct image of the Au (Figure 6.18C-2), Pd (Figure 6.18C-3), and Pt (Figure 6.18C-4) mappings, which clearly confirms the trimetallic triple-layered nanostructures of the Pt$^\wedge$(Au@Pd) particles, with an Au NP in the core, Pd in the inner-layer and Pt-rich clusters in the outermost jagged-surface layers. The outermost layer is composed of Pt instead of the potential PtPd alloy, which was further confirmed by the XPS analysis. Figure 6.18e shows the anodic-scan CV curves for ethanol electrooxidation on Au@Pd, Pt$^\wedge$(Au@Pd), and the reference Pd catalysts. For a better understanding of the observed activity difference, the intrinsic activity and mass activity (MSA) data at -0.1 V were further compared in Figure 6.18f. The intrinsic activity number for Au@Pd (29 A m^{-2} Pd) appeared 6–7 times as high as that for Pd/C (4 A m^{-2} Pd) and Pd NPs (5 A m^{-2} Pd) and about 1.5 times as high as that for Pd black (19 A m^{-2} Pd) and Pt$^\wedge$(Au@Pd) (16 A m^{-2} Pd + Pt), demonstrating a faster ethanol electrooxidation kinetics on the Pd surface of the Au@Pd concave nanocubes. The high intrinsic activity number for ethanol electrooxidation of the present Au@Pd catalyst (Figure 6.18f) would mean a high-density of low-coordinated Pd atoms at the catalyst surface, which is consistent with the concave surface structure of the Au@Pd catalyst exposing high-index Pd facets. Moreover, the enhanced catalytic performance of Au@Pd concave nanocubes may also be due to the modified surface electronic structure of Pd shells by interfacial alloying with the underlying Au cores. The lower intrinsic activity of the trimetallic triple-layered Pt^(Au@Pd) catalyst (Figure 6.18f) could be due to (i) the deposition of Pt resulting in passivation of some active Pd sites and/or significant collapse of the high-index facets of the preformed supporting Au@Pd, and (ii) Pt is known to be intrinsically less active for ethanol electrooxidation than Pd in alkaline media [93,94]. Surprisingly, this trimetallic triple-layered Pt^(Au@Pd) catalyst exhibited the highest activity by MSA (0.46 A mg^{-1}_{Pd+Pt}) (Figure 6.18f), which is 1.2 times that of Au@Pd (0.38 A mg^{-1}_{Pd}) and 2–10 times higher than the other Pd catalysts. Even if the total mass of all the three metals (Pd, Pt, and Au) was taken into account, this Pt^(Au@Pd) catalyst would still show an overall $MSA_{Pd+Pt+Au}$ of 0.38 A mg^{-1} metal, which is 7.6 and 1.7 times higher than the MSA_{Pd} data of the conventional Pd black (0.05 A mg^{-1}_{Pd}) and Pd/C (0.22 A mg^{-1}_{Pd}) catalysts, respectively, which would lead to a Pd saving by 40–85%. Thus, the activity data for ethanol electrooxidation demonstrated that the Au@Pd concave cubes and their derived trimetallic triple-layered Pt^(Au@Pd) structures are highly advantageous compared to conventional Pd black for the anode catalyst. Long-term repeated CV measurements of ethanol electrooxidation were also conducted to compare the electrocatalytic stabilities of the Au@Pd, Pt^(Au@Pd), and Pd-black catalysts. The trimetallic Pt^(Au@Pd) catalyst exhibited outstanding stability, its distinctively high catalytic stability may be attributed to its unique triple-layered structure with jagged Pt-rich surfaces, which could function to stabilize the high-index faceted Pd layers/fabrics with interfacial Pd–Pt interaction in or near the outer surfaces.

6.4
Concluding Remarks and Perspective

This chapter has reviewed recent progess concerning the synthesis of nanoparticles with an open structure. The open surface structure provides abundant active sites, like steps and kinks, whose coordinate number is relatively low. The low coordinate number of the surface sites makes them highly active towards chemical reactions. On the other hand, from a thermodynamic viewpoint, the open surface structure corresponds to high surface energy, resulting in a big challenge for the synthesis of nanoparticles with open surface structure. To overcome the difficult of synthesizing precious metal nanoparticles with high surface energy, kinetic control is indispensable. For the electrochemical route, such as the square-wave method, three major effects can contribute to the successful synthesis of nanoparticles with high surface energy: (i) the dynamic oxygen adsorption/desorption mediated by the square-wave potential; (ii) the significant impediment effect in place-exchange between oxygen and Pt surface atoms on high-index surfaces; and (iii) the decrease in surface energy of the Pt NCs by electrochemical adsorption of hydrogen at low potential and oxygen species at high potential. As for the wet-chemical route, the role played by oxygen species in the electrochemical method is achieved by functional molecules, such as surfactants, capping agents, and additives like amine, halide anion, and so on. Through adsorption on the surface of a nanoparticle, the functional molecules can dramatically change the surface energy from its intrinsic one. Usually the low coordinated sites with high intrinsic energy can be better stabilized by the capping agents, slowing down the growth rate towards the high-energy surface direction and preserving the high energy surface. Etching is another commonly used strategy to create a low-coordinated atom in the wet-chemical route, O_2/Cl^- and Fe^{3+}/Fe^{2+} are most frequently employed. Most of the nanoparticles mentioned above are enclosed by well-defined surface structure, which makes them easy to characterize and visualize esthetically, and also provides the possibility to reveal the structure-functionality of nanoparticle catalysts at a level of surface atomic arrangement. The knowledge about the structure-functionality gained by using precious metal nanocrystals of open surface structure/high surface energy is the basis to rationally design and fabricate practical catalysts with high activity, selectivity and stability. The precious metal nanoparticles of open surface structure/high surface energy can really boost the activity for most electrochemical reactions, like small molecule electrooxidation, oxygen reduction reaction, and so on. It is reasonable to think that the nanoparticles of open surface structure will be widely used in electrocatalysis, both fundamentally and in real applications. In order to achieve real applications, two important points should be addressed: (i) decreasing the size of nanocrystals with open surface structure down to a size comparable to commercial nanocatalysts and (ii) developing techniques for mass production, which certainly need theoretical advances and technical innovations. An electrochemical method could be used to synthesize surfactant-free catalysts, which are ready for use in catalysis. However, growth on the surface of the electrode limits its large scale production. Although the synthesis of HIF-Pt/C with size around 5 nm has been reported, the reaction was fixed at the surface of a glass carbon electrode, which makes it quite difficult to mass produce. As for the wet-chemical route, the nanoparticles with high-index facets synthesized are covered by a large amount of capping agents, which are usually quite difficult to remove for catalytic applications. The present chapter indicates that although significant progress has been made in the preparation of precious metal nanoparticles of high surface energy and thus high catalytic activity, there still remain challenges and there is a long way to go from the fundamental study to real applications.

Acknowledgments

This work was supported by NSFC (21222310, 21073152, 21021002, and 20933004), Foundation for the Author of National Excellent Doctoral Dissertation of China (201126), and Program for New Century Excellent Talents in University (NECT-10-0715 and NECT-11-0301).

References

1 Wolfe, David W. (2001) *Tales from the Underground a Natural History of Subterranean Life*, Perseus Pub., Cambridge, Mass.
2 Appl, M. (2000) *Ullmann's Encyclopedia of Industrial Chemistry*, Wiley-VCH Verlag GmbH.
3 Ryashentseva, M.A. (1998) *Russ. Chem. Rev.*, **67**, 157.
4 Gasteiger, H.A., Kocha, S.S., Sompalli, B., and Wagner, F.T. (2005) *Appl. Catal. B-Environ.*, **56**, 9.
5 Benfield, R.E. (1992) *J. Chem. Soc. Faraday Trans.*, **88**, 1107.
6 Lebedeva, N.P., Koper, M.T.M., Feliu, J.M., and van Santen, R.A. (2002) *J. Phys. Chem. B*, **106**, 12938.
7 Bernasek, S.L. and Somorjai, G.A. (1975) *Surf. Sci.*, **48**, 204.
8 Sun, S.-G., Chen, A.-C., Huang, T.-S., Li, J.-B., and Tian, Z.-W. (1992) *J. Electroanal. Chem.*, **340**, 213.
9 Tian, N., Zhou, Z.-Y., and Sun, S.-G. (2008) *J. Phys. Chem. C*, **112**, 19801.
10 Van Santen, R.A. (2008) *Acc. Chem. Res.*, **42**, 57.
11 Wang, Z.L. (2000) *J. Phys. Chem. B*, **104**, 1153.
12 Wen, Y.-H., Zhang, Y., Zhu, Z.-Z., and Sun, S.-G. (2009) *Chinese Phys. B*, **18**, 4955.
13 Huang, R., Wen, Y.-H., Zhu, Z.-Z., and Sun, S.-G. (2011) *J. Mater. Chem.*, **21**, 11578.
14 Xiong, Y., Wiley, B.J., and Xia, Y. (2007) *Angew. Chem. Int. Ed.*, **46**, 7157.
15 Tian, N., Zhou, Z.Y., Sun, S.G., Ding, Y., and Wang, Z.L. (2007) *Science*, **316**, 732.
16 Xia, Y., Xiong, Y., Lim, B., and Skrabalak, S.E. (2009) *Angew. Chem. Int. Ed.*, **48**, 60.
17 Ahmadi, T.S., Wang, Z.L., Green, T.C., Henglein, A., and El-Sayed, M.A. (1996) *Science*, **272**, 1924.
18 Song, H., Kim, F., Connor, S., Somorjai, G.A., and Yang, P. (2004) *J. Phys. Chem. B*, **109**, 188.
19 Yin, A.-X., Min, X.-Q., Zhang, Y.-W., and Yan, C.-H. (2011) *J. Am. Chem. Soc.*, **133**, 3816.
20 Chen, Y.-X., Chen, S.-P., Zhou, Z.-Y., Tian, N., Jiang, Y.-X., Sun, S.-G., Ding, Y., and Wang, Z.L. (2009) *J. Am. Chem. Soc.*, **131**, 10860.
21 Pangarov, N.A. (1964) *Electrochim. Acta.*, **9**, 721.
22 Mullin, J.W. (1993) *Crystallization*, 3rd edn, Oxford, Butterworth, 172–263.
23 Zhang, K.C. (1987) *Modern Crystallography*, Science Press, Beijing, 76–118.
24 Qian, Yitai (1998) *Crystal Chemistry*, China Science and Technology Press, Hefei, 40–41.
25 Wei, L., Fan, Y.-J., Tian, N., Zhou, Z.-Y., Zhao, X.-Q., Mao, B.-W., and Sun, S.-G. (2011) *J. Phys. Chem. C*, **116**, 2040.
26 Li, Y., Jiang, Y., Chen, M., Liao, H., Huang, R., Zhou, Z., Tian, N., Chen, S., and Sun, S. (2012) *Chem. Commun.*, **48**, 9531.
27 Zhou, Z.-Y., Huang, Z.-Z., Chen, D.-J., Wang, Q., Tian, N., and Sun, S.-G. (2010) *Angew. Chem. Int. Ed.*, **49**, 411.
28 Zhou, Z.-Y., Shang, S.-J., Tian, N., Wu, B.-H., Zheng, N.-F., Xu, B.-B., Chen, C., Wang, H.-H., Xiang, D.-M., and Sun, S.-G. (2012) *Electrochem. Commun.*, **22**, 61.
29 Zhou, Z.-Y., Tian, N., Huang, Z.-Z., Chen, D.-J., and Sun, S.-G. (2009) *Faraday Discuss.*, **140**, 81.
30 Zhou, W.P., Lewera, A., Larsen, R., Masel, R.I., Bagus, P.S., and Wieckowski, A. (2006) *J. Phys. Chem. B*, **110**, 13393.
31 Ge, J.J., Xing, W., Xue, X.Z., Liu, C.P., Lu, T.H., and Liao, J.H. (2007) *J. Phys. Chem. C*, **111**, 17305.
32 Li, Y., Hong, X.M., Collard, D.M., and El-Sayed, M.A. (2000) *Org. Lett.*, **2**, 2385.
33 Farina, V. (2004) *Adv. Synth. Catal.*, **346**, 1553.
34 Tian, N., Zhou, Z.-Y., Yu, N.-F., Wang, L.-Y., and Sun, S.-G. (2010) *J. Am. Chem. Soc.*, **132**, 7580.
35 Huang, X., Zhao, Z., Fan, J., Tan, Y., and Zheng, N. (2011) *J. Am. Chem. Soc.*, **133**, 4718.
36 Personick, M.L., Langille, M.R., Zhang, J., and Mirkin, C.A. (2011) *Nano Lett.*, **11**, 3394.
37 Liu, M. and Guyot-Sionnest, P. (2005) *J. Phys. Chem. B*, **109**, 22192.
38 Jiang, Q., Jiang, Z., Zhang, L., Lin, H., Yang, N., Li, H., Liu, D., Xie, Z., and Tian, Z. (2011) *Nano Res.*, **4**, 612.

39 Chiu, C.Y., Li, Y.J., Ruan, L.Y., Ye, X.C., Murray, C.B., and Huang, Y. (2011) *Nature Chem.*, **3**, 393.
40 Yu, T., Kim, D.Y., Zhang, H., and Xia, Y. (2011) *Angew. Chem. Int. Ed.*, **50**, 2773.
41 Jin, M., Zhang, H., Xie, Z., and Xia, Y. (2011) *Angew. Chem. Int. Ed.*, **50**, 7850.
42 Zhang, J., Zhang, L., Xie, S., Kuang, Q., Han, X., Xie, Z., and Zheng, L. (2011) *Chem. Eur. J.*, **17**, 9915.
43 Zhang, J.A., Langille, M.R., Personick, M.L., Zhang, K., Li, S.Y., and Mirkin, C.A. (2010) *J. Am. Chem. Soc.*, **132**, 14012.
44 Ma, Y., Kuang, Q., Jiang, Z., Xie, Z., Huang, R., and Zheng, L. (2008) *Angew. Chem. Int. Ed.*, **47**, 8901.
45 Hamelin, A. (1996) *J. Electroanal. Chem.*, **407**, 1.
46 Yu, Y., Zhang, Q., Lu, X., and Lee, J.Y. (2010) *J. Phys. Chem. C*, **114**, 11119.
47 Ming, T., Feng, W., Tang, Q., Wang, F., Sun, L., Wang, J., and Yan, C. (2009) *J. Am. Chem. Soc.*, **131**, 16350.
48 Zheng, Y., Tao, J., Liu, H., Zeng, J., Yu, T., Ma, Y., Moran, C., Wu, L., Zhu, Y., Liu, J., and Xia, Y. (2011) *Small*, **7**, 2307.
49 Tran, T.T. and Lu, X. (2011) *J. Phys. Chem. C*, **115**, 3638.
50 Liu, H.B., Pal, U., and Ascencio, J.A. (2008) *J. Phys. Chem. C*, **112**, 19173.
51 Tian, N., Xiao, J., Zhou, Z.-Y., Liu, H.-X., Deng, Y.-J., Huang, L., and Sun., S.-G. (2013) *Faraday Discuss.*, **162**, 77–89.
52 Capon, A. and Parsons, R. (1973) *J. Electroanal. Chem. Interfacial Electrochem.*, **45**, 205.
53 Sun, S.G., Clavilier, J., and Bewick, A. (1988) *J. Electroanal. Chem. Interfacial Electrochem.*, **240**, 147.
54 Chen, Y.X., Heinen, M., Jusys, Z., and Behm, R.J. (2006) *Langmuir*, **22**, 10399.
55 Watanabe, M., Horiuchi, M., and Motoo, S. (1988) *J. Electroanal. Chem.*, **250**, 117.
56 Peng, B., Wang, H.F., Liu, Z.P., and Cai, W.B. (2010) *J. Phys. Chem. C*, **114**, 3102.
57 Chen, Q.S., Zhou, Z.Y., Vidal-Iglesias, F.J., Solla-Gullon, J., Feliu, J.M., and Sun, S.G. (2011) *J. Am. Chem. Soc.*, **133**, 12930.
58 Lopez-Cudero, A., Vidal-Iglesias, F.J., Solla-Gullon, J., Herrero, E., Aldaz, A., and Feliu, J.M. (2009) *J. Electroanal. Chem.*, **637**, 63.
59 Herrero, E., Feliu, J.M., and Aldaz, A. (1994) *J. Electroanal. Chem.*, **368**, 101.
60 Leiva, E., Iwasita, T., Herrero, E., and Feliu, J.M. (1997) *Langmuir*, **13**, 6287.
61 Chen, Q.-S., Zhou, Z.-Y., Vidal-Iglesias, F.J., Solla-Gullón, J., Feliu, J.M., and Sun, S.-G. (2011) *J. Am. Chem. Soc.*, **133**, 12930.
62 Shi-Gang, S., Yan, L., Nan-Hai, L., and Ji-Qian, M. (1994) *J. Electroanal. Chem.*, **370**, 273.
63 Chrzanowski, W. and Wieckowski, A. (1998) *Langmuir*, **14**, 1967.
64 Liu, H.-X., Tian, N., Brandon, M.P., Zhou, Z.-Y., Lin, J.-L., Hardacre, C., Lin, W.-F., and Sun, S.-G. (2012) *ACS Catalysis*, **2**, 708.
65 Del Colle, V., Berna, A., Tremiliosi-Filho, G., Herrero, E., and Feliu, J.M. (2008) *Phys. Chem. Chem. Phys.*, **10**, 3766.
66 Wang, H. and Baltruschat, H. (2007) *J. Phys. Chem. C*, **111**, 7038.
67 Waszczuk, P., Solla-Gullón, J., Kim, H.S., Tong, Y.Y., Montiel, V., Aldaz, A., and Wieckowski, A. (2001) *J. Catal.*, **203**, 1.
68 Obradović, M.D., Tripković, A.V., and Gojković, S.L. (2009) *Electrochim. Acta.*, **55**, 204.
69 Liu, H.-X., Tian, N., Brandon, M.P., Pei, J., Huangfu, Z.-C., Zhan, C., Zhou, Z.-Y., Hardacre, C., Lin, W.-F., and Sun, S.-G. (2012) *Phys. Chem. Chem. Phys.*, **14**, 16415.
70 Wang, R., Wang, C., Cai, W.-B., and Ding, Y. (2010) *Adv. Mater.*, **22**, 1845.
71 Neurock, M., Janik, M., and Wieckowski, A. (2009) *Faraday Discuss.*, **140**, 363.
72 Yang, Y.-Y., Zhou, Z.-Y., and Sun, S.-G. (2001) *J. Electroanal. Chem.*, **500**, 233.
73 Miki, A., Ye, S., and Osawa, M. (2002) *Chem. Commun.*, 1500.
74 Samjeské, G., Miki, A., Ye, S., and Osawa, M. (2006) *J. Phys. Chem. B*, **110**, 16559.
75 Lu, F., Zhang, Y., Zhang, L., Zhang, Y., Wang, J.X., Adzic, R.R., Stach, E.A., and Gang, O. (2011) *J. Am. Chem. Soc*, **133**, 18074–18077.
76 Brankovic, S.R., Wang, J.X., and Adžić, R.R. (2001) *Surf. Sci.*, **474**, L173.
77 Wang, C., Daimon, H., Onodera, T., Koda, T., and Sun, S. (2008) *Angew. Chem. Int. Edit.*, **47**, 3588.
78 Deng, Y.-J., Tian, N., Zhou, Z.-Y., Huang, R., Liu, Z.-L., Xiao, J., and Sun, S.-G. (2012) *Chem. Sci.*, **3**, 1157.

79 Zhang, L., Zhang, J., Kuang, Q., Xie, S., Jiang, Z., Xie, Z., and Zheng, L. (2011) *J. Am. Chem. Soc.*, **133**, 17114–17117.

80 Lu, C.-L., Prasad, K.S., Wu, H.-L., Ho, J.-a.A., and Huang, M.H. (2010) *J. Am. Chem. Soc.*, **132**, 14546.

81 Lu, C.L., Prasad, K.S., Wu, H.L., Ho, J.A., and Huang, M.H. (2010) *J. Am. Chem. Soc.*, **132**, 14546.

82 Kim, D., Lee, Y.W., Lee, S.B., and Han, S.W. (2012) *Angew. Chem. Int. Edit.*, **51**, 159.

83 Liu, Y. and Walker, A.R.H. (2010) *Angew. Chem.*, **122**, 6933.

84 DeSantis, C.J., Peverly, A.A., Peters, D.G., and Skrabalak, S. E. (2011) *Nano Lett.*, **11**, 2164.

85 Lee, Y.W., Kim, M., Kang, S.W., and Han, S.W. (2011) *Angew. Chem.*, **123**, 3528.

86 Lee, Y.W., Kim, M., Kim, Z.H., and Han, S.W. (2009) *J. Am. Chem. Soc.*, **131**, 17036.

87 Yu, Y., Zhang, Q., Liu, B., and Lee, J.Y. (2010) *J. Am. Chem. Soc.*, **132**, 18258.

88 Yu, Y., Zhang, Q.B., Liu, B., and Lee, J.Y. (2010) *J. Am. Chem. Soc.*, **132**, 18258.

89 Wang, F., Li, C.H., Sun, L.D., Wu, H.S., Ming, T., Wang, J.F., Yu, J.C., and Yan, C.H. (2010) *J. Am. Chem. Soc.*, **133**, 1106.

90 Zhang, L., Niu, W., Li, Z., and Xu, G. (2011) *Chem. Commun.*, **47**, 10353.

91 Hu, L. and Xu, G. (2010) *Chem. Soc. Rev.*, **39**, 3275.

92 Zhang, G.-R., Wu, J., and Xu, B.-Q. (2012) *J. Phys. Chem. C*, **116**, 20839.

93 Shen, P.K. and Xu, C. (2006) *Electrochem. Commun.*, **8**, 184.

94 Liang, Z.X., Zhao, T.S., Xu, J.B., and Zhu, L.D. (2009) *Electrochim. Acta.*, **54**, 2203.

7
X-Ray Studies of Strained Catalytic Dealloyed Pt Surfaces
Peter Strasser

7.1
Introduction

Ever since the first observation of an X-ray diffraction pattern by Friedrich, Knipping, and Laue [1] upon directing an X-ray beam through a crystalline solid, X-ray analytics has become an ever more powerful tool to study the geometric structure of matter [2]. Similarly, the discovery of the photoelectric effect has eventually led to the development of sophisticated spectroscopic analysis techniques to gain insights into the chemical state, chemical bonds, and the electronic structure of matter, without which modern chemistry, physics, or materials science would be unthinkable [3–9]. Today, modern X-ray scattering and X-ray spectroscopic techniques are available to the scientific community through public large-scale synchrotron radiation user facilities [10–20].

X-ray based analysis techniques are indispensible to study the nature and behavior of catalysts, that is, functional materials altering the course of a chemical reaction. Catalysts are ubiquitous in chemistry and biology. A specific class of catalysts directs and controls electron transfer at solid/liquid interfaces, where free electrons (electricity) inside conductive electrodes are used to alter the oxidation state of reactant species in the liquid electrode, possibly combined with the making or breaking of chemical bonds. Such catalysts are therefore referred to as electrocatalysts and are critical functional materials to transform electrical energy into chemical energy or vice versa. The former class of electrocatalysts aid the storage of electricity in the form of molecular or solid state chemical bonds, while the latter catalyze the opposite process. Electrochemical galvanic cell devices transforming chemical energy into flowing electricity have been known since Volta's work in 1800. A more specific class of such galvanic cells, referred to originally in 1839 as gas voltaic batteries, typically makes and breaks chemical bonds of molecules rather than solids, and, in today's science jargon, is referred to as fuel cells. A fuel is catalytically oxidized at the fuel cell anode, while an oxidant, mostly oxygen, is catalytically reduced at the fuel cell cathode. Fuel cell catalysis is the branch of science that studies all aspects associated with the electrochemical surface catalysis at the anode and cathode of fuel cells.

Electrocatalysis: Theoretical Foundations and Model Experiments, First Edition.
Edited by Richard C. Alkire, Ludwig A. Kibler, Dieter M. Kolb, and Jacek Lipkowski.
© 2013 Wiley-VCH Verlag GmbH & Co. KGaA. Published 2013 by Wiley-VCH Verlag GmbH & Co. KGaA.

ORR: $O_2 + 4H^+ + 4e^- \rightarrow 2H_2O$

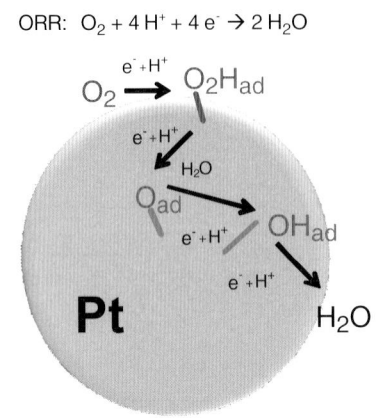

Figure 7.1 An elementary mechanistic model for the four-electron oxygen reduction reaction (ORR).

A prominent type of fuel cell is the so-called polymer electrolyte membrane fuel cell (PEMFC) thanks to its solid polymeric electrolyte [21–28]. As a particularly efficient electrochemical energy conversion technology at relatively low temperatures, PEMFCs have attracted substantial interest over the past two decades [29–32]. The major source of inefficiency of PEMFCs is the slow catalytic rate of the four-electron cathodic oxygen reduction reaction (ORR, see Figure 7.1), which currently relies on the noble metal Pt as the electrocatalyst [25–27]. The limited activity combined with the high cost of Pt have prompted the development of low/non-Pt fuel cell catalysts [33–36]. Estimates dictate that practical ORR PEMFC electrocatalysts must show a stable Pt-normalized mass activity of at least fourfold improvement compared to the state-of-the-art Pt catalyst.

Rational ORR PEMFC electrocatalyst development to meet the activity targets requires first and foremost a comprehensive understanding of the interaction of the catalyst surface and the elementary reaction steps and kinetics. This is why the surface catalysis of the ORR has been extensively studied *ex situ*, and increasingly also *in situ*, and eventually *in operando*. Theoretical computational studies proposed that oxygenated intermediates (such as $-OH_{ad}$) adsorb on the Pt surface too strongly to be efficiently removed and thus result in high oxygenate coverage and slow ORR kinetics. It has therefore been predicted that a slightly weakened binding between the oxygenated surface species and the Pt surface atoms should result in a higher ORR activity. This can be realized by experimental tailoring of the electronic structure of the Pt surface, in particular the d-band center [37–39]. It is further known that the electronic structure can be tuned by tailoring the geometric structure of the top surface layer of an electrocatalyst. *Hence, atomic scale knowledge and control of the geometric arrangement and structure of the surface and near surface atoms has become of the utmost importance to aid in the design of improved electrocatalysts.*

Pt alloys containing a late 3d-transition metal "M" (M = Fe, Co, Ni, etc.) have long been known to be catalytically more active than pure Pt on ORR [29,40–45]. Theoretical studies suggested that there are two major effects responsible for the enhancement: (i) electronic "ligand effects", due to the proximity of transition metals with different electron negativity and thus direct electron interaction, typically operative over one to three atomic layers, and (ii) geometric effects associated with shortened nearest-neighboring Pt–Pt interatomic distances in the Pt alloys. Both effects could induce change in the electronic structure of the Pt surface and, therefore, weakened adsorption of intermediate oxygenated species.

Based on these fundamental effects, innovative structural *core–shell* designs for Pt-rich Pt_3M bimetallic catalysts have been successfully proposed over the past decade [30–32]. First, "Pt monolayer" catalysts consisting of a single Pt monolayer supported on a non-Pt metal substrate like Pd [39,46–51] exhibited significant ORR activity enhancement based on Pt-mass. Second, "Pt-skin" catalysts, formed by Pt surface segregation of Pt_3M alloys upon thermal annealing, exhibit a single pure Pt monolayer on top of a M-enriched second layer [37,52–54]. For both catalyst concepts, the atomic neighborhood of dissimilar atoms in the first and second layers, coupled to their lattice mismatch, ensured that both ligand and strain effects contribute to the observed activity enhancement.

Another important Pt core–shell structure concept is realized in so-called *dealloyed Pt catalysts* [55,56]. Dealloyed Pt catalysts show exceptional ORR activity and, upon careful size control, impressive durability in real PEMFC cathodes. Unlike "Pt skin" and "Pt monolayer" catalysts, dealloyed Pt bimetallic surfaces are characterized by a Pt multilayer surface – the catalyst shell – ranging from 3 to 10 Pt atomic layers. Here, electronic ligand effects cannot affect the surface layer and hence geometric strain effects are the dominant cause of activity enhancement. Synthetically, dealloyed catalysts are easily accessible by selective acid leaching of alloy precursors. The atomic-scale details of the dealloyed core–shell structure formation mechanism are still insufficiently understood; the final morphology and composition of dealloyed catalysts depend sensitively on various macroscopic (dealloying conditions) and microscopic alloy characteristics, such as the sample size (particle size), which is why controlling the catalyst size is so critical when dealing with dealloyed electrocatalysts.

In this chapter, we review recent insights into the geometric bulk and surface structure of strained dealloyed bimetallic Pt catalysts. We address studies on single crystals, polycrystalline thin films, and alloy nanoparticles, highlighting unifying concepts and important differences of dealloyed Pt surfaces of varying length scales (see Figure 7.2).

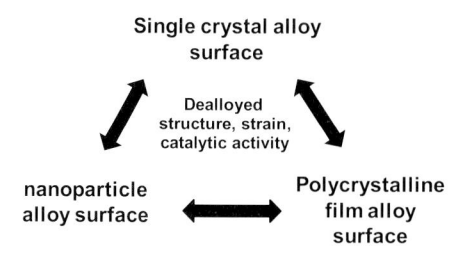

Figure 7.2 Strained dealloyed Pt surfaces studied in this chapter originated from single crystal alloys, polycrystalline film alloys, and nanoparticle alloys.

7.2
Dealloyed Bimetallic Surfaces

Dealloying is a selective chemical or electrochemical leaching process where the less-noble component of a metal alloy is selectively removed from the surface of a bimetallic alloy (see Figure 7.3a) [57–66]. In the corrosion community, dealloying has come to denote the dissolution not only of small quantities of the less noble component from the alloy surface layer – this occurs virtually independently of the initial alloy composition – but also of large quantities from deeper inside the bulk of an alloy [57,58]. When this happens the resulting dealloyed material is strongly enriched in the nobler component near the surface. To achieve bulk dealloying a certain minimum molar ratio of the less noble component is required, sometimes referred to as the parting limit. Above the parting limit, there generally exists an electrochemical potential (dealloying potential) where massive faradic dissolution currents can be observed, associated with the continued loss of the less noble component.

Historically, dealloying has been a subject in the context of corrosion stability of technologically important alloys [67]. An important example is the selective leaching of zinc from brass alloys [68–71] (15 at% Zn) in the presence of oxygen and moisture, resulting in a copper-rich sponge with poor mechanical properties. In this situation, the dealloying process caused undesired performance decline. Dealloying can, however, also be useful in some certain areas regarding its ability

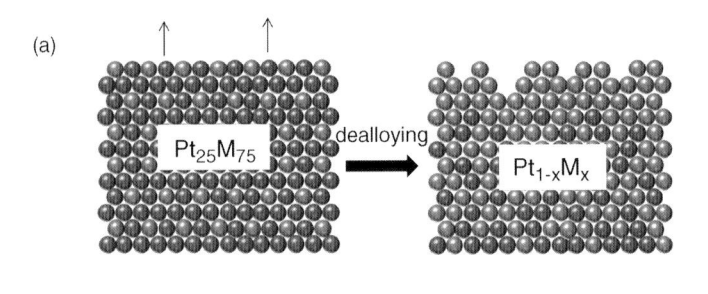

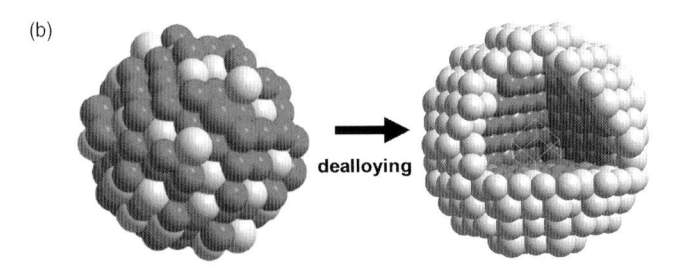

Figure 7.3 Dealloying of Pt poor extended bimetallic surfaces (a) and of Pt poor bimetallic nanoparticles (b) typically results in Pt alloy core–Pt shell architectures with strained surface layers (Reprinted with permission from Ref. [90]).

to create porous structures and, hence, largely enhanced surface areas. A well-known dealloyed catalyst with favorable surface area is Raney nickel, developed by Murray Raney more than 80 years ago [72], which was produced from a macroscopic block of Ni-Al alloy treated with concentrated alkali, dissolving most of the Al out of the alloy and leaving behind a porous structure. The size of these pores was already in the nanoscale region, even though the concept of "nanotechnology" did not arise until several decades later. The resulting nanoporous nickel-rich structure showed a high specific surface area, making it quite useful as a heterogeneous catalyst in a variety of organic syntheses (e.g., hydrogenation reactions).

Inspired by the favorable characteristics of Raney nickel and the rise of nanotechnology, there is increasing interest in synthesizing nanoporous materials, using the dealloying protocol, for applications in catalysis and sensors [58,73–77]. For example, nanoporous gold was synthesized by electrochemical dealloying of silver-gold alloy [78–89]. Other nanoporous materials, such as Pt, Pd, Cu and Cu were also synthesized.

The mechanism of dealloying is fairly well studied and understood in macroscopic samples, thanks to careful electrochemical experimentation coupled to modern Monte Carlo simulations [57,65,73,76,91–93]. Terrace dissolution of the less noble atoms from the topmost surface is thought to be the rate limiting step in dealloying [91]. Terrace vacancies coalesce into vacancy islands with receding perimeters. Meanwhile, the remaining under coordinated noble terrace atoms diffuse to the perimeter steps of the vacancy island, passivate them, and increase their coordination. Overall, dealloying of the less noble component can be perceived as a kinetic competition between chemical dissolution and capillary forces that drive smoothing and passivation by the remaining, more noble component. The competition between the dissolution rate of the less noble metal and the surface diffusion rate of the more noble metal is considered to be a key factor controlling the morphology of the dealloyed material. In bulk alloys, the surface diffusion rate of the noble atoms is slow across the extended surface relative to the dissolution rate of the less noble component; this favors so-called Rayleigh surface instability and leads to the formation of nanoporosity deep into the bulk.

Dealloying mechanisms of nanoscale alloy particles and their confined crystal facets have been studied and understood to a much lesser extent (see Figure 7.3b) [94–99]. This is because virtually all dealloying processes studied to date addressed bulk alloy materials. Since 2007, dealloying of Pt bimetallic alloy nanoparticle catalysts has been investigated by our research group using a variety of lab-based and synchrotron-based X-ray and microscopic methods. Unlike macro bulk materials, dealloying of Pt alloy nanoparticles with small enough particle size can produce non-porous core–shell nanoparticles, because Pt surface diffusion is faster on small particles and helps suppress the formation of pores and voids in the material bulk.

The size dependence of the morphologies of dealloyed particles was recently studied using TEM methods and the results are presented in Figure 7.4. Very small

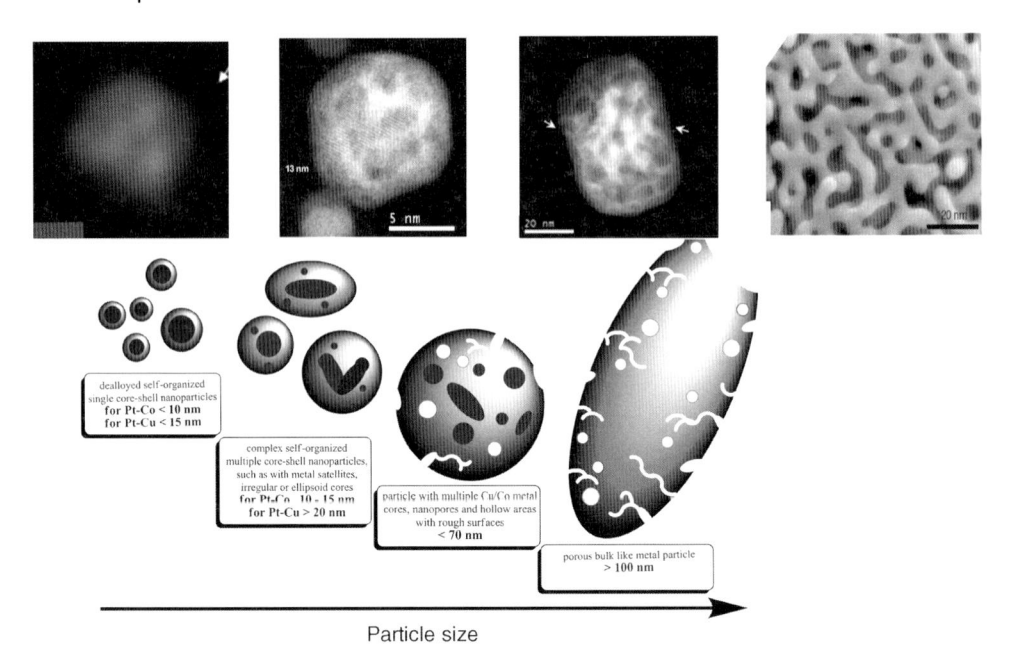

Figure 7.4 The evolution of nanoporosity as a function of the size of the alloy precursor (Reprinted with permission from Ref. [100]).

particles, below about 10 nm, display solid core–shell particle morphologies after a chemical dealloying in acid (acid leaching by free corrosion) or electrochemical dealloying (application of a constant to time variant electrode potential with an upper turning potential well above the critical dealloying potential). With increasing size, dealloyed nanoparticles start exhibiting multiple less noble-metal-enriched cores surrounded by a Pt-enriched shell, and at even larger sizes the dealloyed nanoparticles exhibit corrosion pits, pores, and sometimes voids. At particle sizes approaching macroscale dimensions, the particles showed the nanoporosity structure, which is typical for dealloyed macroscopic samples.

7.3
Dealloyed Strained Pt Core–Shell Model Surfaces

Computational work by Mavrikakis and others has predicted characteristic shifts in the center energy of metal projected d bands (the "d band center") with geometric surface strain [101]. Every metal possesses a characteristic interatomic distance in a pure bulk phase, d_{eq}. Positive lattice strain (tensile strain) with the interatomic distance d being larger than d_{eq} was predicted to shift the center up, thereby generally strengthening adsorbate bonds (see Figure 7.5c) . Negative strain (compressive strain) was predicted to be coupled to a down shift in the d band center energy and would lead to a weakening of adsorbate bonds (Figure 7.5a) [102]. Thus,

7.3 Dealloyed Strained Pt Core–Shell Model Surfaces

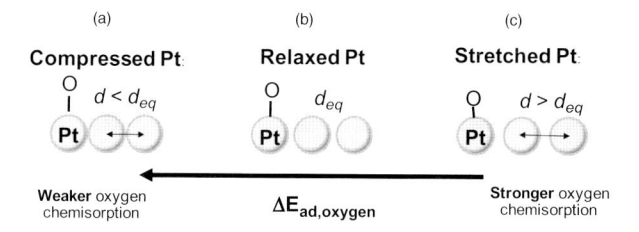

(a) Compressed Pt: O–Pt, $d < d_{eq}$ — Weaker oxygen chemisorption

(b) Relaxed Pt: O–Pt, d_{eq}

(c) Stretched Pt: O–Pt, $d > d_{eq}$ — Stronger oxygen chemisorption

$\Delta E_{ad,oxygen}$

Figure 7.5 (a) Surface lattice strain affects adsorption energies. Compressive strain typically lowers, while tensile strain increases adsorption bond strengths.

tuning the geometric surface lattice strain can be used as a means to tune adsorption and, thereby, catalytic reactivity.

Geometric surface lattice strain typically arises in atomic adlayers of metal A supported on a substrate of metal B, with A and B having dissimilar lattice constants [103–106]. Adlayers of A may also sit on an alloy surface made of A and B as long as their lattice parameters differ. Note that while a bulk terminated substitutional alloy of A and B displays a different lattice constant than either pure A or pure B, that surface would be strained, because lattice strain is typically referred to the lattice constant of the bulk phase with identical composition.

X-ray spectroscopy was used to directly evidence the shift in the d-band center in compressively strained Pt overlayers on a Cu(111) [107]. The inset of Figure 7.6a illustrates a bimetallic model core–shell arrangement of Pt and Cu. Electron diffraction revealed (Figure 7.6a) that compressive strain due to lattice mismatch can be sustained across more than 10 atomic layers of pure Pt, roughly about 3 nm. Figure 7.6b sketches the principles of core level X-ray absorption and X-ray emission spectroscopy used to detect occupied and unoccupied electronic states around the Fermi level. The two X-ray techniques applied at the O1s edge revealed that, for an oxygen-covered 15 ML thick Pt overlayer (∼0% strain), occupied states (peak A in the XES signal in Figure 7.6c) coexisted with unoccupied states (peak C in the XAS signal). A corresponding electronic band scheme is illustrated to the right of the measured data. As the overlayer thickness decreased to 3.5 ML (−2.1% strain) and 2.6 ML (−3.6% strain), the compressive strain increased and the unoccupied states shifted to lower energies, crossed the Fermi energy, and thus became populated as a new XES peak B; concurrently, the occupied states shifted to more negative energies (shifted peak A in top graph). Tuning of electrocatalytic activity on extended single-crystal supported thin metal adlayers was impressively demonstrated in the work of Kibler et al. and Adzic et al. [103,104].

Energetic downshifts associated with increased population of Pt projected electronic 5d states in the valence band of model $PtCu_3$ alloy surfaces were recently probed using a combination of L-edge X ray absorption and high energy X-ray photoemission spectroscopy [107]. In the lattice-contracted alloy state, Pt d band centers were clearly shifted more negative compared to bulk Pt reducing the density of states at the Fermi energy and reducing the white line intensity of the L edge.

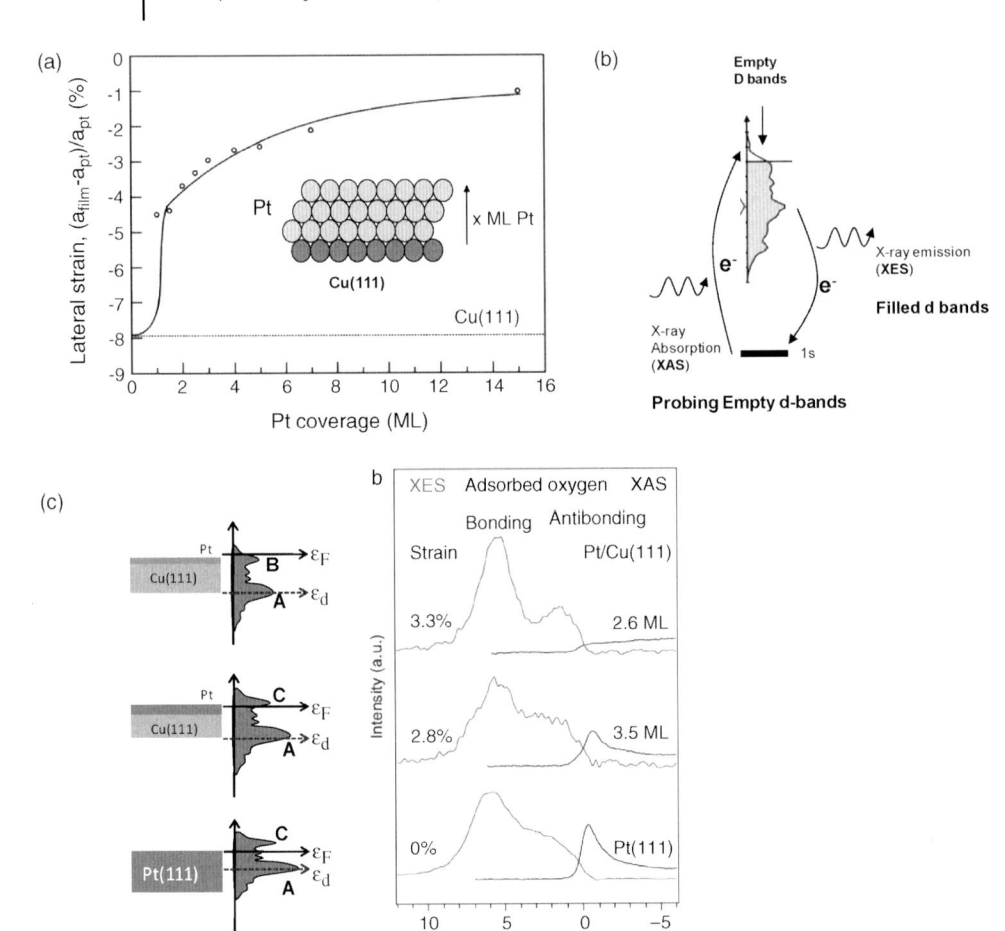

Figure 7.6 (a) Strain in Pt layers deposited on Cu(111), defined as $[(a_{film} - a_{Pt})/a_{Pt} \times 100]$, deduced from LEED patterns as a function of film thickness. The curve is for guidance only. (b) Principles of core level X-ray emission and X-ray absorption spectroscopy (c) In-plane polarized oxygen K-edge XAS and normal emission oxygen K-edge XES of 0.2 ML of oxygen chemisorbed on 3.6 and 2.6 ML films of Pt on Cu(111) and on bare Pt(111). The energy scale is with respect to the oxygen 1s binding energy, which represents the Fermi level. The strain was estimated based on the Pt coverage (Reprinted with permission from Ref. [107]).

7.4
X-Ray Studies of Dealloyed Strained PtCu$_3$(111) Single Crystal Surfaces

Synchrotron based X-ray surface scattering was carried out using a four circle diffractometer at beamline 7.2 of the Stanford Synchrotron Radiation Light source at 16 keV [108]. This surface structure sensitive method was used to follow the dealloying and strain formation in a PtCu$_3$(111) surface. These studies followed earlier excellent studies by Renner *et al.* who applied a similar technique to investigate the early surface passivation stages during the dealloying of AuCu$_3$(111) surfaces [109,110].

Figure 7.7a displays the typical electrochemical response of a PtCu$_3$ surface during an anodic potential scan. Below the Cu Nernst potential, the surface is stable

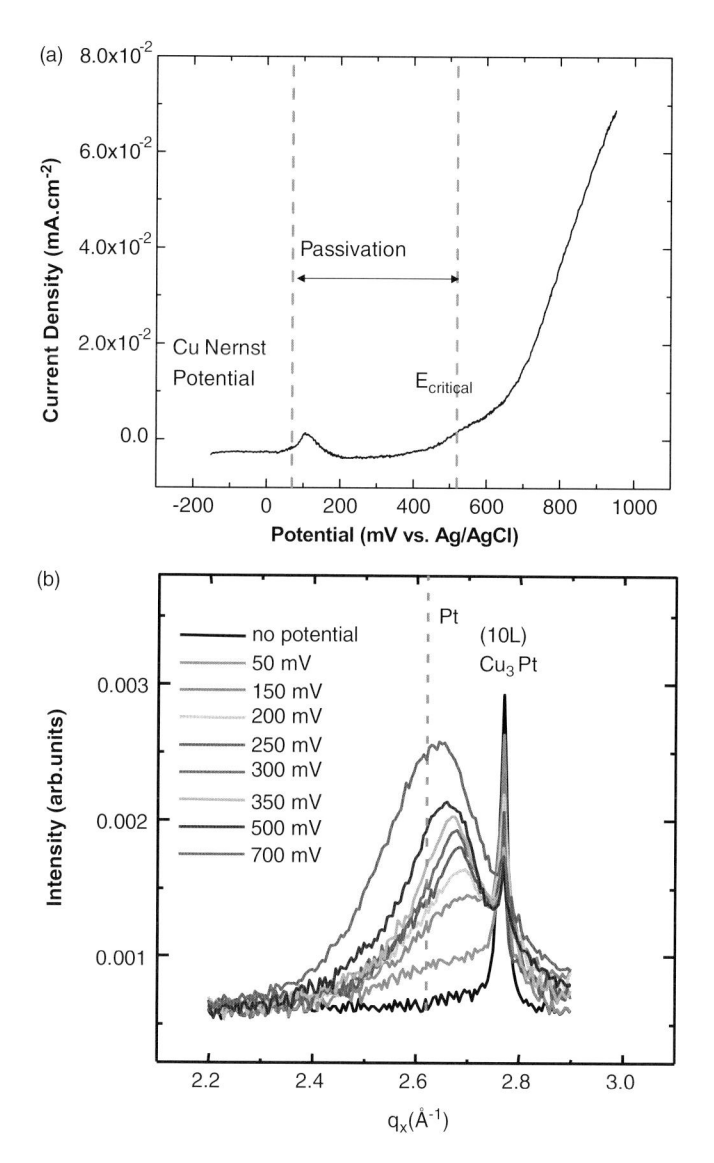

Figure 7.7 (a) Anodic scan of polycrystalline Cu$_3$Pt films showing the critical potential for massive dissolution of Cu from Cu$_3$Pt and the surface passivation below the critical potential (b) Radial in-plane scans at $q_z = 0.2$ Å^{-1} through the (10 L) hex truncation rod after dealloying at the indicated potentials. The diffracted intensity is plotted against q_x, the component of the scattering vector parallel to the surface (Reprinted with permission from Ref. [108]).

and no dissolution occurs. Above the Cu Nernst potential, yet below the critical dealloying potential, a passivation of the surface occurs. This is brought about by the dissolution of Cu from the topmost surface layers, while Pt rapidly diffuses into the surface vacancies and stops further Cu dissolution. After an initial small faradic Cu dissolution wave the surface behaves quasi-stably again. At and beyond the critical dissolution potential large faradic currents are observed, indicating a massive Cu dissolution from the bulk of the alloy. Now Pt surface diffusion is unable to stop the Cu dissolution.

In-plane q_x surface scattering profiles were measured using detector scans parallel to the alloy surface at a constant value of q_z. Such profiles are shown in Figure 7.7b for increasing electrode potentials. Evident is the potential invariant (10 L) Bragg reflexion corresponding to the crystal truncation rod (CTR) of the alloy. At more anodic potentials scattering intensities emerge at smaller values of q_x, indicating the expansion of the surface layers through loss of Cu. At the most positive electrode potential, the peak maximum is located close to the q_x value of pure Pt, indicating an almost complete loss of Cu in and near the surface. The lateral domain size obtained from the peak widths ($2\pi/\Delta q_x$) is between 2.5 and 3.3 nm. This small size suggests that the Pt-rich overlayer consists of nanoscale grains/ligaments after dealloying. The peak intensity increases nearly linearly with the potential, which shows that the Pt-rich overlayer grows in thickness as the potential increases.

Surface lattice strain relative to pure bulk Pt was evaluated from the peak positions in Figure 7.7b according to $S = -(a_{\text{surface}} - a_{\text{Pt}})/a_{\text{Pt}} \times 100\%$ and is plotted in Figure 7.8a. Lattice strain of over -3% was observed at the lowest electrode potentials where the dealloyed Pt layers are still thin and hence the top layer is significantly affected by the bulk alloy phase. At higher electrode potentials the lattice strain drops to about -0.72%.

X-ray surface scattering was subsequently evaluated parallel and perpendicular to the surface at various electrode potentials in two dimensional $q_x - q_z$ plots. After

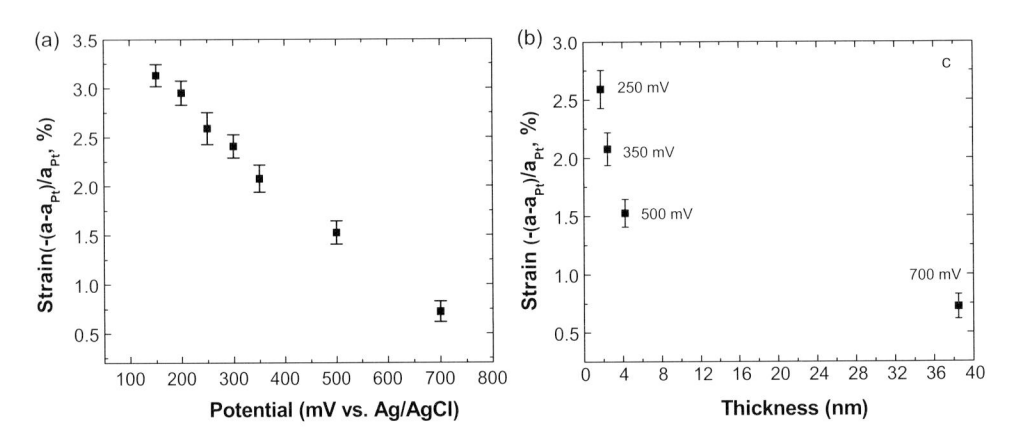

Figure 7.8 (a) q_x-based lattice strain in epitaxial Pt overlayer (compared to pure Pt) as a function of dealloying potentials. (b) q_z-based lattice strain in Pt overlayer (with respect to pure Pt) as a function of the thickness of Pt overlayer (Reprinted with permission from Ref. [108]).

integration of scattering intensity in the x-direction, surface lattice strain and coherence lengths in the surface-normal z-direction were extracted from the resulting intensity–q_z curves. For thin Pt overlayers, scattering coherence lengths were identical to Pt layer thickness. As the Pt overlayer grew thicker with electrode potential, however, the experimental coherence length saturated at the Pt grain size. The Pt overlayer thickness was estimated as 1.7 nm at 250 mV, while it increased dramatically to almost 40 nm at 700 mV. These changes are well correlated with the changes in the surface lattice strain values perpendicular to the surface, as shown in Figure 7.8b.

In conclusion, electrochemical dealloying of PtCu$_3$(111) single crystals is a potential-dependent process. Dealloying above the critical potential results in Cu dissolution from the alloy bulk and yields thick Pt overlayers, ranging from around 2 nm to at least 40 nm at 700 mV. As the overlayer thickness increases, the resulting surface lattice strain decreases. This observation is consistent with experimental strain–thickness relations for the model Pt-Cu core–shell system in Figure 7.6a. Even though detailed electron microscopic information is missing to date, it is likely that due to the macroscopic nature of the dealloying alloys the surface morphology approaches a highly nanoporous labyrinth-type structure, as shown in Figure 7.3.

In order to establish a structure–activity mapping for dealloyed strained Pt surfaces, the catalytic activity for oxygen reduction was measured at 700 mV. The experimental intrinsic, that is, the Pt surface area-normalized ORR activity, was only a factor of 1.3 times higher than for pure Pt, which is a relatively small gain for an alloy. This is consistent with the notion that thick Pt overlayers over Pt-Cu alloy bulks retain only little compressive surface lattice strain in the top layer. Decrease in oxygen adsorption energies can thus not be expected and the catalytic activity is close to that of pure Pt. The highly nanoporous surface does, however, increase the real surface area of the dealloyed surface, and thus lifts the geometric- or mass-based activity of the catalyst somewhat.

7.5
X-Ray Studies of Dealloyed Strained Pt-Cu Polycrystalline Thin Film Surfaces

X-ray scattering techniques were used to establish structure–ORR activity relationships in uniform, polycrystalline sputtered PtCu$_3$ thin films [90]. These studies complemented the single crystal studies from Section 7.4. Synchrotron-based anomalous X-ray diffraction (AXRD) was used to study the structure of the dealloyed Pt-Cu thin film (lattice constant and composition). The measurements were performed on beamline 7-2 at Stanford Synchrotron Radiation Light Source. The synchrotron radiation source provides high incident X-ray intensity and allows variation of the X-ray energy around the absorption edges of individual elements of the alloy. The intensity I of a scattered X-ray beam is proportional to the square of the scattering length density ρ_i of the scattering atoms i according to $I(Q, E) \sim \rho_i^2(E)$ with $\rho_i(E) = n_i f_i(E)$, where n_i is the atomic density of the metal, essentially related to composition, and $f_i(E)$ represents the element-specific atomic X-ray scattering factor

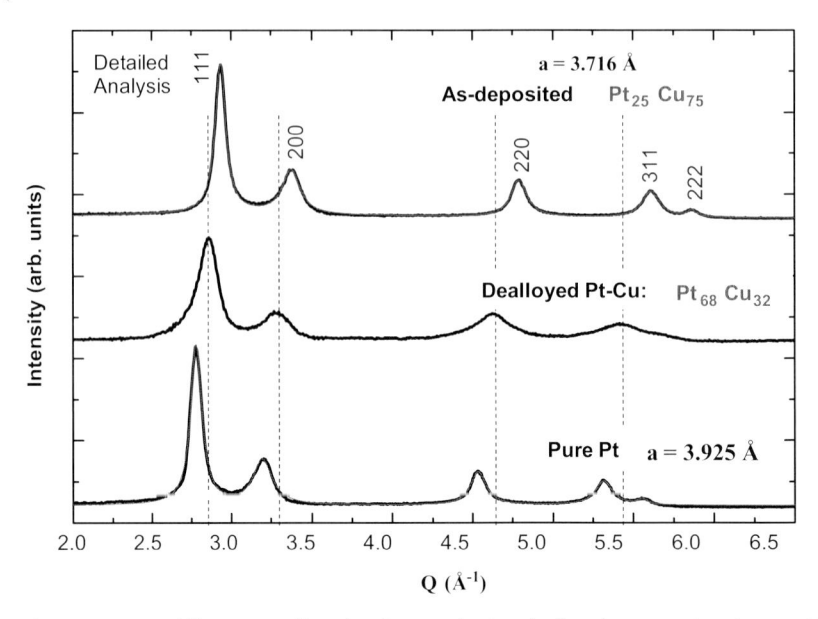

Figure 7.9 X-ray diffraction profiles of as-deposited PtCu$_3$, dealloyed Pt-Cu, and as-deposited Pt film samples on Si(100). (Reprinted with permission from Ref. [90]).

of metal i, which is related to the number of electrons. The X-ray scattering factor is a complex energy dependent quantity; its real part decreases sharply at the characteristic X-ray absorption edges of metal i. This behavior is referred to as "anomalous" and gave this method its name. Other components of the sample, however, show constant f_i near the edge energy. Thus, modeling the variations of the integrated X-ray intensities near the absorption edges of elements enables their accurate compositional analysis, independently of the peak position, that is lattice geometry. This decoupled estimation of phase composition and lattice constant of the scattering phase, in turn, enables the determination of lattice strain in the scattering phase of interest.

For the current X-ray studies the alloy thin films were supported on single crystal Si substrates, while for the electrochemical testing polished glassy carbons served as current collectors and supports. X-ray diffraction profiles of the as-deposited PtCu$_3$, dealloyed Pt-Cu and as-deposited Pt films on Si(100) are shown in Figure 7.9. Both the as-deposited PtCu$_3$ and Pt films show single phase face centered cubic (FCC) structures. The average lattice constants obtained from the best fitting for the as-deposited PtCu3 and Pt film are 3.716Å and 3.921Å, respectively, which is in close agreement with those reported for bulk materials. In contrast, the diffraction peaks of the dealloyed Pt-Cu film were shifted to smaller scattering vectors Q and appeared broadened. The shift suggested an expansion of the lattice due to loss of Cu. This was confirmed by a bulk compositional analysis revealing a drop in Cu from 75 at% to about 32 at%.

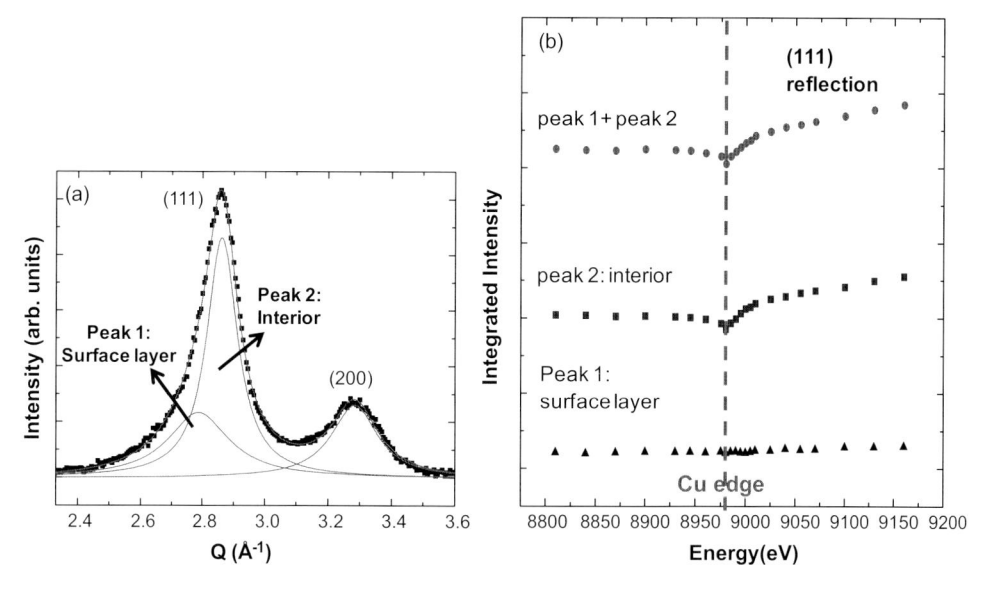

Figure 7.10 (a) Anomalous X-ray diffraction (AXRD) profile of the dealloying PtCu$_3$ catalyst taken at 8810 eV. Fitting (Lorentz function) was achieved using a pure Pt phase (Peak 1) and an alloy phase (Peak 2) (b) Integrated intensity of (111) reflection for the dealloyed Pt-Cu thin film catalyst as a function of X-ray energy (Reprinted with permission from Ref. [90]).

To determine the structure (e.g., lattice constant and composition) of the dealloyed Pt-Cu thin film, AXRD was performed in the energy range of the Cu absorption K edge (8980 eV). As the energy approached the Cu absorption edge the scattering intensity and, hence, the Bragg peak intensity dropped. Figure 7.10a displays a typical diffraction profile and a fit for the asymmetric (111) reflection at an energy near the Cu adsorption edge. The (111) reflection could be fitted using two peaks, Peak 1 and Peak 2. The intensity of these peaks showed characteristic energy dependences, as shown in Figure 7.10b. Peak 1 showed virtually no energy dependence, evidencing that this phase is Cu free. Peak 2, however, decreased in intensity as the Cu edge was approached, indicating the presence of Cu in this phase. XPS analysis showed that the near-surface composition of the dealloyed catalyst was about 9 at% in Cu, so overall Pt$_{91}$Cu$_9$. Thus from the XPS and AXRD, we conclude that peak 1 corresponds to a nearly pure Pt surface layer and peak 2 corresponds to the interior region of the dealloyed Pt-Cu film, still containing Cu. The (111) diffraction peak for the Pt surface layer (peak 1) is centered at $Q = 2.792 \, \text{Å}^{-1}$ (average position from AXRD). The corresponding lattice constant (a_{surface}) is 3.897 Å, which is smaller than that of pure Pt ($a_{\text{Pt}} = 3.923$ Å). The surface region thus has a compressive strain ($S = (a_{\text{surface}} - a_{\text{Pt}})/a_{\text{Pt}} \times 100$) of -0.66% with respect to pure Pt. Quantitative analysis of the AXRD intensity changes near the absorption edge for the overall film, and the two peaks 1 and 2 revealed a composition of Pt$_{26}$Cu$_{74}$ for the pristine alloy film, Pt$_{68}$Cu$_{32}$ for the film after dealloying, Pt$_{59}$Cu$_{41}$ for the interior of the dealloyed film, and Pt$_{100}$ for the surface layer.

From the compressive strain estimate above, one can deduce that the pure Pt surface layers consists of about 4–5 Pt monolayers [107,111]. This means that electronic ligand effects in the surface have become very weak. Neglecting morphology and facet orientation effects on reactivity, the ORR activity enhancement in the dealloyed Pt-Cu thin film can now be attributed to the compressive strain in the Pt surface layer, which modifies the electronic structure of the Pt surface layer such that the d band center shifts to more negative energies.

Since the surface lattice strain is inversely correlated with the Pt surface layer thickness, the smaller thickness of the Pt surface layer in the present dealloyed Pt-Cu thin film (about 1.0 nm) compared to that in the single crystal studies (up to 30 nm), we expect a more ORR active electrocatalytic activity on the dealloyed thin films. In fact, Figure 7.11 shows the experimental ORR activity on the dealloyed Pt-Cu thin films. An activity enhancement of 2.4× was observed, compared to 1.3× on the dealloyed single crystal with its very thick Pt overlayers.

In summary, our results show that dealloyed Pt-Cu thin films consist of a Pt-enriched surface layer and Cu-depleted Pt-Cu alloy interior, and that the Pt surface layer is compressively strained. Due to its thickness, the surface layer compressive strain dominates the ORR activity enhancement in the dealloyed Pt-Cu film, which is consistent with the DFT studies by Norskov [112] and Mavrikakis [101]. We find that the compressive strain (and hence ORR activity) in the Pt surface layer is larger in the films than in the single crystal surfaces.

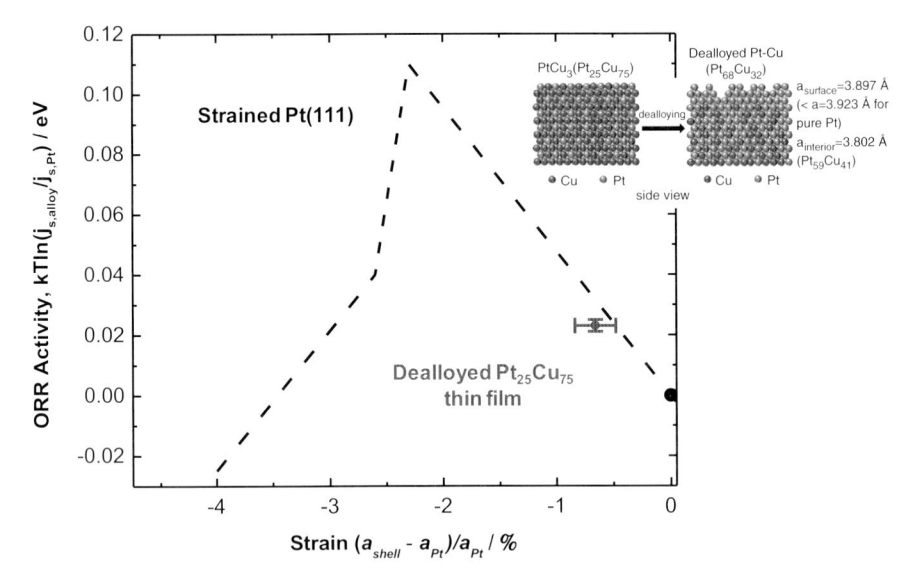

Figure 7.11 Specific (real Pt surface area normalized) ORR activity of the dealloyed Pt-Cu thin film as a function of the extracted surface lattice strain (see data point). Dashed line indicates the theoretical ORR volcano curve for uniform lattice strain in a Pt(111) single crystal surface. The experimental data point fits the computationally predicted value well. The black dot represents the data point for lattice-relaxed pure Pt(111). Inset: Schematic of the dealloyed morphology of the PtCu$_3$ film (Reprinted with permission from Ref. [90]).

7.6
X-Ray Studies of Dealloyed Strained Alloy Nanoparticles

To complement the catalytic activity– surface lattice strain studies on single crystals and polycrystalline thin films, X-ray techniques have been applied to dealloyed bimetallic Pt nanoparticles to evaluate surface lattice strain as a function of synthesis conditions and dealloying conditions [107].

7.6.1
Bragg Brentano Powder X-Ray Diffraction (XRD)

Figure 7.3b illustrated that the preparation of dealloyed surface strained core–shell Pt nanoparticles starts with the synthesis of a uniform, non-noble metal-rich bimetallic Pt-M nanoparticle precursor material, which is subsequently leached in acidic solutions or using electrochemical methods in order to dissolve large portions of the less noble metal from the surface region. A recent review has outlined that dealloying is only one of a variety of methods to prepare such lattice-strained core–shell nanoparticle electrocatalysts [113]. The most widely used method to prepare carbon supported bimetallic nanoparticles involves the co-impregnation of liquid metal salt precursors onto high surface area carbon supports, such as Vulcan XC 72. After drying and a high temperature annealing step in reducing environments, bimetallic nanoparticles are obtained. Figure 7.12 reproduces powder X-ray diffraction patterns of bimetallic $PtCu_3$ nanoparticles on carbon, prepared at three different annealing temperatures. Broad reflections of the material annealed at 600 °C indicate a small mean crystallite size. The sharp reflection around the position of pure Cu indicates the coexistence of a Pt-Cu alloy phase with small

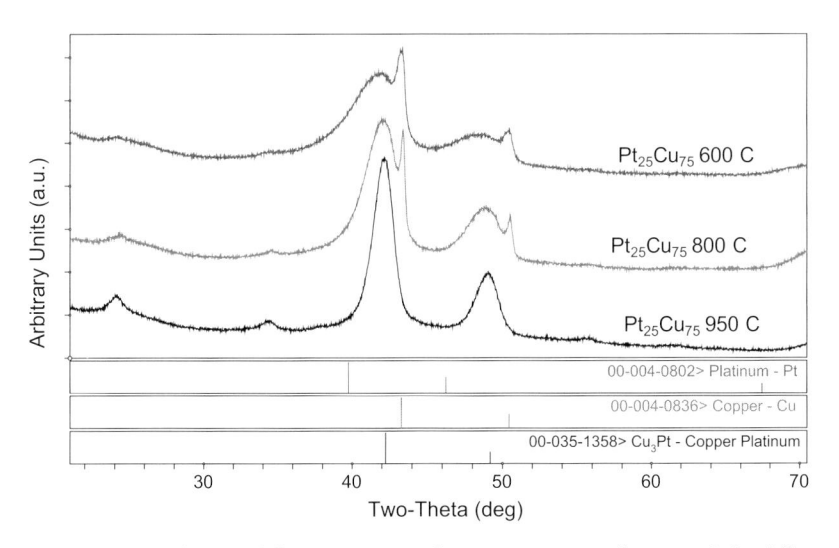

Figure 7.12 Powder X ray diffraction patterns of $Pt_{25}Cu_{75}$ precursor alloys annealed at different temperatures for 7 h (Reprinted with permission from Ref. [114]).

crystallite size and a pure Cu or very Cu-rich crystal phase with relatively large crystallite/particle size. The peak reflections are at 2-theta angles between those of pure Pt and pure Cu, consistent with the formation of a Pt-Cu alloy. Absence of superlattice peaks (for the lower two temperatures) evidence the formation of a solid solution alloy.

The material at 600 °C appears very inhomogeneously alloyed. As the annealing temperature increases the bimetallic becomes more uniformly alloyed. This is coupled to a decrease in the peak widths, suggesting sharply increased crystallite and particle sizes. Finally, the material annealed at 950 °C, displays a single crystalline phase. As more Cu is alloyed into the Pt lattice, the lattice parameter decreases, shifting the diffraction angle of the (111) peak reflection toward more positive 2-theta values, see Figure 7.12.

In Figure 7.12, the maximum annealing temperature seems to control the Cu content of the resulting disordered Pt-Cu phase. Correlation of annealing control parameters, such as heating rate, temperature, and time, with microscopic alloy structure, composition, and particle size has also been studied using *in situ* high temperature XRD [115].

7.6.2
In Situ High Temperature Powder X-Ray Diffraction (XRD)

Figure 7.13a shows experimental *in situ* powder diffraction data. Pt and Cu precursors were impregnated in a ratio of 1:3 onto a high surface area carbon support and then subjected to a "stepped heating protocol". This temperature–time program involved constant temperature plateaus of about 1.5 h at 100 °C temperature intervals when heating from 300 to 800 °C. Figure 7.13a reproduces the lattice parameter *a* versus the annealing temperature for the stepped heating protocol. Black squares denote the structural parameter *a* and the corresponding composition of the scattering crystalline Pt-Cu phase. Thermal lattice expansion is evident in the two lines plotting the lattice parameters of pure Pt and Cu with temperature. The Pt-Cu phase enriches in Cu with temperature. Increasing temperature and time aid the reduction of Cu ions to metallic Cu, which subsequently diffuses and alloys with coexisting Pt or Pt-Cu alloy phases. At 800 °C, the resulting alloy composition is about 33 at% Pt and 67 at% Cu, close to the nominal values of the precursors. The data show that the maximal annealing temperature has a strong influence on the resulting alloy phase composition.

Even though the data in Figure 7.13a provide valuable insights into how temperature controls the resulting phase composition, they cannot provide a time-resolved picture of the annealing process. To achieve this, powder diffraction was applied for an extended period of time at one constant temperature, and these data are shown in Figure 7.13b. The composition trajectory indicated that Cu alloying is a relatively fast process and seems to be essentially completed after about 2 hours (squares). The circles indicate the crystallite/particle size. This curve evidences that annealing at that temperature increases the crystallite size gradually. Hence, from a particle dispersion point of view, it would be desirable to keep the annealing time at this

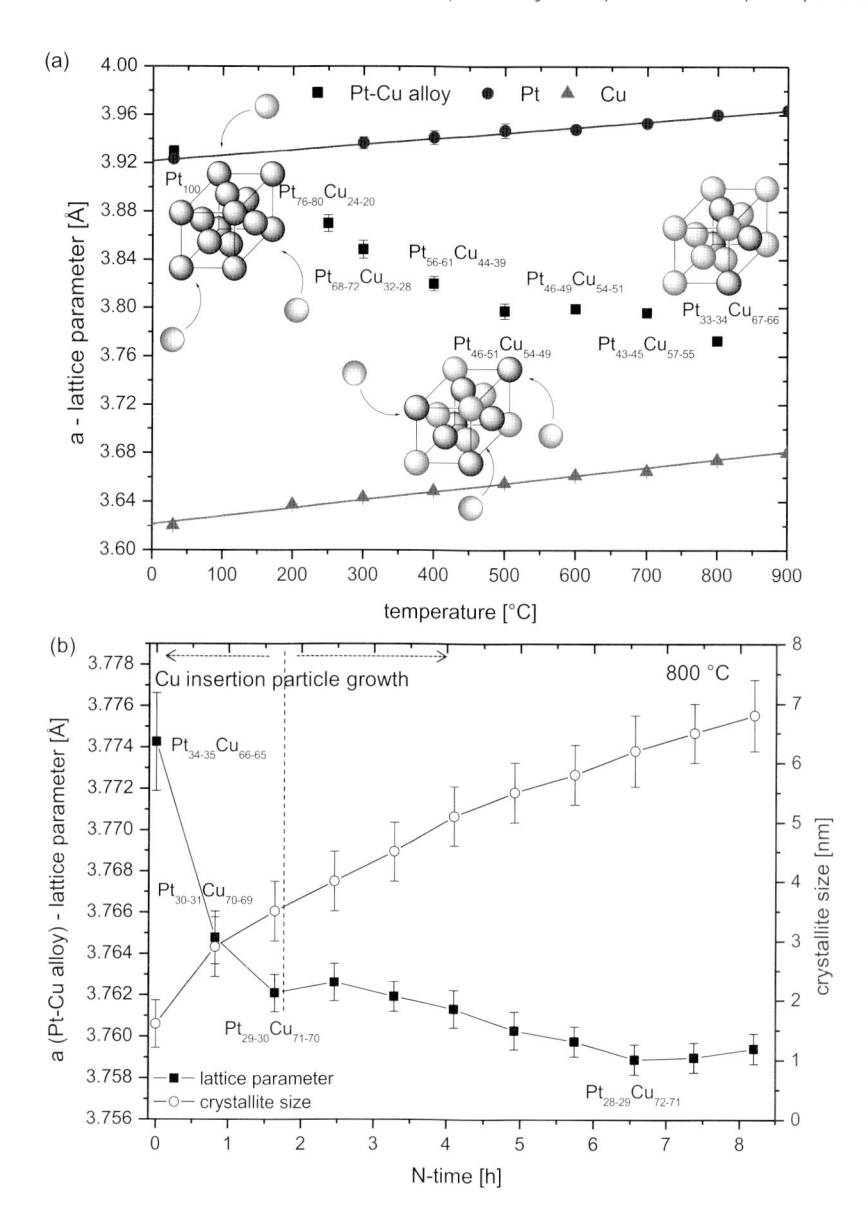

Figure 7.13 (a) Lattice parameters and compositions of disordered Pt-Cu alloy nanoparticles at various temperatures (squares Pt-Cu alloys, circles Pt, triangles Cu). Composition was estimated using a temperature-corrected Vegard's rule (b) Lattice parameter of disordered Pt-Cu alloy and crystallite size versus normalized annealing time at 800 °C. The average crystallite size was determined from the (111) and (200) Pt-Cu reflections via integral breadth method. The y-axis represents a, the lattice parameter; the x-axis represents the normalized annealing time, N-time (Reprinted with permission from Ref. [115]).

temperature as short as possible. A balance between achieving a desired alloy phase composition and keeping the particle size small is thus achieved at an annealing time of about 2 h at 800 °C.

In summary, *in situ* diffraction data suggest that the maximum annealing temperature controls the phase composition and, at 800 °C, is essentially complete after 2 h; prolonged annealing times, however, lead primarily to particle growth and not to more complete alloying. These insights have important consequences for the design of suitable synthesis protocols for Pt alloy nanoparticles.

The subsequent sections will focus on X-ray based analytics to investigate the formation of catalytically active dealloyed Pt core-shell particle architectures. To prepare catalytically active core–shell nanoparticles, the thermally annealed bimetallic Pt alloy nanoparticle precursors were subject to an electrochemical dealloying or a chemical acid leaching treatment. Typical electrochemical dealloying conditions involve several 10s to several 100s potential cycles between 0.05 and 1.0 V/RHE in pH 1 acid electrolyte in oxygen-free conditions, until the voltammetric profile has become time-stable and resembles that of a Pt surface. Depending on the non-noble alloy component metal, dissolution peaks are visible on the very first and subsequent voltammetric cycles [107,116–121].

7.6.3
Synchrotron X-Ray Photoemission Spectroscopy (XPS)

Direct evidence for the structural evolution of a core–shell nanoparticle structure, as illustrated in Figure 7.3b, was obtained from synchrotron-based X-ray photoelectron emission spectroscopy (XPS) at various incident X-ray energies. The inelastic mean free path of photoelectrons depends strongly on their energy, with electrons of about 50 eV being most surface sensitive. Table 7.1 reports Pt at% values of the alloy precursor (before dealloying) and of the dealloyed material at widely varying photoelectron energies. Photoelectron energies of 264 eV and 8 keV probe the material near the surface and deep inside the bulk, respectively. The data evidence high Cu content of 92 at% at the surface of the precursor particles, and a near-nominal 69 at% Cu content in the bulk of the pristine particles. After dealloying, the surface is enriched strongly in Pt (84 at%) while the bulk is enriched to 59 at% Pt. This is consistent with the enrichment of an outer particle shell in Pt while the bulk retains Cu.

Table 7.1 Composition depth profiles for $Pt_{25}Cu_{75}$ nanoparticles annealed at 950 °C before and after electrochemical dealloying. Increasing photon energy correlates with increased probing depth. Pt at% = platinum atomic per cent. (Reprinted with permission from Ref. [107]).

Photoelectron energy (eV)					
	264	616	1133	1480	8000
$Pt_{25}Cu_{75}$ precursor (Pt at%)	8	12	19	18	31
$Pt_{25}Cu_{75}$ dealloyed (Pt at%)	84	68	55	56	59

7.6.4
Anomalous Small Angle X-Ray Scattering (ASAXS)

Additional evidence for the formation of a core–shell particle structure was obtained from anomalous small angle X-ray scattering (ASAXS) experiments. SAXS is an X-ray-based technique which exploits the rich information contained in a scattering profile at very small (<3°) scattering angles near the incident beam. This scattering region can provide useful information on structural characteristics in the colloidal size range, that is, roughly between 1 and 50 nm. In the case of supported nanoparticles, for instance, model fits of SAXS profiles offer information on the morphology and the size distribution of the scattering particles. Due to ambiguities in the choice of suitable structural models for fitting SAXS profiles, structural conclusions should be complemented by other methods, such as electron microscopy. Figure 7.14a illustrates the basic principle of ASAXS, which requires a tunable X-ray source and is usually performed at synchrotron facilities. X-ray scattering using incident X-rays near the Cu K edge and the Pt L3 edge provides element-specific diameters of contiguous scattering domains. If the element-specific diameters are comparable, Pt and Cu are well alloyed without significant surface segregations or enrichment. A smaller Cu diameter indicates the formation of a core–shell structure. Figure 7.14b shows the experimental ASAXS scattering profiles for the pristine Pt-Cu precursor particles as well as the dealloyed active catalysts. After fitting a log normal particle size distribution function to the data of Figure 7.14b, element specific particle size distributions are obtained, Figure 7.14c. The reduced Cu diameter of the dealloyed material suggests the formation of a core–shell structure.

7.6.5
Anomalous Powder X-Ray Diffraction (AXRD)

An analysis of surface lattice strain in dealloyed nanoparticles was performed using AXRD. As outlined above, it permits independent measurements of the average lattice parameter and average composition of the scattering nanoparticles before and after dealloying. Figure 7.15a shows (111) diffraction peak profiles of a $Pt_{25}Cu_{75}$ precursor at several X-ray energies.

As discussed, the scattering power of Cu drops near the Cu absorption edge, and so the diffracted intensity of the alloy shows a characteristic decrease, as indicated by the arrow in Figure 7.15a. Scattering peak positions yield average lattice parameters for the nanoparticles, and fitting energy-dependent scattering intensity to a composition-dependent model provides the chemical composition of the scattering alloy phase. As illustrated in Figure 7.15b, a model for energy-dependent diffraction intensity (black line) was fitted to the experimental integrated (111) peak intensities (circles), which allows estimation of the compositions x_{Pt} and x_{Cu} in the scattering alloy phase.

Correlations between the surface structure and the surface catalytic activity of core–shell particles require knowledge of the surface lattice strain in the top layer of

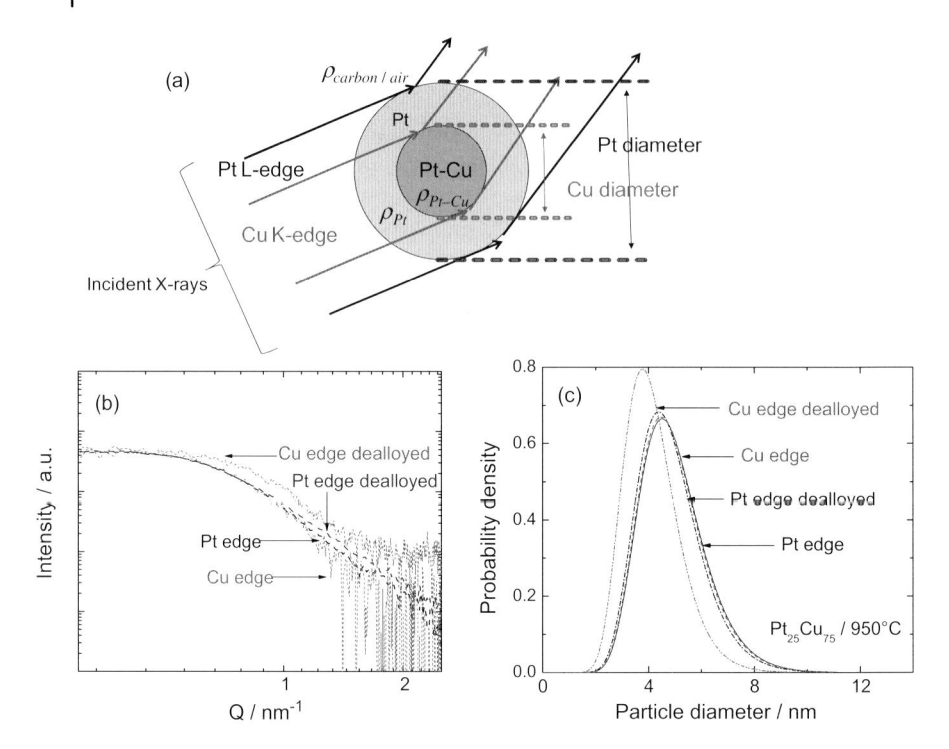

Figure 7.14 Anomalous element-specific small angle X-ray scattering (ASAXS) of core–shell nanoparticles. (a) Principle of ASAXS: Incident X-rays near the Cu K edge (8979 eV) and the Pt L3 edge (11564 eV) probe the diameter of contiguous Cu and Pt domains. Scattered X-ray contrast arises due to a difference in the scattering length density ρ of the shell and core region. Given that monometallic particles are absent, differences in the Pt specific and Cu specific particle diameters evidences the existence of a core–shell structure. (b) Subtracted element specific scattering profiles (Pt: 11450–11551 eV, Cu: 8880–8975 eV) of the pristine PtCu alloy nanoparticle precursor ("Pt edge" and "Cu edge") and the dealloyed catalyst ("Pt edge dealloyed" and "Cu edge dealloyed"). (c) Particle size distribution obtained by fitting the scattering profiles of (b) using an intensity-scattering angle model involving spherical particle form factor and a log normal particle distribution function. The clearly reduced Cu specific diameter after dealloying indicates the formation of a core–shell structure.

the particle shell. Under certain assumptions, the surface strain in the shell can be obtained from the lattice parameter, a_{shell}, in the particle shell. Neglecting strain relaxation in the particle shell and assuming a uniform, pure Pt shell surrounding a Pt-Cu alloy core with lattice parameter a_{core}, the AXRD-derived composition of the scattering phase and the corresponding lattice-parameter data allow the extraction of a_{shell} and the surface lattice strain relative to pure Pt [107]. Figure 7.16 displays the extracted unit cell parameter values in the dealloyed particle shell as a function of the initial Cu content of the precursor particles before dealloying. Clearly, all cell parameters are smaller than that of pure bulk Pt, pointing to compressively strained lattices. High initial Cu contents result in smaller unit cell dimensions. This is consistent with the notion that higher initial Cu contents are associated with higher

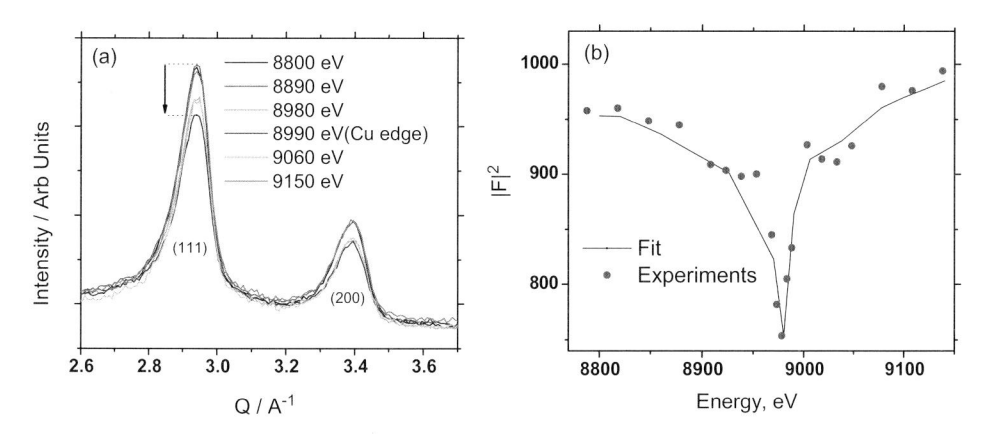

Figure 7.15 AXRD-based structural and phase-composition analysis of Pt–Cu bimetallic nanoparticle precursors and dealloyed nanoparticle catalysts. (a) AXRD intensity profiles of a $Pt_{25}Cu_{75}$ alloy precursor as a function of the scattering vector Q. Diffraction profiles were taken as a function of the X-ray energy across the X-ray absorption edge of Cu at 8990 eV. Scattering intensities of Bragg reflections (shown are the (111) and (200) reflections) decrease (arrow) as the Cu absorption edge energy is approached. The extent of intensity decrease correlates with the Cu content of the scattering phase and can be used to extract alloy phase composition. (b) Fit (black line) of a relation between the square of the scattering amplitude $|F|^2$ and incident X-ray energy to experimental values (circles) in order to extract the molar fractions of Pt and Cu in the scattering phase. AXRD provided the lattice parameter of the scattering phase (structural information) as well as its actual composition (chemical information). (Reprinted with permission from Ref. [107]).

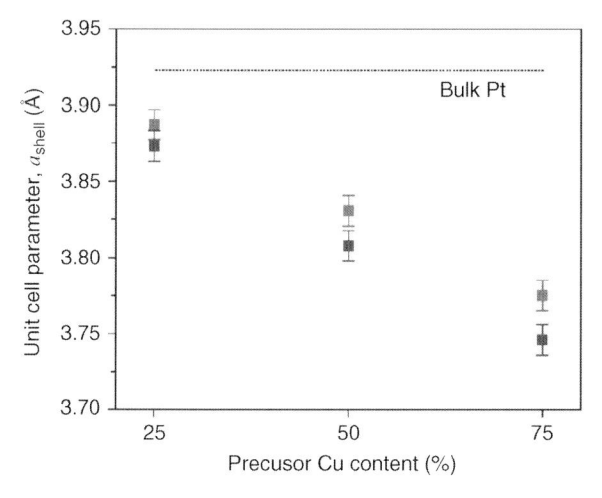

Figure 7.16 Unit cell parameter in the particle shell, a_{shell}, of dealloyed Pt–Cu bimetallic particles, plotted as a function of Cu content of the alloy precursor for two post-synthesis annealing temperatures of 950 °C (lower symbol) and 800 °C (upper symbol). (Reprinted with permission from Ref [107]).

residual Cu contents in the core of the dealloyed particles. Give that the shell is not too thick, higher core Cu contents lead to lattice contraction in the core and, hence, to smaller lattice parameters in the shell. The two different annealing conditions of the precursor particles show little difference in the unit cell parameter values.

Using the surface lattice strain estimates in the three dealloyed Pt core–shell particles, their electrocatalytic ORR activity can be correlated to their structure and a structure–activity relationship can be established. Figure 7.17 attempts to do that. A dimensionless Pt-normalized logarithmic ORR activity at room temperature of dealloyed PtCu nanoparticles, the precursor of which was prepared at 800 and 950 °C, is plotted against the estimate of the lattice strain in the particle shell. The dotted lines are just a guide to the eye. The figure evidences a monotonic increase in ORR activity with surface lattice strain and, hence, via Figure 7.16, with initial Cu content in the precursor. With increasing initial Cu content of the core–shell particle precursor, the final dealloyed catalyst retains more compressive lattice strain in the surface, which changes the electronic band structure of the nanoparticle catalysts so as to weaken the chemisorption of oxygen-containing intermediates. This confirms the validity of earlier computational predictions made for extended surfaces, also for nanoscale strained surfaces. It may well represent the first direct measurement of

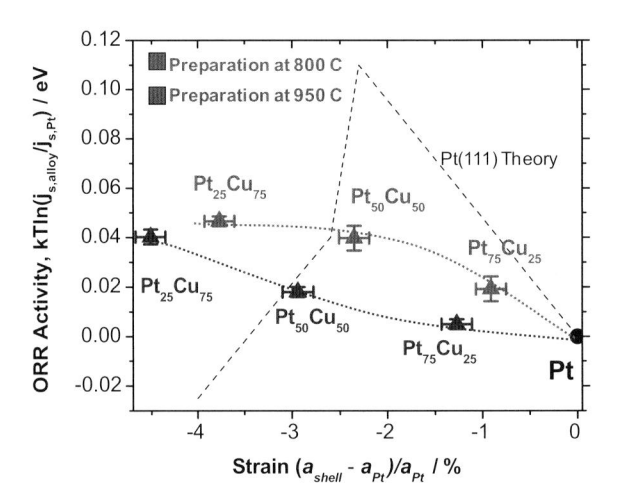

Figure 7.17 Experimental and predicted relationships between electrocatalytic ORR activity and lattice strain. The experimental ORR activity (in units of $kT \ln(j_{s,alloy}/j_{s,Pt})$, $T = 298$ K), based on relative surface area, of two families of dealloyed Pt–Cu bimetallic core–shell nanoparticles plotted as a function of strain in the particle shell (red and blue triangles denote dealloyed Pt–Cu precursors prepared at annealing temperatures of 800 and 950 °C, respectively). The ORR activity is proportional to the logarithm of the ratio of the ORR current density of the dealloyed particles to that of pure Pt nanoparticles; this quantity is the effective difference in ORR activation energies ("reaction driving force"). The experimental curves are upper bounds of the strain in the surface layer. The dashed line is the DFT-predicted, volcano-shaped trend of the ORR activity for a Pt(111) single-crystal slab under isotropic strain. Moderate compressive lattice strain is predicted to enhance the rate of ORR catalysis. (Reprint with permission from Ref [107]).

the effect of geometric lattice strain in nanoscale core–shell catalysts. While the activity seems to be leveling off for the 800 °C core shell material at strains of about −4%, no definitive maximum in either volcano curve could be found.

Also plotted in Figure 7.17 is the predicted ORR activity–strain relation for an ideal Pt(111) single crystal surface. It is the same curve as plotted in Figure 7.11. Starting with a relaxed pure Pt(111) surface, increasing the compressive lattice strain in the surface layer (moving to the left in the graph) increases the ORR activity due to weaker chemisorption of oxygenated reactive intermediates. The strained ideal extended surface exhibits a volcano activity maximum around −2.2% compressive strain, beyond which its ORR activity drops again as adsorption of oxygen becomes rate limiting.

The discrepancy between the ideal computational surface and the real nanoparticles is hardly surprising considering the huge structural difference between an ideal computational single crystal surface and an experimental nanostructured surface. Clearly, strain is a critical controlling parameter for catalytic activity, however, it is not the one and only microscopic parameter affecting activity. Roughness and particle size are additional parameters which may sensitively contribute to the observed catalytic activity. In the present case, however, there is another reason why a volcano maximum and the decreasing leg of the volcano curve cannot be observed. The surface lattice strain estimation of the nanoparticles in Figure 7.17 represents an average strain value that was assumed uniform throughout the Pt particle shells. In reality, however, lattice relaxations occur such that the Pt layers nearest to the alloy core show more, and the outermost layer considerably less strain than the average value would suggest. Hence, the strain values of the nanoparticles represent upper bounds, and the two real 800 and 950 °C activity–strain relations may be shifted somewhat to smaller strain values, bringing them closer to the right leg of the predicted ideal volcano. Furthermore, there may be a maximum sustainable value for the lattice strain which can be physically maintained and measured in such nanoscale structures. Such an upper bound would make the observation of the decreasing high strain leg of the volcano impossible.

7.6.6
High Energy X-Ray Diffraction (HE-XRD) and Atomic Pair Distribution Function (PDF) Analysis

A rarely used, yet powerful X-ray based analytical technique to obtain insight into structural order and disorder, intraparticle coherence lengths, interatomic distances, crystalline phases, and the level of amorphicity of nanoscale materials is the atomic pair distribution function (atomic PDF) analysis [122–129]. For ordinary extended macroscopic well ordered crystals, this information is obtained by traditional (Bragg) X-ray diffraction. It is difficult to use this approach on materials structured at the nanoscale because their diffraction patterns show few, if any, Bragg peaks, and have a pronounced diffuse component associated with order which only extends over short distances. A non-traditional approach based on high-energy X-ray diffraction and atomic PDF data analysis may be used instead. Atomic PDF analyzes diffuse (i.e., non-Bragg type) XRD patterns and obtains important structural information, such as nearest-neighbor atomic distances and coordination numbers for non-crystalline materials.

The frequently used reduced atomic PDF, $G(r)$, gives the number of atoms in a spherical shell of unit thickness at a distance r from a reference atom as [126]:

$$G(r) = 4\pi r[\rho(r) - \rho_o]$$

where $\rho(r)$ and ρ_o are the local and average atomic number densities, respectively, and r is the radial distance. As defined, $G(r)$ is a one-dimensional function that oscillates around zero and shows positive peaks at distances separating pairs of atoms, that is, where the local atomic density exceeds the average. The negative valleys in $G(r)$ correspond to real space vectors lacking atoms at either of their ends. In this respect $G(r)$ resembles the so-called Patterson function that is widely applied in traditional X-ray crystallography. $G(r)$ is the Fourier transform of the experimentally observable total structure function, $S(Q)$ where Q is the scattering vector. As such, the atomic PDF is simply another representation of the experimental XRD data. However, exploring the XRD data in real space is advantageous, especially when studying nanocrystalline materials. This is because the total scattering, including Bragg-like peaks and diffuse (non-Bragg-like) scattering, contributes to the PDF. In this way both the discernible atomic order, manifested in the Bragg-like peaks, and all structural "imperfections" that are responsible for its limited extent, manifested in the intense diffuse component of the diffraction pattern, are reflected in the experimental PDF for a nanocrystalline material [126]. Atomic PDFs are best extracted from X-ray diffraction at very high Q values. To achieve this, very high energy X-ray (small wavelength) radiation from a synchrotron source is used.

Figure 7.18 shows high energy X-diffraction profiles and atomic PDFs of a PtNi$_6$ and a PtNi$_3$ alloy nanoparticle electrocatalyst in their pristine state and after electrochemical dealloying. Using incident synchrotron X-ray radiation above 100 keV, scattering vectors Q larger than 14 Å^{-1} could be sampled. Figure 7.18a impressively illustrates the limited Q range (shaded area) interrogated using conventional Cu Kα ($\lambda = 1.54$ Å) radiation over a typical 2θ range, compared to the use of high energy X-ray radiation ($\lambda = 0.107$ Å). The high energy radiation provides structural information at scattering vectors which would be impossible to reach using conventional Cu radiation. Figure 7.18a can be interpreted just like any conventional Bragg diffraction profile, showing the familiar shift in the major cubic (111) reflections to smaller Qs after dealloying due to a loss of the less noble metal, here Ni. The loss of Ni causes the expansion of the remaining Pt-enriched lattice.

Figure 7.18b displays the atomic PDFs of the four PtNi nanoparticle alloys, combined with that of a pure Pt nanoparticle catalyst. The major interatomic distance at around 2.6 Å corresponding to the Pt–Pt distance in the lattice is evident. The width of this peak increases as the order of the lattice decreases and the integrated intensities of the peaks correlate with the number of atoms in the spherical shell at the distance considered. Both dealloyed PtNi materials show broadened Pt–Pt interatomic distance peaks. In particular, the dealloyed PtNi$_3$

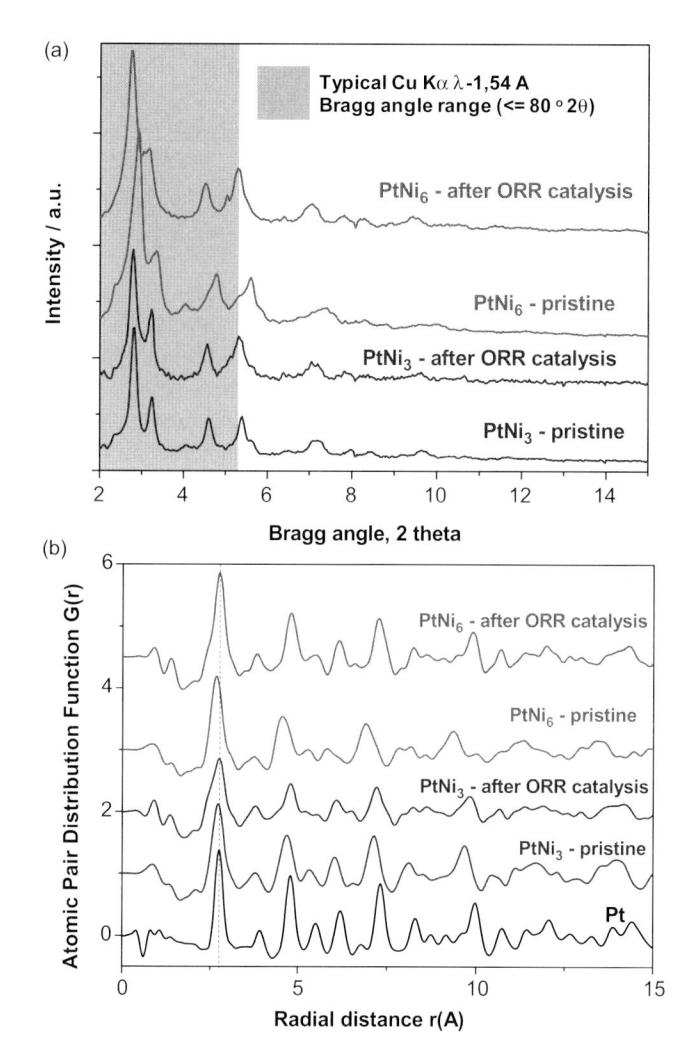

Figure 7.18 (a) High energy X-ray diffraction profiles of pristine and electrochemically dealloyed PtNi$_6$ and PtNi$_3$ nanoparticle alloy. High energy radiation allows the sampling of high scattering vectors Q. The shaded area indicates the limited accessible Q range using Cu Kα radiation. (b) Atomic PDFs, $G(r)$ of the four materials of (a), compared to that of Pt nanoparticles. Positive peaks indicate atomic densities in a spherical shell at distance r exceeding the average atomic density.

materials exhibit a shoulder at smaller distance, consistent with the Ni–Ni distance in metallic Ni. This suggests the formation of segregated Ni domains in the material. Peak intensities gradually decrease for increasing distances and finally vanish, indicating a loss of coherence of the scattered X-rays. For a nanostructured material the distance where PDF peaks vanish indicates the crystallite size or mean particle size. The PDF profiles of Figure 7.18b were fitted using atomic alloy models. While the dealloyed PtNi$_6$ diffraction and PDF pattern was consistent with a structurally ordered Pt$_3$Ni alloy lattice, the dealloyed PtNi$_3$ PDF profile could be best fitted using a solid solution model with Ni-rich segregated domains.

7.7
Conclusions

This chapter has explored X-ray-based analytical techniques to characterize dealloyed lattice-strained surfaces of macroscopic and nanoscale dimension with regard to their geometric and electronic structure.

The geometric arrangement of the atoms in the topmost surface layer is a critical factor determining the catalytic properties of the surface. At a given crystallographic orientation and uniform chemical composition, the lattice parameter of the surface atoms remains an important determinant of the electronic structure, chemical bonding and catalytic activity of the surface. Deviations from the thermodynamically most stable surface lattice parameter represent surface lattice strain. Strain can be used to tune the electronic and catalytic properties of surfaces.

Lattice strain in monometallic surfaces can arise due to their nanometer-scale dimension or their structural imperfections (defects). Surface lattice strain of a much larger magnitude can be introduced in a controlled way in overlayers of metal A on top of a dissimilar metal B. Unlike direct orbital hybridization (ligand) effects between A and B, strain effects in overlayers are long-range and may persist across up to 10 atomic metal layers. This is why core–shell material architectures can be used to control surface lattice strain and catalytic activity. Dealloying of non-noble metal-rich precursors is a suitable synthetic technique to arrive at strained core–shell structures in extended or nanoscale materials.

A variety of X-ray methods, such as conventional lab-based XRD, synchrotron-based grazing angle surface XRD, *in situ* high temperature XRD, synchrotron-based AXRD and ASAXS, photoelectron emission spectroscopy, X-ray absorption, X-ray emission, high energy XRD, and atomic PDF have been used to characterize the surface strain and structure of core–shell electrocatalyst materials for the oxygen reduction reaction.

Macroscopic single crystal alloy surfaces dealloy deep into the bulk and, hence, the surface strain is very small and leads to limited activity enhancements for the oxygen reduction reaction. Polycrystalline thin film alloys dealloyed on the order of up to 40 nm and, hence, the remaining surface strain in the surface remains low, for example −0.66% compressive strain for a PtCu$_3$ film. Nanoscale alloys dealloy very differently to extended surfaces. If the initial size is below the critical size where porosity can occur, a solid core–shell nanoparticle forms and retains a sizeable magnitude of surface lattice strain of up to −4–5% compressive strain. Compressive surface strain correlates with the experimentally observed ORR activity. Hence, the ORR activity of dealloyed solid core–shell particles ranges up to factors of 10× [95,96,100,116,130–133], while the ORR activity of extended surfaces (single crystals and thick films) remains at factors of 1–3× [90,108,134].

Thanks to the advanced electron characterization techniques, such as aberration-corrected electron microscopy and spectroscopy, real space images of complex core–shell fine structures of dealloyed nanoparticles can today be investigated at the atomic scale [130,132,133]. X-ray studies combined with electron microscopy enable a comprehensive understanding of structure–activity–stability relationships of nanoscale electrocatalyst materials. Future work in the area of core–shell catalyst materials for fuel cells will concentrate on understanding and optimizing the core–shell fine structures of dealloyed nanoparticles through the control of particle size, composition, shape, and the dealloying conditions.

The future for X-ray-based analytical techniques in material science and catalysis is bright and will continue to yield new insights into the surface structure of functional materials.

Acknowledgments

The author thanks Professor Anders Nilsson, Dr. Michael F. Toney, Professor Ruizhi Yang, Professor Valeri Petkov and, in particular, Dr. Shirlaine Koh for their help and support in synchrotron-based X-ray scattering and spectroscopy.

The author also thanks Dr. Chengfei Yu, Dr. Ratndeep Srivastava, Dr. Prasanna Mani, Dr. Zengcai Liu, Dr. Mehtap Oezaslan, and Dr. Xenia Tuaev for their support over the past years.

This work was supported by U.S. DOE through a past BES award via subcontract through SSRL and a current EERE award DE-EE0000458 via subcontract through General Motors P.S. acknowledges financial support through the Cluster of Excellence in Catalysis (UniCat) funded by DFG and managed by TU Berlin.

References

1 Friedrich, W., Knipping, P., and Laue, M.v. (1912) in Proceedings of the Royal Bavarian Academy of Science (The Royal Bavarian Academy of Science - communicated by A Sommerfeld, Munich; June 8, 1912).

2 Frenkel, A.I., Rodriguez, J.A., and Chen, J.G.G. (2012) Synchrotron techniques for in situ catalytic studies: capabilities, challenges, and opportunities. *ACS Catalysis*, **2**, 2269–2280.

3 Siegbahn, H., Lundholm, M., Arbman, M., and Holmberg, S. (1983) On the measurement of binding-energies in liquid Esca and the relation to electrochemical half-cell emfs. *Phys. Scr.*, **27**, 241–244.

4 Fellnerfeldegg, H. et al. (1974) New developments in ESCA-instrumentation. *J. Electron Spectrosc. Relat. Phenom.*, **5**, 643–689.

5 Siegbahn, K. (1970) Electron spectroscopy for chemical analysis (ESCA). *Philos. Trans. R. Soc. Ser. A*, **268**, 33–57.

6 Siegbahn, K., Gelius, U., Siegbahn, H., and Olson, E. (1970) Angular distribution of electrons in ESCA spectra from a single crystal. *Phys. Lett. A*, **32**, 221–222.

7 Nordberg, R., Albridge, R.G., Bergmark, T., Ericson, U., Hedman, J., Nordling, C., Siegbahn, K., and Lindberg, B.J. (1968) Molecular spectroscopy by means of ESCA – charge distribution in nitrogen compounds. *Ark. Kemi*, **28** (3), 257.

8 Fahlman, A. and Siegbahn, K. (1966) ESCA method using monochromatic X-rays and a permanent magnet spectrograph. *Ark. Fysik*, **32** (2–3), 111.

9 Fahlman, A., Hagstrom, S., Hamrin, K., Nordberg, R., Nordling, C., and Siegbahn, K. (1966) An apparatus for ESCA method. *Ark. Fysik*, **31** (6), 479.

10 Kim, S.G. and Lee, S.J. (2013) Quantitative visualization of a gas diffusion layer in a polymer electrolyte fuel cell using synchrotron X-ray imaging techniques. *J. Synchrotron Radiat.*, **20**, 286–292.

11 Nam, K.W. et al. (2013) Combining in situ synchrotron X-ray diffraction and absorption techniques with transmission electron microscopy to study the origin of thermal instability in overcharged cathode materials for lithium-ion batteries. *Adv. Funct. Mater.*, **23**, 1047–1063.

12 Hackett, M.J. et al. (2011) Synchrotron techniques and research: objectives at the K-edge (novel investigations of the role of sulfur in stroke pathogenesis using synchrotron based K-edge X-ray absorption spectroscopy). *Stroke*, **42**, E597–E597.

13 Krumrey, M., Gleber, G., Scholze, F., and Wernecke, J. (2011) Synchrotron radiation-based X-ray reflection and scattering techniques for dimensional nanometrology. *Meas. Sci. Technol.*, **22**, 129901.

14 Luo, L. and Zhang, S.Z. (2010) Applications of synchrotron-based X-ray techniques in environmental science. *Sci. China-Chem.*, **53**, 2529–2538.

15 Ludwig, W. et al. (2010) Characterization of polycrystalline materials using synchrotron X-ray imaging and diffraction techniques. *JOM*, **62**, 22–28.

16 Boscherini, F. et al. (2010) X-ray Techniques for Advanced Materials, Nanostructures and Thin Films: from Laboratory Sources to Synchrotron Radiation Proceedings of the EMRS 2009 Spring Meeting - Symposium R Strasbourg, France, June 8-June 12 2009. *Nuclear Instruments & Methods in Physics Research Section B-Beam Interactions with Materials and Atoms* **268**,V-V.

17 Buonassisi, T., Vyvenko, O.F., Istratov, A. A., Weber, E.R., and Schindler, R. (2002) Application of X-ray synchrotron techniques to the characterization of the chemical nature and recombination activity of grown-in and process-induced defects and impurities in solar cells. MRS 2002 Spring Meeting, San Francisco, CA, 1–5 April 2002, Symposium F, Defect and Impurity Engineered Semiconductors and Devices III, **719**, pp. 179–184.

18 Roberts, K.J. (1993) The application of synchrotron X-ray techniques to problems in crystal science and engineering. *J. Cryst. Growth*, **130**, 657–681.

19 Robinson, J. and Walsh, F.C. (1993) In-situ synchrotron-radiation X-ray techniques for studies of corrosion and protection. *Corros. Sci.*, **35**, 791–800.

20 North, A.N. et al. (1988) Small-angle X-ray-scattering studies of heterogeneous systems using synchrotron radiation techniques. *Nucl. Instrum. Meth. B*, **34**, 188–202.

21 Busby, R.L. (2005) *Hydrogen and Fuel Cells*, Pennwell Corporation, Tulsa.

22 Kordesch, K. and Simader, G. (1996) *Fuel Cells and Their Applications*, VCH, Weinheim.

23 Larminie, J. and Dicks, A. (2000) *Fuel Cell Systems Explained*, Wiley, New York.

24 Li, X. (2006) *Principles of Fuel Cells*, Taylor - Francis, New York.

25 O'Hayre, R., Cha, S.-W., Colella, W., and Prinz, F.B. (2006) *Fuel Cell Fundamentals*, Wiley, New York.

26 Hoogers, G. (ed.) (2003) *Fuel Cell Technology Handbook*, CRC Press.

27 Koper, M. (ed.) (2009) *Fuel Cell Catalysis - A surface Science Approach*, Wiley, New Jersey.

28 Vielstich, W., Gasteiger, H.A., and Yokokawa, H. (eds) (2009) *Handbook of Fuel Cells:Advances in Electrocatalysis, Materials, Diagnostics and Durability*, vols. 5 & 6, John Wiley & Sons Ltd., Chichester, West Sussex, UK.

29 Gasteiger, H.A., Kocha, S.S., Sompalli, B., and Wagner, F.T. (2005) Activity benchmarks and requirements for Pt, Pt-alloy, and non-Pt oxygen reduction catalysts for PEMFCs. *Appl. Catal. B - Environ.*, **56**, 9–35.

30 Gasteiger, H.A. and Markovic, N.M. (2009) Just a dream or future reality? *Science*, **324**, 48–49.

31 Wagner, F.T., Lakshmanan, B., and Mathias, M.F. (2010) Electrochemistry and the future of the automobile. *J. Phy. Chem. Lett.*, **1**, 2204–2219.

32 Debe, M.K. (2012) Electrocatalyst approaches and challenges for automotive fuel cells. *Nature*, **486**, 43–51.

33 Bashyam, R. and Zelenay, P. (2006) A class of non-precious metal composite catalysts for fuel cells. *Nature*, **443**, 63–66.

34 Lefevre, M., Proietti, E., Jaouen, F., and Dodelet, J.P. (2009) Iron-based catalysts with improved oxygen reduction activity in polymer electrolyte fuel cells. *Science*, **324**, 71–74.

35 Wu, G., More, K.L., Johnston, C.M., and Zelenay, P. (2011) High-performance electrocatalysts for oxygen reduction derived from polyaniline, iron, and cobalt. *Science*, **332**, 443–447.

36 Li, Y. et al. (2012) An oxygen reduction electrocatalyst based on carbon nanotube-graphene complexes. *Nat. Nanotechnol.*, **7**, 394–400.

37 Stamenkovic, V. et al. (2006) Changing the activity of electrocatalysts for oxygen reduction by tuning the surface electronic structure. *Angew. Chem. - Int. Ed.*, **45**, 2897–2901.

38 Greeley, J. et al. (2009) Alloys of platinum and early transition metals as oxygen reduction electrocatalysts. *Nature Chem.*, **1**, 552–556.

39 Zhang, J.L., Vukmirovic, M.B., Xu, Y., Mavrikakis, M., and Adzic, R.R. (2005) Controlling the catalytic activity of platinum-monolayer electrocatalysts for oxygen reduction with different substrates. *Angew. Chem. - Int. Ed.*, **44**, 2132–2135.

40 Toda, T., Igarashi, H., Uchida, H., and Watanabe, M. (1999) Enhancement of the electroreduction of oxygen on Pt alloys with Fe, Ni, and Co. *J. Electrochem. Soc.*, **146**, 3750–3756.

41 Min, M.-k., Cho, J., Cho, K., and Kim, H. (2000) Particle size and alloying effects of Pt-based alloy catalysts for fuel cell applications. *Electrochim. Acta*, **45**, 4211–4217.

42 Paulus, U.A. et al. (2002) Oxygen reduction on carbon-supported Pt-Ni and Pt-Co alloy catalysts. *J. Phys. Chem. B*, **106**, 4181–4191.

43 Salgado, J.R.C., Antolini, E., and Gonzalez, E.R. (2004) Structure and activity of carbon-supported Pt-Co electrocatalysts for oxygen reduction. *J. Phys. Chem. B*, **108**, 17767–17774.

44 Yang, H., Alonso-Vante, N., Leger, J.M., and Lamy, C. (2004) Tailoring, structure, and activity of carbon-supported nanosized Pt-Cr alloy electrocatalysts for oxygen reduction in pure and methanol-containing electrolytes. *J. Phys. Chem. B*, **108**, 1938–1947.

45 Yang, H., Vogel, W., Lamy, C., and Alonso-Vante, N. (2004) Structure and electrocatalytic activity of carbon-supported Pt-Ni alloy nanoparticles toward the oxygen reduction reaction. *J. Phys. Chem. B*, **108**, 11024–11034.

46 Zhang, J. et al. (2004) Platinum monolayer electrocatalysts for O-2 reduction: Pt monolayer on Pd(111) and on carbon-supported Pd nanoparticles. *J. Phys. Chem. B*, **108**, 10955–10964.

47 Zhang, J. et al. (2005) Platinum monolayer on nonnoble metal-noble metal core-shell nanoparticle electrocatalysts for O-2 reduction. *J. Phys. Chem. B*, **109**, 22701–22704.

48 Zhang, J.L. et al. (2005) Mixed-metal Pt monolayer electrocatalysts for enhanced oxygen reduction kinetics. *J. Am. Chem. Soc.*, **127**, 12480–12481.

49 Adzic, R.R. et al. (2007) Platinum monolayer fuel cell electrocatalysts. *Top. Catal.*, **46**, 249–262.

50 Zhang, J., Sasaki, K., Sutter, E., and Adzic, R.R. (2007) Stabilization of platinum oxygen-reduction electrocatalysts using gold clusters. *Science*, **315**, 220–222.

51 Wang, J.X. et al. (2009) Oxygen reduction on well-defined core-shell nanocatalysts: particle size, facet, and pt shell thickness effects. *J. Am. Chem. Soc.*, **131**, 17298–17302.

52 Stamenkovic, V.R., Mun, B.S., Mayrhofer, K.J.J., Ross, P.N., and Markovic, N.M. (2006) Effect of surface composition on electronic structure, stability, and

electrocatalytic properties of Pt-transition metal alloys: Pt-skin versus Pt-skeleton surfaces. *J. Am. Chem. Soc.*, **128**, 8813–8819.

53 Stamenkovic, V.R. *et al.* (2007) Improved oxygen reduction activity on Pt3Ni(111) via increased surface site availability. *Science*, **315**, 493–497.

54 Stamenkovic, V.R. *et al.* (2007) Trends in electrocatalysis on extended and nanoscale Pt-bimetallic alloy surfaces. *Nat. Mater.*, **6**, 241–247.

55 Oezaslan, M., Heggen, M., and Strasser, P. (2011) Dealloyed Pt nanoparticle electrocatalysts for PEMFC cathodes: core–shell fine structure and size-dependent morphology. 242nd National Meeting of the American Chemical Society, Denver, CO, p. 290.

56 Mani, P., Srivastava, R., and Strasser, P. (2008) Dealloyed Pt-Cu core-shell nanoparticle electrocatalysts for use in PEM fuel cell cathodes. *J. Phys. Chem. C*, **112**, 2770–2778.

57 Erlebacher, J. (2004) An atomistic description of dealloying - Porosity evolution, the critical potential, and rate-limiting behavior. *J. Electrochem. Soc.*, **151**, C614–C626.

58 Erlebacher, J., Aziz, M.J., Karma, A., Dimitrov, N., and Sieradzki, K. (2001) Evolution of nanoporosity in dealloying. *Nature*, **410**, 450–453.

59 Erlebacher, J. and Sieradzki, K. (2003) Pattern formation during dealloying. *Scr. Mater.*, **49**, 991–996.

60 Lichtenstein, J. (2003) Dealloying. *Mater. Performance*, **42**, 66–67.

61 Paffett, M.T. and Gottesfeld, S. (1988) Dealloying of Pt alloy fuel-cell electrocatalysts. *J. Electrochem. Soc.*, **135**, C348–C348.

62 Pugh, D.V., Dursun, A., and Corcoran, S.G. (2005) Electrochemical and morphological characterization of Pt-Cu dealloying. *J. Electrochem. Soc.*, **152**, B455–B459.

63 Renner, F.U., Grunder, Y., Lyman, P.F., and Zegenhagen, J. (2007) In-situ X-ray diffraction study of the initial dealloying of Cu3Au(001) and Cu0.83Pd0.17(001). *Thin Solid Films*, **515**, 5574–5580.

64 Schofield, E.J., Ingham, B., Turnbull, A., Toney, M.F., and Ryan, M.P. (2008) Strain development in nanoporous metallic foils formed by dealloying. *Appl. Phys. Lett.*, **92**, 043118.

65 Sieradzki, K. *et al.* (2002) The dealloying critical potential. *J. Electrochem. Soc.*, **149**, B370–B377.

66 Wagner, K., Brankovic, S.R., Dimitrov, N., and Sieradzki, K. (1997) Dealloying below the critical potential. *J. Electrochem. Soc.*, **144**, 3545–3555.

67 Bardal, E. (2004) *Corrosion and Protection*, Springer.

68 Sieradzki, K., Kim, J.S., Cole, A.T., and Newman, R.C. (1987) The Relationship between dealloying and transgranular stress-corrosion cracking of Cu-Zn and Cu-Al alloys. *J. Electrochem. Soc.*, **134**, 1635–1639.

69 Vanorden, A.C. and Handwerker, C.A. (1985) Dealloying of Cu-Zn and Cu-Au alloys in acidic solutions. *J. Met.*, **37**, A84-A84.

70 Parthasarathi, A. and Polan, N.W. (1982) Stress-corrosion of Cu-Zn and Cu-Zn-Ni alloys - the role of dealloying. *Metall. Mater. Trans. A*, **13**, 2027–2033.

71 Heidersb, R.H. and Verink, E.D. (1969) Clarification of mechanisms of dealloying phenomena in Cu-Zn alloys. *Corrosion*, **25**, 18.

72 Raney, M. (1927)

73 Ding, Y. and Erlebacher, J. (2003) Nanoporous metals with controlled multimodal pore size distribution. *J. Am. Chem. Soc.*, **125**, 7772–7773.

74 Ding, Y., Chen, M.W., and Erlebacher, J. (2004) Metallic mesoporous nanocomposites for electrocatalysis. *J. Am. Chem. Soc.*, **126**, 6876–6877.

75 Ding, Y., Kim, Y.J., and Erlebacher, J. (2004) Nanoporous gold leaf: "Ancient technology"/advanced material. *Adv. Mater. (Weinheim, Ger.)*, **16**, 1897.

76 Snyder, J., Asanithi, P., Dalton, A.B., and Erlebacher, J. (2008) Stabilized Nanoporous Metals by Dealloying Ternary Alloy Precursors. *Adv. Mater. (Weinheim, Ger.)*, **20**, 4883–4886.

77 Snyder, J., Fujita, T., Chen, M.W., and Erlebacher, J. (2010) Oxygen reduction in

nanoporous metal-ionic liquid composite electrocatalysts. *Nat. Mater.*, **9**, 904–907.

78 Dou, R. and Derby, B. (2009) The Strength of Gold Nanowires and Nanoporous Gold. *Fractal Geometry and Stochastics IV*, **61**, 185–190.

79 Dou, R. and Derby, B. (2010) Strain gradients and the strength of nanoporous gold. *J. Mater. Res.*, **25**, 746–753.

80 Fang, C., Bandaru, N.M., Ellis, A.V., and Voelcker, N.H. (2012) Electrochemical fabrication of nanoporous gold. *J. Mater. Chem.*, **22**, 2952–2957.

81 Fujita, T. et al. (2012) Atomic origins of the high catalytic activity of nanoporous gold. *Nat. Mater.*, **11**, 775–780.

82 Hodge, A.M., Hayes, J.R., Caro, J.A., Biener, J., and Hamza, A.V. (2006) Characterization and mechanical behavior of nanoporous gold. *Adv. Eng. Mater.*, **8**, 853–857.

83 Ji, C.X. and Searson, P.C. (2002) Fabrication of nanoporous gold nanowires. *Appl. Phys. Lett.*, **81**, 4437–4439.

84 Nagle, L.C. and Rohan, J.F. (2011) Nanoporous gold anode catalyst for direct borohydride fuel cell. *Int. J. Hydrogen Energy*, **36**, 10319–10326.

85 Wang, D. and Schaaf, P. (2012) Nanoporous gold nanoparticles. *J. Mater. Chem.*, **22**, 5344–5348.

86 Wittstock, A., Biener, J., and Baumer, M. (2010) Nanoporous gold: a new material for catalytic and sensor applications. *Phys. Chem. Chem. Phys.*, **12**, 12919–12930.

87 Wittstock, A. et al. (2009) Nanoporous Au: an unsupported pure gold catalyst? *J. Phys. Chem. C*, **113**, 5593–5600.

88 Zhou, H., Jin, L., and Xu, W. (2007) New approach to fabricate nanoporous gold film. *Chin. Chem. Lett.*, **18**, 365–368.

89 Zielasek, V. et al. (2006) Gold catalysts: nanoporous gold foams. *Angew. Chem. Int. Edit.*, **45**, 8241–8244.

90 Yang, R., Leisch, J., Strasser, P., and Toney, M.F. (2010) Structure of dealloyed $PtCu_3$ thin films and catalytic activity for oxygen reduction. *Chem. Mater.*, **22**, 4712–4720.

91 Snyder, J. and Erlebacher, J. (2010) Kinetics of crystal etching limited by terrace dissolution. *J. Electrochem. Soc.*, **157**, C125–C130.

92 Ding, Y., Chen, M., and Erlebacher, J. (2004) Metallic mesoporous nanocomposites for electrocatalysis. *J. Am. Chem. Soc.*, **126**, 6876–6877.

93 Erlebacher, J., Aziz, M.J., Karma, A., Dimitrov, N., and Sieradzki, K. (2001) Evolution of nanoporosity in dealloying. *Nature*, **410**, 450–453.

94 Rudi, S., Tuaev, X., and Strasser, P. (2012) Electrocatalytic oxygen reduction on dealloyed $Pt_{1-x}Ni_x$ alloy nanoparticle electrocatalysts. *Electrocatalysis*, **3**, 265–273.

95 Hasche, F., Oezaslan, M., and Strasser, P. (2012) Activity, structure and degradation of dealloyed $PtNi_3$ nanoparticle electrocatalyst for the oxygen reduction reaction in PEMFC. *J. Electrochem. Soc.*, **159**, B25–B34.

96 Oezaslan, M. and Strasser, P. (2011) Activity of dealloyed $PtCo_3$ and $PtCu_3$ nanoparticle electrocatalyst for oxygen reduction reaction in polymer electrolyte membrane fuel cell. *J. Power Sources*, **196**, 5240–5249.

97 Koh, S. and Strasser, P. (2010) Dealloyed Pt nanoparticle fuel cell electrocatalysts: stability and aging study of catalyst powders, thin films, and inks. *J. Electrochem. Soc.*, **157**, B585–B591.

98 Oezaslan, M., Hasché, F., and Strasser, P. (2010) Structure–activity relationship of dealloyed $PtCo_3$ and $PtCu_3$ nanoparticle electrocatalyst for oxygen reduction reaction in PEMFC. *ECS Trans.*, **33** (1), 333–341.

99 Koh, S., Hahn, N., Yu, C.F., and Strasser, P. (2008) Effects of composition and annealing conditions on catalytic activities of dealloyed Pt-Cu nanoparticle electrocatalysts for PEMFC. *J. Electrochem. Soc.*, **155**, B1281–B1288.

100 Oezaslan, M., Heggen, M., and Strasser, P. (2012) Size-dependent morphology of dealloyed bimetallic catalysts: linking the nano to the macro scale. *J. Am. Chem. Soc.*, **134**, 514–524.

101 Mavrikakis, M., Hammer, B., and Norskov, J.K. (1998) Effect of strain on the reactivity of metal surfaces. *Phys. Rev. Lett.*, **81**, 2819–2822.

102 Hammer, B. and Norskov, J.K. (2000) Theoretical surface science and catalysis - calculations and concepts. *Adv. Catal.*, **45**, 71–129.

103 Zhang, J., Vukmirovic, M.B., Xu, Y., Mavrikakis, M., and Adzic, R.R. (2005) Controlling the catalytic activity of platinum-monolayer electrocatalysts for oxygen reduction with different substrates. *Angew. Chem. Int. Edit.*, **44**, 2132–2135.

104 Kibler, L.A., El-Aziz, A.M., Hoyer, R., and Kolb, D.M. (2005) Tuning reaction rates by lateral strain in a palladium monolayer. *Angew. Chem. Int. Ed.*, **44**, 2080–2084.

105 Kibler, L.A. and Kolb, D.M. (2003) Physical electrochemistry: recent developments. *Z. Phys. Chem.*, **217**, 1265–1279.

106 Greeley, J., Kibler, L., El-Aziz, A.M., Kolb, D.M., and Nørskov, J.K. (2006) Hydrogen evolution over bimetallic systems: understanding the trends. *ChemPhysChem*, **7**, 1032–1035.

107 Strasser, P. et al. (2010) Lattice-strain control of the activity in dealloyed core-shell fuel cell catalysts. *Nature Chem.*, **2**, 454–460.

108 Yang, R., Strasser, P., and Toney, M.F. (2011) Dealloying of Cu3Pt (111) studied by surface X-ray scattering. *J. Phys. Chem. C*, **115**, 9074–9080.

109 Renner, F.U. et al. (2008) In situ x-ray diffraction study of the initial dealloying and passivation of Cu3 Au (111) during anodic dissolution. *Phys. Rev. B*, **77**, 235433.

110 Renner, F.U. et al. (2006) Initial Corrosion observed on the atomic scale. *Nature*, **430**, 707–710.

111 Fusy, J., Meneaucourt, J., Alnot, M., Huguet, C., and Ehrhardt, J.J. (1996) Growth and reactivity of evaporated platinum films on Cu(111): a study by AES, RHEED and adsorption of carbon monoxide and xenon. *Appl. Surf. Sci.*, **93**, 211–220.

112 Stamenkovic, V. et al. (2006) Changing the activity of electrocatalysts for oxygen reduction by tuning the surface electronic structure. *Angew. Chem. Int. Edit.*, **45**, 2897–2901.

113 Oezaslan, M., Hasche, F., and Strasser, P. (2013) Pt core-shell catalyst architectures for oxygen fuel cell electrodes. *J. Phys. Chem. Lett.*, Sept 12 online, DOI: 10.1021/jz4014135.

114 Koh, S. and Strasser, P. (2007) Electrocatalysis on bimetallic surfaces: modifying catalytic reactivity for oxygen reduction by voltammetric surface dealloying. *J. Am. Chem. Soc.*, **129**, 12624–12625.

115 Oezaslan, M., Hasché, F., and Strasser, P. (2011) In situ observation of bimetallic alloy nanoparticle formation and growth using high-temperature XRD. *Chem. Mater.*, **23**, 2159–2165.

116 Oezaslan, M., Hasche, F., and Strasser, P. (2012) Oxygen electroreduction on PtCo3, PtCo and Pt3Co alloy nanoparticles for alkaline and acidic PEM fuel cells. *J. Electrochem. Soc.*, **159**, B394–B405.

117 Oezaslan, M., Hasche, F., and Strasser, P. (2012) PtCu$_3$, PtCu and Pt$_3$Cu alloy nanoparticle electrocatalysts for oxygen reduction reaction in alkaline and acidic media. *J. Electrochem. Soc.*, **159**, B444–B454.

118 Srivastava, R., Mani, P., and Strasser, P. (2009) In-situ voltammetric de-alloying of fuel cell catalyst electrode layer: a combined scanning electron microscope/electron probe micro analysis study. *J. Power Sourc.*, **190**, 40–47.

119 Strasser, P., Koha, S., and Greeley, J. (2008) Voltammetric surface dealloying of Pt bimetallic nanoparticles: an experimental and DFT computational analysis. *Phys. Chem. Chem. Phys.*, **10**, 3670–3683.

120 Srivastava, R., Mani, P., Hahn, N., and Strasser, P. (2007) Efficient oxygen reduction fuel cell electrocatalysis on voltammetrically dealloyed Pt—Cu—Co nanoparticles. *Angew. Chem. Int. Edit.*, **46**, 8988–8991.

121 Koh, S. and Strasser, P. (2007) Electrocatalysis on bimetallic surfaces: modifying catalytic reactivity for oxygen reduction by voltammetric surface dealloying. *J. Am. Chem. Soc.*, **129**, 12624–12625.

122 Welborn, M., Tang, W.J., Ryu, J., Petkov, V., and Henkelman, G. (2011) A combined density functional and X-ray

diffraction study of Pt nanoparticle structure. *J. Chem. Phys.*, **135**, 014503.

123 Petkov, V. and Shastri, S.D. (2010) Element-specific structure of materials with intrinsic disorder by high-energy resonant X-ray diffraction and differential atomic pair-distribution functions: a study of PtPd nanosized catalysts. *Phys. Rev. B*, **81**, 165428-1–165428-8.

124 Petkov, V. *et al.* (2009) Atomic-scale structure of biogenic materials by total X-ray diffraction: a study of bacterial and fungal MnO(x). *ACS Nano*, **3**, 441–445.

125 Petkov, V., Cozzoli, P.D., Buonsanti, R., Cingolani, R., and Ren, Y. (2009) Size, shape, and internal atomic ordering of nanocrystals by atomic pair distribution functions: a comparative study of gamma-Fe_2O_3 nanosized spheres and tetrapods. *J. Am. Chem. Soc.*, **131**, 14264.

126 Petkov, V. (2008) Nanostructure by high-energy X-ray diffraction. *Materials Today*, **11**, 28–38.

127 Petkov, V., Ohta, T., Hou, Y., and Ren, Y. (2007) Atomic-scale structure of nanocrystals by high-energy X-ray diffraction and atomic pair distribution function analysis: study of Fe_xPd_{100-x} (x=0 26, 28, 48) nanoparticles. *J. Phys. Chem. C*, **111**, 714–720.

128 Petkov, V. *et al.* (2005) Three-dimensional structure of nanocomposites from atomic pair distribution function analysis: study of polyaniline and (polyaniline)(0.5)V_2O_5 center dot 1.0H_2O. *J. Am. Chem. Soc.*, **127**, 8805–8812.

129 Luo, J. *et al.* (2005) Phase properties of carbon-supported gold-platinum nanoparticles with different bimetallic compositions. *Chem. Mater.*, **17**, 3086–3091.

130 Gan, L., Heggen, M., O'Malley, R., Theobald, B., and Strasser, P. (2013) Understanding and controlling nanoporosity formation for improving the stability of bimetallic fuel cell catalysts. *Nano Lett.*, **13**, 1131–1138.

131 Oezaslan, M., Hasché, F., and Strasser, P. (2012) $PtCu_3$, PtCu and Pt_3Cu alloy nanoparticle electrocatalysts for oxygen reduction reaction in alkaline and acidic media. *J. Electrochem. Soc.*, **159**, B444–B454.

132 Gan, L., Heggen, M., Rudi, S., and Strasser, P. (2012) Core-shell compositional fine structures of dealloyed Pt_xNi_{1-x} nanoparticles and their impact on oxygen reduction catalysis. *Nano Lett.*, **12**, 5423–5430.

133 Cui, C. *et al.* (2012) Octahedral PtNi nanoparticle catalysts: exceptional oxygen reduction activity by tuning the alloy particle surface composition. *Nano Lett.*, **12**, 5885–5889.

134 Yang, R.Z., Bian, W.Y., Strasser, P., and Toney, M.F. (2013) Dealloyed $PdCu_3$ thin film electrocatalysts for oxygen reduction reaction. *J. Power Sources*, **222**, 169–176.

Index

a
angle energy 29
anodic biofilms 148
– characterization 149–156
– cyclic voltammetry 154, 155
– electrochemical and biological studies 149–156
– Raman microscopy 155, 156
– techniques/methods 150–154
anomalous powder X-ray diffraction (AXRD) 277–281
– Pt–Cu bimetallic nanoparticle precursors, structural and phase composition analysis 279
anomalous small angle X-ray scattering (ASAXS) 277
– of core–shell nanoparticles 278
atomic hypothesis 2
atomic pair distribution function (PDF) analysis 281–284
atomistic simulations 35
Au-modified Pt THH NCs 241–243
– active Pt sites 241
– adatoms 241
– Au-decorated anodes 243
– Au-decorated THH Pt NCs 243
– backward scans, cyclic voltammograms
– – characterizing formic acid 241, 242
– CO_{ads} adsorption 241
– CO poisoning 243
– CO stripping experiments 243
– coverage 241, 243
– formic acid electrooxidation 241
– – in situ FTIR spectra, unmodified THH Pt NC electrode 242, 243
– forward scans, cyclic voltammograms
– – for Au-decorated THH Pt NC electrodes 241, 242
– gold decoration 241

– IR absorption frequency 243
– – Stark effect 243
– linearly bonded CO 243
– nanoparticles 241
– peak potential 241, 242
– Pt oxide formation 241
– third body effect 243
– voltammetric profile 241
– – graphs 242

b
BES. see bioelectrochemical systems (BES)
biaxial modulus, M 174
bimetallic NCs, shape-controlled synthesis 235–254
– alloy NCs 245–254
– – core–shell structured NCs 248–254
– – HOH PdAu alloy 247
– – THH PdPt alloy 245
– surface modification 236–245
– – Au-modified Pt THH NCs 241
– – Bi-modified THH Pt NCs 237
– – electrochemical approaches to modify metal 236, 237
– – – direct electrodeposition 236
– – – illustration 237
– – – underpotential deposition (UPD) 236, 237
– – Pt-modified Au prisms, high-index facets 243
– – Ru-modified THH Pt NCs 239, 240
– – vs. bimetallic alloy 236
– surface-modified nanoparticle 236
Bi-modified THH Pt NCs 237–239
– as-synthesized THH Pt nanocrystal 237
– – comparing voltammograms with different Bi coverage 237, 238
– – modification 237
– Bi coverages 237

Electrocatalysis: Theoretical Foundations and Model Experiments, First Edition.
Edited by Richard C. Alkire, Ludwig A. Kibler, Dieter M. Kolb, and Jacek Lipkowski.
© 2013 Wiley-VCH Verlag GmbH & Co. KGaA. Published 2013 by Wiley-VCH Verlag GmbH & Co. KGaA.

– Bi-decorated Pt nanospheres 238
– – steady-state current density, formic acid oxidation 238
– – vs. Bi-decorated THH Pt NCs 238
– Bi decoration 238, 239
– – effects 239
– Bi modification 237
– – catalytic activity 237, 238
– – effects 237, 238
– current densities 239
– CV profiles 238
– formic acid electrooxidation 237
– oxidation, via CO pathway 237
– preparation via Bi irreversible adsorption 237
– third-body effect 237
bioelectrochemical systems (BES)
– hydrogen peroxide, cell voltage behavior of 140
– sketches 141
biofilm, fed-batch reactor and photograph 150
Boltzmann factor 36
bond energy 29
Born–Oppenheimer approximation 16, 26
Butler–Volmer (BV) equation 79

c
cantilever-bending experiments 198
cathodic biological energy dissipation 144
clamping constraints 194
coordination number (CN) 222
core-shell structured NCs 248–254
– Au core 251
– AuPd alloy 250
– Au@Pd catalyst 253, 254
– – high-index Pd facets 254
– Au@Pd concave nanocubes 254
– – catalytic performance 254
– Au–Pd core@shell heterostructures 248
– Au–Pd core–shell NCs 248, 250
– – tetrahexahedral structure 248, 249
– catalytic processes advantages 248
– concave TOH Pd@Au core–shell NCs 251
– – SEM image 252
– convex Au@Pd NCs 250
– – morphologic advantage 250
– convex polyhedral Au@Pd core–shell NCs 250
– – elemental mapping, Au and Pd 249, 250
– convex polyhedral Au@Pd NCs, peak current density 249, 250
– core–shell structure 250
– Cu UPD layer 250
– CV measurements, ethanol electrooxidation
– – comparing electrocatalytic stabilities 254
– ethanol electrooxidation 248, 249
– HAADF-STEM-EDS 249, 250
– jagged-surface layers 253
– low-coordinated Pd atoms 254
– Miller indices, NCs 251
– multi-component heterogeneous seed-mediated growth route 252
– oxidation activities, of different NCs 249, 250
– Pd@Au core–shell nanocrystal, HAADF-STEM image 252
– Pd layer, heteroepitaxial growth 251
– Pd shell, examples 251
– polyhedral Au@Pd NCs 251
– – concave HOH NCs with (hkl) facets 251
– – concave trisoctahedral (TSH) NCs, (hhl) facets 251
– – model 252
– – THH NCs with facets 251
– Pt^(Au@Pd) catalyst 252, 253, 254
– – CV curves, ethanol electrooxidation 253
– – HAADF-STEM image 253
– – trimetallic 252
– Pt-on-(Au@Pd) trimetallic nanostructures 252
– sub-30nm Au@Pd concave nanocubes 252
– Suzuki coupling reaction, high-index-faceted Pd nanoshells 251
– THH Au–Pd core–shell nanocrystals 248
– – SEM images 249
– – TEM images 249
– THH nanocrystal (NCs) 248
– – electro-catalytic activity 249
– TOH Pd@Au core–shell NCs 252
– – ECL density 252
– – electrocatalytic activity 252
– trimetallic triple-layered Pt^(Au@Pd) catalyst 254
– – image 253
– XPS analysis 253
– Z-contrast, atomic number contrast 253
coupled cluster methods 13
coupled proton electron transfer (CPET) process 47, 56, 57, 63, 65
covalent bond energy 28, 29
covalent bond interactions 27, 28, 32
crystal truncation rod (CTR) 268
cyclic cantilever-bending experiments 199
cytochromes signals 156

d

dealloyed PtNi$_6$/PtNi$_3$ nanoparticle alloy
– high energy X-ray diffraction profiles of 283
dealloyed Pt surfaces, strained catalytic
– adsorption energies, surface lattice strain affects 265
– alloy nanoparticles 273
–– anomalous powder X-ray diffraction (AXRD) 277–281
–– anomalous small angle X-ray scattering (ASAXS) 277
–– atomic pair distribution function (PDF) analysis 281–284
–– Bragg Brentano powder X-ray diffraction (XRD) 273–274
–– high energy X-ray diffraction (HE-XRD) 281–284
–– *in situ* high temperature powder X-ray diffraction (XRD) 274–276
–– synchrotron X-ray photoemission spectroscopy (XPS) 276
– alloy precursor
–– nanoporosity, evolution of 264
– bimetallic surfaces 262–264
– core–shell model surfaces 264–266
– nanoporosity, evolution of 264
– ORR PEMFC electrocatalyst development 260
– Pt-Cu polycrystalline thin film surfaces 269–272
– PtCu$_3$(111) surface 266–268
– Pt monolayer catalysts 261
– X-ray diffraction 259
dealloying PtCu$_3$ catalyst
– anomalous X-ray diffraction (AXRD) profile of 271
– lattice parameters and compositions 275
– ORR activity of 272, 280
density functional theory (DFT) 2, 14, 34, 55, 60, 66, 75, 79, 87, 185, 187, 191, 196, 197, 213
– calculations 48, 49, 60
diffusion rates 1, 78, 94
direct electron transfer (DET) 144, 145, 148
– transmembrane complex 146
dynamical system 83, 84
– Chapman–Kolmogorov equation 83
– cluster approximation 84
– dynamic QCA 84
– kinetic modeling 84
– Markovian process 83
– voltammetry, relation 84
dynamic electro-chemo-mechanical analysis 199–203
– experimental results 202, 203
– experimental set-ups 201
– thin-film Au electrode 202
– thin-film Pd electrode 203

e

elastic solid surfaces, free energy 167–170
– absolute values, strain 168
– bulk phases 167
– capillary forces 168
– charge neutral surfaces 169
– consequence 170
– constitutive assumption 168
–– equation 168
– density 168
– elastic deformation 170
– electrochemical surface mechanics 167
– energy minimization, approach 169
– Faraday constant 169
– free energy densities 167
– general constitutive equation 169
– Helmholtz-type, free energy density function 168
– Lagrangian coordinates 168
– laws of electrostatics 169
– net free energy 167
– polarizable solid metallic electrode B 167
– strain-free reference configuration 168
– superficial excess 168
– superficial free energy density 167, 168
– thermodynamic analysis, surface 168
– thermodynamic description 167
electrical energy 1
electroactive microbial biofilms
– techniques and methods 150–154
electrocapillary coupling, at equilibrium 177
– cantilever-bending experiment 181–183
– coupling coefficient 179–181
–– for adsorption 185, 186
–– for Langmuir isotherm 186, 187
– during electrosorption 184, 185
– Lippmann equation 179–181
– Maxwell relations 183, 184
– outline–polarizable/nonpolarizable electrodes 177–179
– for zero charge and work function 190, 191
electrocapillary coupling coefficients 193
– for adsorption 185, 186
– define 195
– empirical data 193–198
– for Langmuir isotherm 186, 187
– values of apparent coefficients
–– hydrogen adsorption/oxygen species 197

–– for nominally capacitive processes 196
electrocapillary maximum 180–183
electrocatalysis 85–91
– chronoamperometric experiment 92
–– expression for current transient 92
–– nucleation-and-growth (N&G) model 92, 93
– kinetic modeling simulations, of CO oxidation on Pt(111) 93
– KMC simulations, for model PtRu alloy surfaces 93, 94
– Langmuir–Hinshelwood-type mechanism 92
– rate of CO oxidation
–– using mean-field approximation 92
– stripping voltammetry, for CO oxidation on 93, 94
electrochemical energy conversion process 141
electrochemical route
– nanoparticles synthesis, effects 254, 255
electrochemical surface reactions, lattice-gas modeling 76–79
electrochemical systems 2
electrochemistry 1–3, 79
– electronic structure models 22
electrode, Lagrangian area 174
electrodeposition 85, 90, 91, 236, 245
– interaction model for Cu and sulfate 90, 91
electrode potential 1, 2, 25, 26, 53–55, 54, 56, 59, 79, 115, 116, 131, 169, 178, 180, 182, 195, 203
– current density $vs.$ 182, 203
– influence of 53
– modeling 25, 26
– variation 195
electrode surface modeling 23
– cluster $vs.$ slab 23, 24
electron correlation, in electrochemistry 14
electron energy-loss spectroscopy (EELS) 49
electronic coupling 105, 114, 118, 123
electronic structure methods 4, 6, 19, 22, 26, 27, 34, 35, 37, 39, 40
electronic structure modeling 3, 6–26
electron transfer mechanisms 103–106
– Fowler Nordheim tunneling models 108
– Hopping models 111–115
–– rate constants for electron transfer 112, 113
–– weak/strong coupling limits 114
– Kuznetsov Ulstrup model
–– for two-step electron transfer across molecular junctions 114, 115
– rate of electron transfer 105

– Simmons model 108
– tunneling 106–109
–– coherent 108, 109
–– current 110, 111, 117
–– resonant 109–111 (See also scanning tunneling microscopy (STM))
electro-oxidation 227, 228, 231, 240, 248, 250, 254
– of formic acid (see formic acid oxidation)
electrosorption 76, 79, 85–91, 88, 185, 187, 193, 213, 214
– bisulfate and OH butterflies on Pt(111) 88
– chloride on Ag(100) 88
–– electrosorption valency, modeled as 88
–– off-lattice model 88
– DFT calculations 87, 88
– Frumkin isotherm 86, 87
– MC simulation of H-upd
–– and bromide adsorption, on Pt(100) electrode 89
–– Pt(111) region and Pt surfaces 90
–– sharpness 90
– mean-field prediction 87
– mutual depolarization 88
– potential 211
– potential-dependent bromide
–– coverage on Ag(100) electrode and MC simulation 87
–– QCA isotherm 87
– Pt(100) electrode, bromide adsorption 88
– underpotential deposition of hydrogen (H-upd) 86
– voltammograms of Pt(111) 85, 86
electrostatic embedding (EE) schemes 38
electrostatic interactions 32
Eley–Rideal-type (ER-type) reaction 47
energy profiles
– entropy corrections 18–20
– of LH-type mechanisms 50
enthalpy 20, 52, 208, 210
entropy 18, 19, 20, 21, 39, 75, 77, 186, 208
$Escherichia\ coli$ 148
exopolymeric substances (EPS) 146, 152
explicit solvation model 24
Eyring's equation 55

f

face centered cubic (FCC) 270
Faraday current 204
Fermi energy 191, 192
fluid droplet scales 176
fluids, capillary equations 175–177
formic acid oxidation, on Pt(111) 59, 60

- DFT calculations 60
- electrode potential 63–65
- Eley–Rideal mechanisms 63–65
- explicit solvation model 61–63
- gas phase reactions 60, 61
- kinetic rate model 65, 66
- overview 59
- proposed mechanisms 60

free energy 22, 25, 51, 52, 105, 167, 173, 180, 192

g

gas–solid interfaces 2
generalized gradient approximation (GGA) 15, 60
Geobacter-based biofilm electrode 143
Geobacter-dominated anodic biofilm 154
- cyclic voltammogram 155
Geobacter dominated biofilm 155
geometry optimization 16, 17
Gibbs adsorption equation, adsorption equation 165, 170
Gibbs free energy 20, 21, 52, 143, 167, 171
Gokhshteins measurement scheme 180
Gokshthein's theory 193
gold 234, 235
- capping agents 234
-- CTAB 234
-- CTAC 234
-- PDDA, organic quaternary ammonium salt 234
- concave cube nanoparticles 234
-- SEM image 235
- concave trisoctahedral (TOH) Au NCs 235
-- CV trace 235
-- FESEM image 235
- electrode, cantilever-bending experiment 182
- high-index facets nanoparticles 234
-- illustration 235
-- synthesis 234
- shape-controlled synthesis 234
- wet-chemical method 235
-- well-shaped metal NCs synthesis 235

h

Haber–Bosch process 221
Haiss' concept 193
Hartree–Fock equations 13
Hellmann–Feynman forces 194
heterogeneous catalyst 221
high energy X-ray diffraction (HE-XRD) 281–284

high-resolution transmission electron microscopy (HRTEM) 224–226, 229, 231, 232, 234, 245, 251
Hohenberg–Kohn variational theorem 14
HOH PdAu alloy 247, 248
- as-prepared hexoctahedral Au–Pd alloy NCs 248
- concave 48-facet polyhedron 247
- Cu UPD 247
- element mapping analysis 247
- hexoctahedral (HOH), NCs 247
-- inductively coupled plasma atomic emission spectroscopy 248
-- SEM images, different orientations 247
- high-index facets, formation 248
- nanospheres 248
- octadecyl trimethyl ammoniumchloride (OTAC) 247
Hooke's law 28
hydrogen bonding 34
hydrogen desorption potential 188
hydrogen evolution reaction (HER) 2
hydrogen peroxide, cell voltage behavior 140

i

implicit solvation model 24
inductively coupled plasma atomic emission spectroscopy (ICP-AES) 248
in situ FTIR spectroscopic studies 228
interatomic interactions 27
ionic charge 1

j

Jacob's ladder, illustrating hierarchy of approximations 15

k

Core-shell structured NCs
- comparing electrocatalytic stabilities 254
- CV measurements, ethanol electrooxidation 254
Kohn–Sham (KS) orbitals 16
Kolid *vs.* fluid, theory of capillarity 167

l

Langmuir–Hinshelwood-type (LH-type) reaction 47
Langmuir model 186
lattice-gas model 76
- adsorption energies, depending on electrode potential 79
- advantages 76
- Butler–Volmer (BV) equation 79

– DFT calculations 79
– diffusion 78
– electrosorption valency 79
– rate constants for adsorption/desorption 78
– stepped Pt(111) surface indicating lattice sites 77
Lennard-Jones potential 33
ligand effect 187, 194, 213, 236
Lippmann equation 179, 180
localized density approximation (LDA) 15, 16

m

Marcus theory 105
Maxwell relation 183, 186, 187
mechanical embedding (ME) schemes 38
mechanically controlled break junctions (MCBJs) 101
mechanically modulated catalysis 203, 204
– Butler–Volmer kinetics 206, 207
– capacitive/pseudocapacitive accumulation processes 204
– current–strain coupling coefficient 205
– Faraday current density 205
– Heyrowsky reaction 207–212
– – enthalpy diagram 208
– pseudo-capacitive current density 204, 205
– superficial charge density 204
mediated electron transfer (MET) 144, 146, 147
– mechanisms based on secondary metabolites 147
– on primary metabolites 147, 148
– transmembrane complex for 146
MEP. see minimum energy pathways (MEP)
MET. see mediated electron transfer (MET)
meta-generalized gradient approximation (MGGA) 15
metal–electrolyte interfaces 39
Metropolis algorithm 44, 84, 85
MFCs. see microbial fuel cells (MFCs)
MGGA. see meta-generalized gradient approximation (MGGA)
microbial bioelectrocatalysis 141
– to microbial bioelectrochemical systems 137–156
microbial bioelectrocatalysts 141
microbial bioelectrochemical systems 137–156
– anodic acetate oxidation, by *Geobacteraceae* 142, 143
– anodic biofilms, characterizing 149–156
– – cyclic voltammetry, use of 154, 155
– – Raman microscopy 155, 156
– – techniques 150–154
– cathodic electron transfer mechanisms 148
– cathodic hydrogen evolution reaction 143, 144
– chronoamperometric curve 150
– direct electron transfer (DET) 144, 145
– – on EPS matrix 146
– – on microbial nanowires 145, 146
– – sketch of 145
– – on trans-membrane electron transfer proteins 145
– energetic considerations 141, 142
– mediated electron transfer 146, 147
– – on primary metabolites 147, 148
– – on secondary metabolites 147
– metabolic networks, establishment of 149
– microbial electron transfer mechanisms 144
– microbial fuel cells 137–139
– microbial interactions
– – ecological networks 148
– – interspecies electron transfer 148, 149
– – scavenging of redox-shuttles 148
– microorganisms catalyze electrochemical reactions 140, 141
– strength, through diversity 139, 140
microbial electrolysis cells (MEC) 139
microbial fuel cells (MFCs) 137–139
– number of publications 138
– sketch of 138
microbial interactions
– ecological networks 148
– interspecies electron transfer 148, 149
– scavenging of redox-shuttles 148
minimum energy pathways (MEP) 16–18
modern electronic structure theory 6
– basis sets 10
– Born–Oppenheimer approximation 8, 9
– configuration interaction 13
– coupled cluster methods 13
– density functional theory 14
– – energy functional 14, 15
– – exchange–correlation functionals 15, 16
– – Hohenberg–Kohn theorems 14
– electron correlation 14
– – methods 12
– enforcing Pauli principle 11, 12
– perturbation theory 13
– plane waves 11
– quantum mechanical foundations 6, 8
– single-electron Hamiltonians 9, 10
– Slater/Gaussian type orbitals 10
modern surface science techniques 2

molecular dynamics (MD) 36
– Ehrenfest 35
– simulations 36
– study of surface-induced pressure, case study 177
molecular modeling, in electrochemistry 39
molecular simulations 27
– angles 29, 30
– covalent bond interactions 27, 28
– covalent bond strength 28, 29
– cross terms 31, 32
– electrostatics 32
– energy terms, and force field parameters 27
– hydrogen bonds 34
– inversion 31
– Lennard-Jones potential 33
– non-covalent interactions 32
– reactive *vs.* non-reactive 28
– resonance 32
– torsion 30, 31
– under/over coordination 31
– Van der Waals contribution 33
Møller–Plesset (MP) perturbation theory 13
Mo nonmetallic nanocrystals 224–235
– electrochemical route 224–230
– – palladium 228–230
– – platinum 224–228
– shape-controlled synthesis 224–235
– wet-chemical route 230–235
– – gold 234, 235
– – palladium 232–234
– – platinum 230–232
Monte Carlo simulations 35, 36, 41, 45, 76, 87, 88, 90, 95, 96
– first reaction method 85
– importance sampling 84
– Metropolis algorithm 84, 85
Morse potential 29
motorways, for electrons 145
multiscale modeling 3–5
– according to time and length scales 4
– increases as DFT calculations 67
– QM/MM modeling 37

n

nanoparticles 221
– alloy 236
– catalysts, based on 221
– core–shell structure 236
– metal nanoparticles 222, 223
– – electrochemical sqaure wave method 223
– – strategies 223
– – thermodynamics 223
– nanocatalysts 222
– Pt nanoparticle catalysts 222
Newton–Raphson methods 17
non-covalent interactions 32
non-noble metal catalysts 222
nonpolarizable electrode 178
– $AgNO_3$ solution 178
nudged elastic band (NEB) method 18

o

oxygen reduction reaction (ORR) 46, 47, 53, 54, 55, 67, 139, 222, 231, 232
– electrochemical mechanism 54
– energy profiles, of LH-type mechanisms 50
– four-electron, elementary mechanisticmodel 260
– free energy contributions 52, 53
– H_2O_2 dissociation 50, 53
– mechanisms 49
– OOH-dissociation 52
– potential-dependent rate constants 58
– on Pt(111) 46
– single rate processes, kinetic rates analysis 56
– solvation effects 52
– at zero potential 54

p

palladium (Pd) 228–230, 232–234
– as-synthesized concave nanocubes 232
– catalytic applications 228
– commercial Pd black 229
– – cyclic voltammograms 229
– concave hexoctahedral Pd NCs 229, 230
– – SEM image 229
– concave Pd cubic nanoparticles, directly synthesized 233
– – angle of concave structure, tuning of 234
– – SEM image 233
– electrochemical square-wave method 229
– – Pd NCs, synthesis of 229
– high-index surface formation 234
– Pd concave nanocubes 232
– – electrocatalytic activity 232
– – Miller indices determination 232
– – performance 232, 233
– – SEM image 233
– – TEM image 233
– Pd nanocrystals 233
– – catalytic activities 234
– – rapid reduction 233
– Pd system
– – *vs.* Pt system 232

– reduction kinetics 232
– THH Pd NCs 228
–– cyclic voltammograms 229
–– dominant facets 228, 229
–– SEM image 229
– THH polyhedron model 229
– trapezohedral Pd NCs 229
–– SEM image 229
parametrization, and validation 34, 35
Patterson function 282
PEMFC. see proton exchange membrane fuel cells (PEMFC)
perturbation theory 13
photoexcitation 105
platinum (Pt) 224–228, 230–232
– commercial Pt black 230
– concave Pt cubic nanoparticles, synthesis by Xia's group 231
–– concave Pt THH nanoparticles 231
– concave Pt NCs, high-index facets 230
–– sluggish reduction rate 231
–– SME image 231
–– structural advantages 230
–– structural model 231, 232
– concave Pt THH nanoparticles 230
–– HRTEM images 231
– ethanol electrooxidation 227, 228
–– comparing catalytic activity of, NHH Pt NCs, Pt nanotubes, Pt/C 227, 228
– fivefold twinned Pt nanorods 226
–– HRSEM image 225
– formation of high-index facets, crucial factor 224
– formic acid and ethanol electrooxidation 230, 231
–– CV curves, concave Pt NCs, Pt black, and Pt/C (E-TEK) 230, 231
– formic acid and methanol electrooxidation 227, 228
–– TPH Pt NCs, and commercial Pt/C catalysts 227, 228
– high-index faceted Pt NCs 225–228
–– carbon black (HIF-Pt/C) 226
–– fcc metals 277
–– HRTEM image 225
– kinetic current densities, specific activity 231, 232
–– Pt concave cubes, Pt cubes, cuboctahedra 231, 232
– low-energy facets 226
– methylamine 230
–– Pt octapod NCs, role in formation of 230
– nanocrystals (NCs) 224

– oxidation current density 227, 228
– platinum-based catalysts 221
– platinum high-index planes
–– vs. low-index planes 223
– potential dependent steady-state current density 227, 228
–– of ethanol electrooxidation 227
–– THH Pt NCs vs. Pt nanospheres 227
– properties 224
– Pt_{35} cluster model 48, 49
– Pt/C (E-TEK) 230
– Pt electrocatalysts 226
– Pt octapod NCs 230
– Pt pyrophosphato complex 231
– Pt trapezohedral (TPH) NCs 225, 226
–– SEM image 225
– shape transformation 226
–– HRTEM image 225
–– Pt cubic nanoparticles 226
– square-wave method 225, 226
–– fivefold twinned Pt nanorods, synthesis 226
–– periodic oxygen adsorption/desorption, indused by 226
–– Pt trapezohedral (TPH) NCs, synthesis 225, 226
– synthesis methods, high-index facets
–– deep eutectic solvents (DESs) 225
–– electrochemical method 224
–– one-step method 224
–– thermodynamic control method 224
– tetrahexahedral(THH) PtNCs 224
–– catalytic activity 227
––– vs. polycrystalline Pt nanospheres, catalytic activity 227
–– high-magnification SEM image 225
–– surface structure determination 224
–– TEM images 225
– wet-chemical method 230
–– amine, capping agent 230
–– concave Pt NCs, synthesis of 230
Poisson–Boltzmann equation 51
Poisson number 189
polarizable electrodes 204
polycrystalline Cu_3Pt films
– anodic scan of 267
polymer electrolyte (or proton-exchange) membrane fuel cells (PEMFCs) 46
porphyrin structures 115
potential energy surface (PES) 18
potential of zero charge (pzc) 180
potential–strain coupling coefficients 194
principle of detailed balance 77

proton exchange membrane fuel cells
  (PEMFC) 222
– electrocatalysts 222
– membrane electrode assembly (MEA) 222
– PEMFC-powered automobile 222, 223
– Pt electrocatalysts
– – Pt loading 222
– research and development 222
$Pt_{25}Cu_{75}$ nanoparticles 276
$Pt_{25}Cu_{75}$ precursor alloys
– powder X ray diffraction patterns 273
$PtCu_3$(111) single crystals
– electrochemical dealloying of 269
Pt-modified Au prisms 243–245
– Ag monolayer 244
– – repeated galvanic reaction 244
– Ag UPD 244
– – effects 244
– – onset of Ag^+ UPD, various Au crystalline facets 244
– facet-dependent catalytic behavior 245
– high-magnification SEM image, stressing typical TDP profiles 243
– polarization curves, hydrogen evolution/oxidation reactions 244, 245
– Pt monolayer lattice 244
– SEM image, as-prepared Au TDP NPs 244
– truncated ditetragonal Au prisms (TDP) 243
– – illustration 244
– voltammetry curve, Au (TDP)–Pt (monolayer) sample 244
Pt poor extended bimetallic surfaces
– dealloying of 262

q
QM/MM modeling 4, 37
– methods for coupling 38
– schematic representation of simulation 37

r
Raman spectra 155, 156
Raney nickel 263
random phase methods (RPM) 16
reaction energies, and rates 21, 22
resonance 32
resting states (RS) 16
Ru-modified THH Pt NCs 239, 240
– bare THH Pt NCs 239
– bi-functional effects 239
– cleanTHH Pt NC electrode 239
– CVs with various Ru coverage 239, 240
– methanol oxidation 239
– PtRu alloy 239

– – catalytic activity 239
– removal of CO 239
– Ru benefits 239
– Ru coverage 239
– – effects on oxygen adsorption/desorption currents 239
– – evaluation 239
– Ru decoration 239
– – enhancement effect 239
– – Pt nanoparticles 239
– Ru-induced enhancement factor 239
– Ru-modified Pt black catalyst 240

s
SAED. *see* selected-area electron diffraction (SAED)
sampling, and analysis 39
scanning electrochemical microscopy (SECM) 99, 100, 153
scanning tunneling microscopy (STM) 99, 100
– characterization of molecule
– – electrical properties, general modes 101
– – self-assembled monolayer 102
– determining single molecule conductance
– – I(s) technique 102
– – I(t) method 102–104
– electrochemical studies, for single molecule (*see* single molecule electrochemical studies, with STM)
– single molecular junction 102
SCF optimization 38
Schrödinger equation 6–9, 12, 13, 26
SECM. *see* scanning electrochemical microscopy (SECM)
selected-area electron diffraction (SAED) 224, 226, 229, 230, 232, 234, 245
shape-controlled synthesis 224–235
– electrochemical route 224–230
– wet-chemical route 230–235
Shewanella oneidensis 145, 149
single crystal alloys
– strained dealloyed Pt surfaces studied 261
single molecule electrical measurements 101–103
single molecule electrochemical studies, with STM 115
– adsorbed iron complexes 115–118
– – FePP, *in situ* electrochemical STM images of 116
– – FePP *vs.* simulations 117
– – porphyrin structures 115

- oligo(phenylene ethynylene) derivates 130, 131
-- molecular conductance 131
- osmium and cobalt metal complexes 122–125
- perylene tetracarboxylic diimides 128, 129
-- single molecule conductance values 129
-- tunneling current *vs.* electrochemical potential 130
- pyrroloTTF (pTTF) 125–128
-- aniline heptamer 127
-- molecular structure 126
-- single molecule conductance *vs.* overpotential 127
- viologens 118–122
-- current–distance decay curves 120
-- electrochemical switching 118
-- single molecule conductance *vs.* electrochemical overpotential data 120
-- STM image 119
solid–electrolyte interface 2
solids, capillary equations 175–177
solid surface deforming 170–172
- atomic structure surface 170
- deformation 170
-- two states 170
- elastic strain 171
- increasing the physical area, two non-equivalent ways of 171
- Lagrangian area 172
- Lagrangian coordinates 171
- physical area 170
- 'plastically' deformed state 172
- Shuttleworth equation 172
- specifying surface area, two conventions 172
- surface excess energy 171
- thermodynamic potentials 171
- undeformed state 170
solid *vs.* fluid 167
- capillarity 167
- Gibbs free energy 167
- Gibbsian thermodynamics 167
- shear stress 167
- substitution, stress and strain 167
- theory of capillarity 167
- thermodynamics
-- heterogeneous equilibria 167
-- potentials 167
solvent modeling
- explicit *vs.* implicit 24, 25
static system 79–83
- Frumkin isotherm 80

- hard hexagon model 82
-- cyclic voltammetry 83
-- disorder–order phase transitions 83
-- isotherms of derivatives 82
- hard-square model 81
-- solutions in MFA 82
- Langmuir isotherm 80, 81
- lateral interactions 80
- mean-field approximation 80, 81
- modeling of static lattice gas system 79
-- energetic interactions 79
- one-dimensional lattice gas, expression 81
- partition function 80
- quasi-chemical approximation 81
STM. *see* scanning tunneling microscopy (STM)
Stoneys equation 174
strictly localized bond orbitals (SLBOs) 38
superficial charge density 204
surface-induced mean pressure, atomistic computation 178
surface stress, surface tension 164–166
- capillary phenomena 164
- changing physical area of surface, two modes 165
- Gibbs explanation 164
- negative surface stress 166
- net free excess free energy 165
- *positive* surface stress, metal 166
- Shuttleworth's analysis 166
-- consequences 166
- solid mechanics 164
- strain term, projection 164
- *strengthened* bonding 166
- surface atoms 166
- surface stress microscopic origin
-- schematic illustration 166
- surface tension of solid 164
- tangential stress 164, 165
-- elastic strain 164, 165
surface–vacuum interfaces 2

t

thermal desorption spectroscopy 49
thermodynamic state functions 20, 21
THH PdPt alloy 245, 246
- alloy effect 246
- alloy THH Pd–Pt NCs 245
-- atomic composition 245
-- different alloy composition 245
-- peak current densities, j_P 246
-- peak potentials, E_P 246
-- SEM image 246

– formic acid electrooxidation 246
– – current–potential curves 246
– Pd catalysts 246
– Pd tetrahexahedral NCs (THH Pd NCs) 245
– programmed potential electrodeposition 245
– THH $Pd_{0.90}$–$Pt_{0.10}$ NCs 245
– – SEM image 246
– – XPS test 246
thought experiment in electrowetting, case study 172–175
– *ab initio* computer simulation 174
– conclusion 174
– elastic strain changes 174
– electrode surface, Langrangian areas 173
– experiment setup 172
– free-energy change, dependence 173
– Lagrangian frame 174
– net free energy, system 172
– physical surface area 172
– – non-equivalent ways, of varying it 173
– relevant capillary parameter 173
– second process, elastic strain 173, 174
– Stoneys equation 174
– – equilibrium curvature 174
– Young equation 173
torsional angle 30
torsional deformation 30

torsional energy 30
transition states (TS) 4, 9, 16, 26, 35, 48, 49, 60, 163
– searches 17, 18

u

underpotential deposition (UPD) 190, 237, 241, 244
– Cu 247
– hydrogen 187
– surface stress variation during 195
UPD. *see* under-potential deposition (UPD)
UV/Vis absorption spectra 233

v

van der Waals interactions 33
vibrational modes 26, 29

x

X-ray diffraction (XRD)
– Bragg Brentano powder 273–274
– *in situ* high temperature powder 274–276
X-ray photoemission spectroscopy (XPS)
– synchrotron-based 276

y

Young–Laplace equation 175–178, 181